PLANTS OF
ARIZONA

PLANTS OF ARIZONA

Second Edition

ANNE ORTH EPPLE
UPDATED BY JOHN F. WIENS
PHOTOGRAPHY BY LEWIS E. EPPLE

GUILFORD, CONNECTICUT
HELENA, MONTANA
AN IMPRINT OF GLOBE PEQUOT PRESS

FALCONGUIDES®

Previously published by Falcon Publishing, Inc.

FalconGuides is an imprint of Globe Pequot Press.
Falcon, FalconGuides, and Outfit Your Mind are registered trademarks of Morris Book Publishing, LLC.

Photos by Lewis E. Epple unless otherwise credited.
Illustration on page 447 by D. D. Dowden
Text design: Sheryl P. Kober
Layout: Sue Murray
Project editor: Gregory Hyman
Map by Daniel Lloyd © Morris Book Publishing, LLC.

ISBN 978-0-7627-7035-9
Library of Congress Cataloging-in-Publication Data is available on file.

Printed in USA
10 9 8 7 6 5 4 3 2

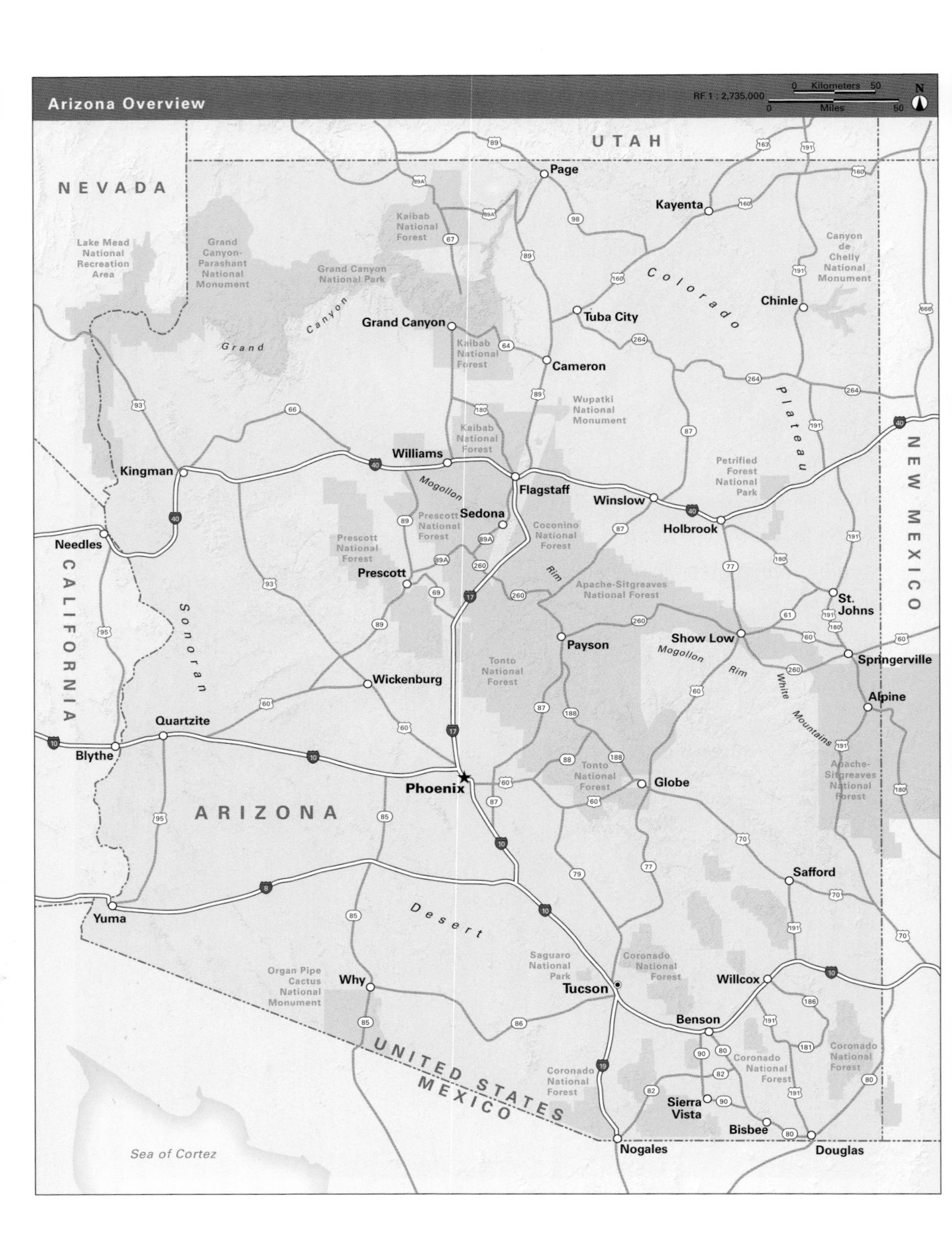
Arizona Overview
RF 1 : 2,735,000
Kilometers
Miles
UTAH
NEVADA
NEW MEXICO
CALIFORNIA
ARIZONA
UNITED STATES
MEXICO
Page
Kayenta
Lake Mead National Recreation Area
Grand Canyon-Parashant National Monument
Kaibab National Forest
Grand Canyon National Park
Canyon de Chelly National Monument
Colorado
Chinle
Tuba City
Grand Canyon
Cameron
Grand Canyon
Wupatki National Monument
Plateau
Williams
Kingman
Flagstaff
Winslow
Holbrook
Petrified Forest National Park
Mogollon
Sedona
Prescott National Forest
Coconino National Forest
Needles
Prescott
Rim
Apache-Sitgreaves National Forest
St. Johns
Sonoran
Payson
Show Low
Mogollon Rim
Springerville
White Mountains
Alpine
Tonto National Forest
Wickenburg
Quartzite
Blythe
Phoenix
Globe
Safford
Yuma
Desert
Organ Pipe Cactus National Monument
Why
Saguaro National Park
Tucson
Coronado National Forest
Willcox
Benson
Sierra Vista
Bisbee
Nogales
Douglas
Sea of Cortez

CONTENTS

ACKNOWLEDGMENTS

The author wishes to thank the following for their invaluable assistance in identifying the more difficult plant subjects:

Dick Anderson, Chiricahua National Monument
Arizona-Sonora Desert Museum Botany Department, Tucson
Michael Bencic, Wupatki National Monument
Dr. Susann Biddulph
Jane Cole, Desert Botanical Garden, Phoenix
Dr. Carol Crosswhite, Boyce Thompson Southwestern Arboretum
Robert Dyson, Alpine Ranger District
Karen Foster, Saguaro National Park West
Peggy Lu Gladhill, Alpine Ranger District
Wendy Hodgson, Desert Botanical Garden, Phoenix
Philip D. Jenkins, The University of Arizona, Tucson
Ron Kearns, Kofa Mountain Wildlife Refuge
Les Landrum, Arizona State University, Tempe
Dr. Charles Mason, The University of Arizona, Tucson
Sandy McMahan, Saguaro National Park West
Vince Ordonez, Apache-Sitgreaves National Forest
Barbara Phillips, Coconino National Forest
Dr. Donald J. Pinkava, Arizona State University, Tempe
Patrick Quirk, Desert Botanical Garden, Phoenix
Roy Simpson, Chiricahua National Monument
Kathy Warren, Grand Canyon National Park

INTRODUCTION

After retiring to Arizona my husband and I searched for a comprehensive field guide to Arizona plants—wildflowers, cacti, trees, shrubs, vines, and ferns. When we found none, we decided to write our own. And so, we planned this compact guide for hikers, campers, rangers and other amateur botanists who are interested in recognizing, as well as learning more about, Arizona's plants. We now present a new edition with updated plant information thanks to the help of John Wiens. This edition also reorganizes the plants to aid in visual indentification; plant photos now accompany their corresponding descriptions and can be found by division, color, and family more easily.

While writing this guide we traveled throughout the state photographing and carefully recording the measurements, elevations, and habitats of our plant subjects. These statistics were compared with the numerous references used to compile this guide. Where references proved inadequate, we queried plant experts. We are grateful for their interest in our extensive project, and for sharing their wealth of knowledge with us.

Arizona boasts over 3,000 species of plants—over 900 of which are pictured in this field guide. Grasses and most weeds are not included. It is impossible, too, for a book of this size to picture all of the species within a given genus. For this reason the **comments** section under each plant indicates the number of species found in Arizona for that particular genus.

The plants in any genus usually resemble each other; however, each species differs in some respect. Where a genus contains a large number of species, the amateur often finds it impossible to name the specific plant. In these instances, using a general name such as goldenrod, or thistle, or evening primrose will often suffice. On the other hand, dedicated botanists frequently spend years deliberating over a species, often changing the genus or species name several times before agreeing. These professionals deserve our greatest respect.

For the convenience of the amateur botanist, plants with conspicuous flowers, the first species described in this book, are arranged by flower color (then listed according to family name in Latin, then by genus and species) into the following seven groups:

White to Cream
Yellow
Orange
Red
Pink to Purple
Blue
Green to Brown

Plants whose flowers have minute or absent petals have been grouped in with the green to brown section. It is also important to note that there are sometimes great color variations among flowers of the same species. In those cases, the colors in which the plant's flowers may appear are listed in the species description.

Appearing next are the cacti. Although they are true flowering plants, cacti have such a distinctive appearance that we chose to group them separately. They can be tree- or shrublike, cylindrical or round, or have jointed stems or pad stems.

Conifers, a more primitive, nonflowering group, follow the true flowering plants and cacti. The ferns, the most primitive plants included in this guide, appear last and are also arranged by family and genera.

Each species is listed by its common name (e.g., Quaking Aspen); an alternate common name (e.g., Golden Aspen); its scientific name (e.g., *Populus tremuloides*); in parentheses, any other scientific name that may be commonly used; and the family name in English and Latin (e.g., Willow Family [Salicacae])—in that order. Common names, while fun to use, vary from place to place. Scientific names, on the other hand, are standard throughout the world. The scientific names in this guide, as well as the order of the families are based on *The Flora of North America* and *Seinet.* Of course, changes are ongoing and authorities often disagree on names.

We have tried to keep the scientific jargon to a minimum in the text. A glossary is included for easy reference.

The appearance of any plant varies according to the time of year and the elevation (not to mention natural factors such as the amount of rainfall). Therefore, as an aid to the reader, we indicate the location as well as the specific day of the month that the photo was taken.

With each succeeding generation more and more of our desert, woodland, and forest plants are destroyed. Many of the rarest species in our state are becoming extinct. We must protect Arizona's flora for future generations to enjoy by staying on established trails when hiking and by not picking or vandalizing our plant life.

Through the photographs and text in this guide, we hope the plants of Arizona—the common as well as the uncommon—take on a new meaning for you.

ARIZONA'S LIFE ZONES

The diversity of plant life in Arizona spans six main life zones, from a low point of 70 feet near Yuma to a high point of 12,670 feet in the San Francisco Peaks near Flagstaff. Although certain species characterize specific zones, others frequently overlap several zones. The contour of the land—canyons or mountain slopes—also influences the variety of flora within a zone. Rainfall, too, greatly affects life within a given environment. Thus, life zones are merely guides, rather than well-defined territories. Climbing 1,000 feet approximates traveling northward about 300 miles and results in a temperature drop of about five degrees.

The **Lower Sonoran Zone** is situated below 4,500 feet. Here, creosote, jojoba, paloverde, mesquite, ironwood, saltbush, bursage, and cacti abound. The plants in this zone endure high temperatures and low precipitation. Spring annuals survive as seeds, some for decades. When temperatures are just right and rainfall in the fall and winter (January and February) is sufficient, the seeds germinate. These so-called *ephemerals* ("of very short duration") develop quickly, burst into blossom, then soon go to seed. In contrast, paloverde, mesquite, ironwood, and bursage survive desert conditions because of their reduced leaf area. Others, such as jojoba and creosote, have specialized leaves for desert survival, while the ocotillo and brittlebush shed their leaves entirely during drought conditions. Still others, such as the cacti and certain bushes, have vestigial leaves or have modified their leaves to spines over evolutionary time.

The **Upper Sonoran Zone** ranges from 4,500 to 6,500 feet. At these elevations rainfall is more plentiful. Here, grasslands and sagebrush, as well as woodlands of oak, juniper, and pinyon pine are found. Here, too, lie large areas of chaparral with thickets of manzanita.

Between 6,500 and 8,000 feet lies the **Transition Zone.** Abundant rainfall at these altitudes produces huge stands of ponderosa pines. Scattered among these pines grow junipers, Gambel oaks, and Douglas firs.

The **Canadian Zone,** at 8,000 to 9,500 feet, is the province of cool, moist, fir forests. Douglas fir dominates this zone, with a mixture of blue and Engelmann spruce, quaking aspen, and white or subalpine fir.

Within the **Hudsonian Zone** (9,500 to 11,500 feet) grow spruce, fir, and bristlecone pines. These trees are usually stunted due to a short growing season; often they are twisted from the windy conditions found at these higher elevations.

The **Alpine Zone,** at above 11,500 feet, is represented on the San Francisco Peaks by sedges, lichens, grasses, and alpine wildflowers. It is above the timberline.

Each of the above zones is rich in wildflowers—some species are unique to a certain zone; others are present in several zones.

PLANT FAMILY DESCRIPTIONS

Flowering Plants: Dicots

The largest group of families of flowering plants (angiosperms) are the dicots. As the name suggests, each seed has two cotyledons (seed leaves). Other characteristics of this group of plants is that principal veins of the leaves branch from a midrib or the base, forming a network; they are not parallel. The sepals and petals are usually in twos, fours, or fives, and there is a single cylindrical vascular bundle within the stem.

Acanthus Family

Acanthaceae. Represented in Arizona by herbs or shrubs. *Flowers:* Irregular, 4 to 5 united petals, often 2-lipped (2-lobed upper and 3-lobed lower); 5-part calyx, 2 to 4 stamens. Pistil with long style and usually 2 stigmas. Superior ovary, 2-celled. *Fruit:* 2-celled capsule. *Leaves:* Opposite, simple, occasionally lobed.

Maple Family

Aceraceae. Recent studies show that this family of trees and large shrubs may better fit into the Soapberry Family (Sapindaceae) but will be kept separate here. *Flowers:* Small, either 4 or 5 petals or petalless; 4 or 5 sepals, 4 to 12 stamens, 1 pistil, 1 style with 2 stigmas. Superior ovary, 2-celled. *Fruit:* 2-winged key ("samara," or seed case) united at base. *Leaves:* Opposite, single, palmately or pinnately compound.

Amaranth Family

Amaranthaceae. Weedy herbs. *Flowers:* Small, inconspicuous, petalless; 3 to 5 sepals, 3 to 5 stamens opposite the sepals; in spikes or in dense clusters. Superior ovary, 1-celled. *Fruit:* Small, dry, 1-seeded. *Leaves:* Alternate or opposite, simple.

Cashew Family

Anacardiaceae. Represented in Arizona by shrubs. *Flowers:* Regular, small, with generally 5 petals, 5 sepals, 5 stamens, 1 pistil. Superior ovary, 1-celled. *Fruit:* Small, 1-seeded drupe. *Leaves:* Alternate; simple or compound.

Carrot Family

Apiaceae (Umbelliferae). Represented in Arizona by herbs. *Flowers:* Regular, with 5 petals, 5 sepals, 5 stamens, 2 styles. Usually small and in umbrella-shaped clusters. Inferior ovary, 2-celled. *Fruit:* 2-ribbed or winged, 1-seeded pods with aromatic oil ducts. *Leaves:* Alternate or basal, pinnately compound; usually finely cut; strong-smelling.

Dogbane Family

Apocynaceae. Represented in Arizona by herbs and slightly woody plants. *Flowers:* Regular, usually bell-shaped or funnel-shaped; generally 5 sepals, 5-lobed joined petals; occasionally a 4-lobed corolla; 5 stamens, 2 pistils; twisted in bud. Superior ovary, 2 with single style. *Fruit:* Usually paired, dry pods. *Leaves:* Alternate, opposite, or whorled; simple.

Milkweed Family

Asclepiadaceae. Current research shows this family, represented in Arizona by herbs and slightly woody plants, may belong within the dogbane family (Apocynaceae), although we have kept it separate in this publication. *Flowers:* Regular, hourglass-shaped; 5 separate sepals; 5 joined petals, generally swept backward; with central column, 5 stamens. In roundish clusters or umbels. 2 superior ovaries, each 1-celled. *Fruit:* Dry pod with seeds attached to tufts of hairs. *Leaves:* Mostly opposite, occasionally whorled or alternate; simple. Normally produce a milky sap.

Sunflower Family

Asteraceae (Compositae). Represented in Arizona by herbs or shrubs. *Flowers:* Regular or irregular, in heads surrounded by bracts, scales, or bristles. Individual florets are tiny to small, compound of outer straplike, flat ray flowers (as in the "petals" of the daisy) and tiny to small, tubular disk flowers (as at the center of the daisy). Some are rayless, with 5 united petals forming tube. Usually 5-stamened. Inferior ovary, 1-seeded. *Fruit:* An achene. *Leaves:* Alternate or opposite, occasionally whorled; simple, lobed, or dissected.

Barberry Family

Berberidaceae. Represented in Arizona by small, woody plants and shrubs. *Flowers:* Regular, usually tiny; 6 petals in 2 circles, 6 sepals in 2 circles; 6 stamens. Superior ovary, 1-celled. *Fruit:* Berry or capsule. *Leaves:* Alternate, simple, or compound, often spiny.

Birch Family

Betulaceae. Represented in Arizona by trees and shrubs. *Flowers:* Male flowers in long, slender catkins; female flowers in short spikes or conelike clusters; petalless, 1 pistil. Inferior ovary, 2-celled. *Fruit:* 1-seeded nutlet, often winged or conelike. *Leaves:* Alternate, simple.

Bignonia Family

Bignoniaceae. Represented in Arizona by shrubs or small trees. *Flowers:* Irregular, tubular or funnel-shaped; 2-lipped (2-lobed upper lip and 3-lobed lower lip); usually 4 stamens, in 2 sets; 5-lobed calyx; 1 style, usually 2-lobed stigma. Superior ovary, 2-celled. *Fruit:* Dry, elongated, 2-valved capsule. *Leaves:* Opposite, simple, or compound.

Forget-me-not Family

Boraginaceae. Represented in Arizona by herbs and slightly shrubby plants. *Flowers:* Regular; 5 petals united into trumpetlike tube flared at top, with 5 lobes, 5 sepals, 5 calyx lobes, 5 stamens. Mostly small, on 1-sided, rolled up, coiled spike that gradually unfolds. Superior ovary. *Fruit:* Divides into 4 1-seeded nutlets. *Leaves:* Mostly alternate, simple.

Mustard Family

Brassicaceae (Cruciferae). Represented in Arizona by herbs and slightly woody plants. *Flowers:* Regular, small; 4 petals arranged like a cross, 4 sepals; 6 stamens: 4 long, 2 shorter; and 1 pistil. Style with 1 or 2 stigmas. Superior ovary, 2-celled. *Fruit:* Mostly a dry capsule with 2 cells. *Leaves:* Alternate; simple or pinnately divided; bitter tasting.

Torch Wood Family

Burseraceae. Represented in Arizona by two species of shrubs or small trees. *Flowers:* Regular, tiny, with 3 to 5 petals, 3 to 5 calyx lobes, 6 to 10 stamens, 1 pistil. Ovary 3-celled. *Fruit:* Drupelike. *Leaves:* Alternate, pinnately compound.

Cactus Family

Cactaceae. Represented in Arizona by mostly succulent, spiny, leafless plants with special organs (areoles) from which spines, stems, or flowers occur. *Flowers:* Mostly regular, with numerous petals and sepals forming funnel-shaped receptacle; many stamens; 1 style with several stigma lobes. Inferior ovary, 1-celled. *Fruit:* Large or small, fleshy or dry berry. *Leaves:* Leafless. Plants have varying numbers of stems and joints and various shapes.

Bellflower Family

Campanulaceae. Represented in Arizona chiefly by herbs. *Flowers:* Regular or irregular, mostly bell-shaped with 5 flaring lobes and 5 stamens; or tubular and 2-lipped with stamens joined. 1 style with 2 to 5 stigmas; 5 sepals. Inferior ovary, 1- to 5-celled. *Fruit:* Dry pod (capsule). *Leaves:* Alternate or basal, simple.

Hemp Family

Cannabaceae. Represented by a twining herb in Arizona. *Flowers:* Regular, tiny, inconspicuous; in loose clusters. Superior ovary, 1- to 2-celled. *Fruit:* Surrounded by bracts. *Leaves:* Alternate or opposite, simple and palmately lobed.

Honeysuckle Family

Caprifoliaceae. Herbs, vines, shrubs, and small trees. *Flowers:* Regular or irregular; bell-, funnel-, or tube-shaped; 4- to 5-lobed corolla, 4- to 5-toothed or lobed calyx; 4 to 5 stamens, 1 style, 1 to 5 stigmas. Inferior ovary, 2- to 5-celled. *Fruit:* Berry, drupe, or capsule. *Leaves:* Opposite, simple, or compound.

Pink Family

Caryophyllaceae. Represented in Arizona by herbs and slightly woody plants. *Flowers:* Regular, 4 to 5 petals, 4 to 5 sepals; 8 to 10 stamens, 1 pistil, 2 to 5 styles. Superior ovary, 1-celled. *Fruit:* Dry; many-seeded capsule or 1-seeded achene. *Leaves:* Opposite or whorled; simple.

Casuarina Family

Casuarinaceae. An Australian tree, not native to Arizona. *Flowers:* Tiny, petalless, 1 pistil. Superior ovary. *Fruit:* Conelike, hard, woody. *Leaves:* Scalelike.

Bittersweet Family

Celastraceae. Represented in Arizona by woody plants and shrubs. *Flowers:* Regular, small; 4 to 6 calyx lobes and 4 to 6 petals; as many or twice number of stamens as petals; 1 style. Superior or half inferior ovary, 2- to 5-celled. *Fruit:* Dry or fleshy, depending on genus. *Leaves:* Alternate or opposite; simple; often reduced to scales.

Goosefoot Family

Chenopodiaceae. Herbs and shrubs. *Flowers:* Small, inconspicuous, when present; 1 to 5 perianth segments, stamens same number or fewer than perianth segments, 2 or 3 styles. Superior ovary, 1-celled. *Fruit:* Achene or a small, often bladdery, 1-seeded fruit that does not split. *Leaves:* Mostly alternate, simple.

Cleome Family

Cleomaceae. Represented in Arizona by herbaceous or woody plants. *Flowers:* Regular or nearly regular, 4 petals, 4 sepals, 6 or more stamens. Superior ovary, 1-celled. *Fruit:* Mostly a 2-valved capsule. *Leaves:* Alternate, simple, or palmately compound.

Morning Glory Family

Convolvulaceae. Represented in Arizona by herbs and slightly woody plants. *Flowers:* Regular, with 4 or 5 united petals forming funnel-shaped or long, tubular corolla; 5 sepals, 5 stamens; 1 or 2 styles, 1 or 2 stigmas, usually twisted buds. Superior ovary, mostly 2-celled. *Fruit:* Generally a capsule. *Leaves:* Alternate, simple.

Dogwood Family

Cornaceae. Represented in Arizona by 1 shrub. *Flowers:* Regular, small, with 4 petals, 4 stamens, 1 pistil, 1 style. Inferior ovary, 1- or 2-celled. *Fruit:* A drupe. *Leaves:* Opposite, simple.

Orpine Family

Crassulaceae. Represented in Arizona by herbs. *Flowers:* Regular, usually starlike; 4, 5, or no petals (when present, petals sometimes unite into a tube); 4 to 5 sepals, 4 or 5 stamens or twice as many; 3 or more pistils. Superior ovary. *Fruit:* Dry, 1-celled capsule. *Leaves:* Alternate, opposite, or basal, simple; succulent.

Crossosoma Family

Crossosomataceae. Represented in Arizona by shrubs. *Flowers:* 5 petals, 5 sepals, numerous stamens, 2 to 5 pistils. *Fruit:* Several-seeded follicle. *Leaves:* Alternate, simple.

Gourd Family

Cucurbitaceae. Herbs. *Flowers:* Regular, small and inconspicuous; or large and funnel-shaped; or of 5 separate petals; calyx joined to 5-lobed ovary; 3 or 5 stamens (2 pairs united), 1 style, usually 3 stigmas. Inferior ovary, 1- to 4-celled. *Fruit:* Varying; generally berrylike or gourdlike. *Leaves:* Alternate, simple or compound.

Dodder Family

Cuscutaceae. Represented in Arizona by parasitic herbs. Closely related to the Morning Glory Family (Convolvulaceae), in which some taxonomists place them. *Flowers:* Regular, with 4 or 5 united petals forming funnel-shaped corolla; 5 sepals, 5 stamens. Superior ovary, mostly 2-celled. *Fruit:* Generally a capsule. *Leaves:* Alternate, simple, reduced to scales.

Teasel Family

Dipsacaceae. A weedy herb. *Flowers:* Small, in bracted heads; tubular, 4- or 5-lobed; 5 sepals, 4 stamens, threadlike style. Inferior ovary, 1-celled. *Fruit:* Dry achene. *Leaves:* Opposite, simple.

Oleaster Family

Elaeagnaceae. Represented in Arizona by shrubs and small trees. *Flowers:* Regular, small, 4-toothed perianth, 4 to 8 stamens, long style. Superior ovary, 1-celled. *Fruit:* Dry nut or achene. *Leaves:* Alternate or opposite, simple, covered with minute, silvery or light brownish scales.

Heather Family

Ericaceae. Herbs, saprophytes, shrubs, and small trees. *Flowers:* Regular or nearly regular, bell-shaped, urn-shaped, or tubular, 4 to 5 lobes or petals, 4- to 5-lobed calyx, 8 to 12 stamens, 1 pistil, 1 style, 1 stigma. Superior or inferior ovary, 4- to 10-celled. *Fruit:* Dry or juicy. *Leaves:* Mostly alternate or basal; simple or scalelike. Some taxonomists separate Chimaphila, Monese, Orthila, and Pyrola into the Pyrola Family *(Pyrolaceae).*

Spurge Family

Euphorbiaceae. Represented in Arizona by herbs, shrubs, and small trees. *Flowers:* Usually small and inconspicuous, but bracts around and beneath them are often colorful. Flowers with or without petals or sepals; 1 stamen or many. Superior ovary, usually 3-celled. *Fruit:* Mainly 3-lobed capsule. *Leaves:* Alternate, opposite or whorled; simple or compound.

Pea Family

Fabaceae (Leguminosae). Represented in Arizona by herbs, shrubs, and small trees. *Flowers:* Variable, mostly irregular. Many are pealike, with upper petal or *banner*, 2 side petals called *wings*, and 2 lower petals that form the *keel*. Some are without petals, with pistils and stamens forming dense balls, spikes or long, hanging clusters. 10 to 20 stamens; 1 pistil. Superior ovary, 1-celled. *Fruit:* 2-valved pod called a *legume*. *Leaves:* Alternate, compound (some are twice or thrice compound), 3-leaved, rarely single.

Beech Family

Fagaceae. Represented in Arizona by oak trees. *Flowers:* Male flowers in catkins; 5-lobed perianth; 5 to 10 stamens. Female flowers solitary or in small clusters; 6-lobed perianth, 3 styles. Inferior ovary, 1-celled. *Fruit:* Acorns on oak trees. *Leaves:* Alternate, simple, lobed or toothed; deciduous or evergreen.

Ocotillo Family

Fouquieriaceae. Represented in Arizona by a thorny shrub. *Flowers:* Regular; 5 united petals forming 5-lobed tubular corolla; 5 sepals, 10 to 17 protruding stamens. Superior ovary. *Fruit:* 3-celled capsule. *Leaves:* Alternate, simple, in clusters in axils of spines.

Fumitory Family

Fumariaceae. Represented in Arizona by herbs. *Flowers:* Irregular; 4 petals (in 2 pairs) somewhat joined, with 1 or 2 petals often forming a spur or hood; 4 to 6 stamens, 2 tiny sepals, 1 pistil. Superior ovary. *Fruit:* Mainly a long, dry, 1-celled capsule. *Leaves:* Alternate; compoundly dissected.

Silktassel Family

Garryaceae. Represented in Arizona by a small tree and a shrub. *Flowers:* Regular, small, in dense, catkinlike spikes. Inferior ovary. *Fruit:* Berrylike. *Leaves:* Opposite, simple, thick, evergreen.

Gentian Family

Gentianaceae. Herbs. *Flowers:* Regular, tubular or bell-shaped; corolla is usually 5-lobed, sometimes 4-lobed (rarely 12-lobed); 4 to 12 stamens alternating with corolla lobes and of same number; same number of sepals as petals: 1 style, 2-lobed stigma. Superior ovary, 1-celled. *Fruit:* Many-seeded capsule. *Leaves:* Generally opposite, simple.

Geranium Family

Geraniaceae. Represented in Arizona by herbs. *Flowers:* Generally regular; usually with 5 separate petals; 5 sepals; 2 or 3 times as many stamens as petals; elongated style. Superior ovary, 3- to 5-celled. *Fruit:* Dry, 1-seeded pod. *Leaves:* Alternate, opposite, or basal; mostly simple, variously lobed.

Currant Family

Grossulariaceae. Represented in Arizona by shrubs (often spiny). *Flowers:* Regular, with 4 or 5 petals, 4 or 5 sepals, 4 to 10 (or more) stamens, and 1 pistil. Inferior ovary, 2-celled. *Fruit:* Capsule or berry. *Leaves:* Simple, alternate (often in fascicles), deciduous.

Waterleaf Family

Hydrophyllaceae. Represented in Arizona by herbs or shrubs. *Flowers:* Regular, often bell- or funnel-shaped; in coils; 5 untied petals forming 5 lobes; 5 united sepals, 5 stamens; 1 or 2 styles, 2 stigmas. Superior ovary, mostly 1-celled. *Fruit:* Many-seeded capsule. *Leaves:* Mostly alternate, often in basal rosettes, simple or pinnately compound.

Hydrangea Family

Hydrangeaceae. Represented in Arizona by shrubs. *Flowers*: 4 petals, inflorescences in cymes or heads. *Fruit*: A many-seeded capsule, usually fleshy. *Leaves*: Opposite, simple, smooth to slightly hairy.

St. John's Wort Family

Hypericaceae (Guttiferae). Represented in Arizona by 2 perennial herbs. *Flowers:* Regular, with 4 or 5 petals, 4 or 5 sepals, numerous stamens arranged in 3 or 5 distinct clusters, 2 to 5 styles. Superior ovary, 3-celled. *Fruit:* A capsule. *Leaves:* Opposite, simple, dotted with glands.

Walnut Family

Juglandaceae. Represented in Arizona by one species of tree. *Flowers:* Male flowers have several stamens in drooping catkins; female flowers have 3- to 5-lobed calyx, solitary or in small clusters. Inferior ovary, 1-celled. *Fruit:* Round, hard-shelled nut. *Leaves:* Alternate, pinnately compound.

Ratany Family

Krameriaceae. Represented in Arizona by low shrubs. *Flowers:* Irregular; 5 petals; upper 3 long-clawed; other 2 reduced and glandlike; 3 or 4 stamens; 1 pistil. *Fruit:* Roundish, spiny. *Leaves:* Alternate, simple.

Mint Family

Lamiaceae (Labiatae). Represented in Arizona by herbs and shrubs. *Flowers:* Irregular, with 5-lobed corolla, usually 2-lipped; upper lip 2-lobed, lower lip 3-lobed. 4 stamens, 2 longer than others. Superior ovary, 2-celled, 4-lobed. *Fruit:* 4 1-seeded nutlets. *Leaves:* Opposite, rarely whorled, with fragrant oil glands; simple; 4-angled or square stems.

Bladderwort Family

Lentibulariaceae. Represented in Arizona by an aquatic herb. *Flowers:* Irregular, prominently 2-lipped; 2 stamens, spurred. Superior ovary, 1-celled. *Fruit:* 1-celled capsule. *Leaves:* Have threadlike segments; submerged in water.

Flax Family

Linaceae. Represented in Arizona by herbs. *Flowers:* Regular, with 4 or 5 petals, 4 or 5 sepals; usually 4 or 5 stamens, 1 pistil, 2 to 5 styles. Superior ovary, 2- to 5-celled. *Fruit:* 4- to 10-valved capsule. *Leaves:* Usually alternate, rarely opposite or whorled; simple.

Stick Leaf Family

Loasaceae. Herbs or slightly woody plants. *Flowers:* Regular, with 5 petals, 5 sepals and numerous stamens (outer stamens are often petallike). Inferior ovary, 1-celled. *Fruit:* Dry pod (capsule). *Leaves:* Alternate or pinnately cleft, simple, often with barbed hairs.

Malpighia Family

Malpighiaceae. Represented in Arizona by woody vines. *Flowers:* Small, 5-petaled; with 5 sepals, 10 stamens, 3 styles. Superior ovary. *Fruit:* 3 nutlike segments, sometimes winged. *Leaves:* Opposite, simple.

Mallow Family

Malvaceae. Represented in Arizona by herbs and shrubs. *Flowers:* Regular, usually 5 broad petals and 5 sepals; numerous stamens united to form center column which surrounds style. Petals of buds twist. Superior ovary, 2- to many-celled. *Fruit:* Dry capsule splitting into several parts. *Leaves:* Alternate, simple; usually deeply lobed, cut, or dissected.

Melia Family

Meliaceae. Represented in Arizona by one tree species. *Flowers:* Showy, with 4 to 5 united petals, 4 to 5 calyx lobes, 8 to 10 stamens, 1 pistil. Superior ovary, mostly 2- to 5-celled. *Fruit:* Fleshy, single-seeded capsule. *Leaves:* Alternate, compound.

Unicorn Plant Family

Martyniaceae (Pedaliaceae). Annual herbs. *Flowers:* Irregular with 2 lips and lobes; large, funnel-shaped; 4 to 5 calyx lobes; 4 stamens. Superior ovary, 1-celled. *Fruit:* Capsule with curved beak. *Leaves:* Alternate or opposite, simple, sticky, hairy.

Mulberry Family

Moraceae. Represented by two shrub or tree species. *Flowers:* Regular, tiny, inconspicuous; in catkins. Superior ovary, 1- to 2-celled. *Fruit:* Fleshy. *Leaves:* Alternate or opposite, simple, often palmately lobed.

Myrsine Family

Myrsinaceae. Represented in Arizona by herbs. This family is closely related to the Primrose Family (Primulaceae). *Flowers:* Regular; mostly funnel-shaped or tubular and 4- or 5-lobed; with 5 sepals, generally 5 stamens (if corolla is 5-lobed), 1 pistil with single style and stigma. Superior ovary, 1-celled. *Fruit:* Dry pod (capsule). *Leaves:* Alternate, opposite, whorled, or in basal rosette; simple.

Four O'clock Family

Nyctaginaceae. Represented in Arizona by herbs. *Flowers:* Regular or near regular, petalless, colorful, and tubular; calyx flares into 4 or 5 lobes, usually 5 stamens but varies from 3 to 10, 1 pistil with long style and buttonlike stigma. Superior ovary, 1-celled. *Fruit:* 1-seeded, nutlike achene. *Leaves:* Mostly opposite, simple.

Olive Family

Oleaceae. Herbs, shrubs, and trees. *Flowers:* Regular; with tubular corolla, with 4 to 6 lobes, or 4 separate petals, or petalless; 2 to 4 stamens, 1 style. Superior ovary, 2-celled. *Fruit:* Varying; drupe, winged samara, or capsule. *Leaves:* Alternate or opposite; simple or pinnate.

Evening Primrose Family

Onagraceae. Represented in Arizona by herbs and small shrubs. *Flowers:* Parts are usually regular and in fours: 4 petals, 4 sepals, 4 or 8 stamens, 4-lobed stigma, single style. Inferior ovary, 2- to 4-celled. *Fruit:* Usually a dry, 4-celled capsule. *Leaves:* Alternate, opposite, whorled or basal, simple.

Broomrape Family

Orobanchaceae. Represented in Arizona by root-parasitic herbs with fleshy stems and lacking chlorophyll. *Flowers:* Irregular, tubular; 5 united petals forming 2-lobed upper lip and 3-lobed lower lip; 4 stamens in pairs, 2 to 5 sepals, 1 style. Superior ovary, 1-celled. *Fruit:* Many-seeded, chambered capsule. *Leaves:* Alternate, simple, scalelike.

Oxalis Family

Oxalidaceae. Herbs. *Flowers:* Regular, 5 petals, 5 sepals, 10 stamens, 5 styles. Superior ovary, 5-celled. *Fruit:* Capsule seed pod. *Leaves:* Alternate, compound, often divided into 3 heart-shaped leaflets.

Poppy Family

Papaveraceae. Represented in Arizona by herbs. *Flowers:* Regular or irregular; 4 to 6 petals, often wrinkled; usually 2 sepals, 4 to 6 or more stamens; bulky pistil, style short or absent. Superior ovary, 1-celled. *Fruit:* Dry capsule. *Leaves:* Alternate or basal; simple or divided.

Pokeberry Family

Phytolaccaceae. Represented in Arizona by herbs. *Flowers:* Regular, small, petalless; 4 or 5 perianth segments, 4 to 12 stamens, 1 or several styles. Superior ovary. *Fruit:* Juicy berry. *Leaves:* Alternate, simple.

Plantain Family

Plantaginaceae. Ephemeral herbs. *Flowers:* Regular, small; calyx and corolla are 4-divided or 4-lobed; 2 to 4 stamens. Superior ovary, 2- to 4-celled. *Fruit:* A capsule. *Leaves:* Basal.

Plane Tree Family

Platanaceae. Trees. *Flowers:* Minute sepals and petals in a dense ball-like cluster; 3 to 7 stamens, 5 to 9 pistils. *Fruit:* 4-angled nutlet crowded in ball. *Leaves:* Alternate, simple, palmately lobed.

Phlox Family

Polemoniaceae. Represented in Arizona by herbs and slightly woody plants. *Flowers:* Mostly regular, tubular or funnel-shaped; with 5 corolla lobes or teeth, 5 sepals and 5 stamens alternating with corolla lobes; 1 style usually 3-lobed. Superior ovary, mostly 3-celled. *Fruit:* Usually 3-celled capsule. *Leaves:* Alternate or opposite; simple or compound.

Milkwort Family

Polygalaceae. Represented in Arizona by herbs and slightly shrubby plants. *Flowers:* Irregular; 3 petals, often united at base; 5 sepals, sometimes united and colored; 6 to 8 stamens; 1 style with 1 to 4 stigmas. Superior ovary, mostly 1-celled. *Fruit:* Mostly a flat, 1- or 2-celled capsule. *Leaves:* Alternate, opposite, or whorled; simple.

Buckwheat Family

Polygonaceae. Represented in Arizona by herbs and woody plants. *Flowers:* Regular, tiny, petalless; 2 to 6 petallike sepals, 2 to 9 stamens; 2 to 3 styles, 4- to 5-lobed calyx. One pistil with 2 to 4 stigmas. Superior ovary, usually 1-celled. *Fruit:* Usually 3-sided nutlet or achene. *Leaves:* Usually alternate or in rosettes; simple.

Purslane Family

Portulacaceae. Herbs. *Flowers:* Regular or near regular, 4 or 5 or more petals; generally 2 sepals; usually 5 stamens, but occasionally 10; 1 pistil, several styles. Mostly superior ovary, 1-celled. *Fruit:* Dry, 3-valved pod (capsule). *Leaves:* Alternate or opposite; simple, toothless, thick, and succulent.

Primrose Family

Primulaceae. Herbs. *Flowers:* Regular; mostly funnel-shaped or tubular and 4- or 5-lobed; with 5 sepals, generally 5 stamens (if corolla is 5-lobed), 1 pistil with single style and stigma. Superior ovary, 1-celled. *Fruit:* Dry pod (capsule). *Leaves:* Alternate, opposite, whorled, or in basal rosette; simple.

Buttercup Family

Ranunculaceae. Represented in Arizona by herbs and slightly shrubby plants. *Flowers:* Regular or irregular; generally 3 to 5 petals, but sometimes petalless; spurred or hooded; 3 to 15 sepals, often petallike; numerous stamens; 1 pistil or many. Superior ovary, 1-celled. *Fruit:* 1-celled achene, pod, or berry. *Leaves:* Alternate or opposite and occasionally basal; simple or compound.

Buckthorn Family

Rhamnaceae. Represented in Arizona by shrubs and small trees. *Flowers:* Regular, small, with 4 or 5 petals or petalless; 4 or 5 stamens; 4- or 5-lobed calyx. Superior or partly inferior ovary, 2- to 4-celled. *Fruit:* A drupe or capsule. *Leaves:* Alternate or opposite, simple.

Rose Family

Rosaceae. Represented in Arizona by herbs, shrubs, and trees. *Flowers:* Regular, with 4 to 5 petals, 4 to 5 sepals, 10 or more stamens; in clusters or solitary; 1 to many pistils. Superior or inferior ovary, 1- to 5-celled. *Fruit:* Pod, achene, drupe, or pome. *Leaves:* Alternate; simple or compound, or basal.

Madder Family

Rubiaceae. Represented in Arizona by herbs and shrubs. *Flowers:* Regular or nearly regular, with 4- to 5-lobed united corolla, 4 to 5 sepals, 4 to 5 stamens, 1 style. Inferior ovary, mostly 2-celled. *Fruit:* Capsule, berry, or drupe. *Leaves:* Opposite or whorled, simple. Square stems.

Rue Family

Rutaceae. Represented in Arizona by shrubs, small trees, and nearly herbaceous plants. *Flowers:* Generally regular, with 4 or 5 petals, 4 or 5 sepals, 8 or 10 stamens (usually double the number of petals); 1 pistil. Superior ovary, 2- to 5-celled. *Fruit:* Varying, depending on genus. *Leaves:* Alternate or opposite, simple or compound.

Willow Family

Salicaceae. Trees or large shrubs. *Flowers:* Minute, without sepals or petals; in catkins. Male and female on separate trees, appearing with or before the leaves. Ovary 1-celled. *Fruit:* Small, 2- to 4-valved splitting capsule; seeds surrounded by silky hairs. *Leaves:*

Simple, alternate, deciduous. Identifying a particular species of willow is very difficult. Leaves within a species vary, and some species hybridize.

Sandalwood Family

Santalaceae. Represented by a root parasite in Arizona. *Flowers:* Petalless, 4- to 5-lobed, with bell-shaped calyx, 5 stamens. Inferior ovary. *Fruit:* Nutlike. *Leaves:* Alternate, simple.

Soapberry Family

Sapindaceae. Represented in Arizona by a vine, shrub, and small tree. *Flowers:* Regular or nearly regular; 4 to 5 petals or petalless; 5 to 10 stamens; 1 style. Superior ovary, 2- to 4-celled. *Fruit:* Berrylike or dry, winged capsule. *Leaves:* alternate, simple, or pinnately compound.

Sapote Family

Sapotaceae. Represented in Arizona by a large shrub. *Flowers:* Regular and small, with 5 petals, 5 sepals, 10 stamens. Superior ovary. *Fruit:* 1-seeded drupe. *Leaves:* Alternate, simple, leathery, and thick.

Lizard Tail Family

Saururaceae. Represented in Arizona by one herb. *Flowers:* Regular; calyx and corolla absent; white petallike bracts; in dense spike; 3, 6, or 8 stamens, 3 or 4 pistils. Ovary 1-celled. *Fruit:* Succulent capsule. *Leaves:* Alternate, mostly basal, simple.

Saxifrage Family

Saxifragaceae. Represented in Arizona by herbs and shrubs. *Flowers:* Regular, with 4 or 5 petals, 4 or 5 sepals, 4 to 10 (or more) stamens, and 1 pistil. Inferior ovary, 2-celled. *Fruit:* Capsule or berry. *Leaves:* Alternate, opposite, or basal, mostly simple.

Figwort Family

Scrophulariaceae. Represented in Arizona by herbs and shrubs. *Flowers:* Generally irregular, with 5 united petals; tubular; usually 2-lipped, with upper lip 2-lobed (sometimes hooked) and lower lip 3-lobed; often with 3 sacs. Usually 4 stamens, 2 shorter than the others; forked or unforked style. Superior ovary, mostly 2-celled. *Fruit:* Generally a 2-celled capsule. *Leaves:* Alternate, opposite, or basal; simple or pinnate. Some plant taxonomists now split genera and place them into Orobanchaceae, Phrymaceae, and Plantaginaceae.

Simarouba Family

Simaroubaceae. Represented in Arizona by a tree and a shrub. *Flowers:* Generally small, with 3- to 8-lobed calyx; either 3 to 8 petals or petalless; 2 to 5 simple pistils. *Fruit:* Winged or drupelike. *Leaves:* Alternate, pinnately compound; or reduced to deciduous scales.

Jojoba Family

Simmondsiaceae. Represented in Arizona by a shrub. *Flowers:* Of 1 sex, petalless; with 4 to 6 sepals, 10 to 12 stamens, 3 styles. Superior ovary, 3-celled. *Fruit:* Large, acorn-shaped capsule. *Leaves:* Opposite, simple, thick, leathery.

Nightshade Family

Solanaceae. Represented in Arizona by herbs and shrubs. *Flowers:* Regular, bell- or funnel-shaped; 5 united lobes folded lengthwise in bud; 5 stamens, 1 style, 1- or 2-lobed stigma. Superior ovary, 2-celled. *Fruit:* Dry capsule or berry, often very poisonous. *Leaves:* Alternate, simple or pinnately compound; lobed or dissected.

Tamarix Family

Tamaricaceae. Large shrub or small tree. *Flowers:* Regular, very small, with 4 or 5 petals, 4 or 5 sepals, 4 or more stamens, 3 to 5 styles. Superior ovary, 1-celled. *Fruit:* 3- to 5-valved capsule. *Leaves:* Alternate, small, scalelike.

Elm Family

Ulmaceae. Represented by three species of woody shrubs and trees in Arizona. *Flowers:* Inconspicuous, petalless, 4 or 5 sepals, generally 4 to 5 stamens. Superior ovary, 1-celled. *Fruit:* Flat, winged, waferlike fruit, or a drupe. *Leaves:* Alternate, simple, lop-sided at base. Some plant taxonomists feel the genus *Celtis* belongs in the Hemp Family (Cannabaceae).

Nettle Family

Urticaceae. Represented by herbs in Arizona. *Flowers:* Petalless; male flowers with 4- to 5-parted calyx, female flowers tubular or with 3- to 5-parted calyx; 4 to 5 stamens. Superior ovary, 1-celled. *Fruit:* Dry, 1-seeded achene. *Leaves:* Alternate or opposite; simple.

Valerian Family

Valerianaceae. Herbs. *Flowers:* Irregular and tubelike, with 4 to 5 lobes, generally 3 stamens, 1 style. Inferior ovary, 3-celled. *Fruit:* Achenelike, 1-seeded. *Leaves:* Opposite.

Vervain Family

Verbenaceae. Represented in Arizona by herbs or shrubs. *Flowers:* Usually irregular, tubular or funnel-shaped; with 4 to 5 lobes or 2-lipped; generally 4 stamens (2 shorter), 5 sepals, 1 style, 1 to 2 stigmas. Superior ovary, 2-, 4-, or 5-celled. *Fruit:* A drupe or a fruit with 2 to 4 nutlets. *Leaves:* Opposite or whorled; simple or compound.

Violet Family

Violaceae. Represented in Arizona by herbs. *Flowers:* Irregular; 5 petals, 4 arranged in pairs, lower petal larger and spurred; 5 sepals, 5 stamens. Superior ovary, 3-celled. *Fruit:* 3-celled, many-seeded capsule. *Leaves:* Alternate or basal; simple or occasionally lobed.

Mistletoe Family

Viscaceae. Parasites on shrubs and trees. *Flowers:* Perianth is calyxlike; small, 2- to 4-lobed or toothed. Inferior ovary, 1-celled. *Fruit:* Berry. *Leaves:* Opposite. Jointed stems.

Grape Family

Vitaceae. Represented in Arizona by woody vines. *Flowers:* Regular, very small, with 4 or 5 petals, 4 or 5 calyx lobes, 4 or 5 stamens; 1 pistil, 1 style or none. Superior ovary, 2-celled. *Fruit:* Berrylike. *Leaves:* Alternate, simple or compound.

Caltrop Family

Zygophyllaceae. Represented in Arizona by herbs and shrubs. *Flowers:* Regular, usually 5 petals and 5 sepals; generally 10 stamens, 1 pistil, simple style. Superior ovary, 2- to 12-celled. *Fruit:* Usually a chambered capsule. *Leaves:* Alternate or opposite; compound or pinnately dissected.

Flowering Plants: Monocots

We have separated the monocot class family descriptions from the dicot class family descriptions. Monocots, as the name suggests, have only one cotyledon (seed leaf) in each seed. They are more recently evolved from the dicot class, which have two cotyledons in each seed. Other characteristics of this most highly evolved group of plants is that they have parallel veins in the leaves, sepals and petals are in threes or multiples of three, and vascular bundles are scattered within the stems.

Agave Family

Agavaceae. Herbs or somewhat woody at base. *Flowers:* Regular or nearly regular; 3 petals and 3 sepals, which look alike and are united below into a tube; 6 stamens. Inferior ovary, 3-celled. *Fruit:* A capsule. *Leaves:* Grasslike or succulent, narrow; forming basal rosette of rigid leaves. Some taxonomists argue that this family is invalid and that the genus *Agave* belongs in the Asparagus Family (Asparagaceae), *Dasylirion* belongs in the Lily Family (Liliaceae), and *Nolina* belongs in the Beargrass Family (Ruscaceae).

Water-Plantain Family

Alismataceae. Aquatic or semi-aquatic herbs. *Flowers:* Regular, white, 3 petals, 3 sepals, 6 or more stamens, 6 or more pistils. Ovary 1-celled, usually 1-seeded. *Fruit:* Achenelike, compressed. *Leaves:* Mostly basal, with bases of leaves sheathing stems.

Palm Family

Arecaceae (Palmae). Arizona's only native palm species is the California fan palm, Washingtonia filifera. *Flowers:* Small, in large, drooping clusters; 6 perianth segments; usually 6 stamens, 1 style, 1 stigma. Superior ovary, 3-celled. *Fruit:* Small drupe. *Leaves:* Palmate; leaf segments radiate outward from a single point.

Spiderwort Family

Commelinaceae. Herbs. *Flowers:* Regular or irregular, 3 colored petals, 3 green sepals, 6 stamens, single style (2- or 3-lobed). Superior ovary, 3-celled. *Fruit:* 3-celled capsule. *Leaves:* Alternate, thick, linear; parallel-veined, leaf base clasps stem.

Iris Family

Iridaceae. Herbs. *Flowers:* Regular or nearly regular, flower parts in threes; either 3 petallike sepals stand out or down and 3 petals stand up, or all 6 parts are of same size and color and form a flat circle; 2 bracts, 3 stamens, single, 3-cleft style. Inferior ovary, 3-celled. *Fruit:* 3-celled, many-seeded capsule. *Leaves:* Long, narrow, toothless, sword-shaped.

Lily Family

Liliaceae. Represented by herbs in Arizona. *Flowers:* Regular or near regular. Typically a flower cup with 6 sections; 3 petals, 3 sepals, 6 stamens (3 frequently without anthers); most commonly with a long pistil with 3-lobed stigma. Superior ovary, 3-celled. *Fruit:* 3-celled capsule or berry. *Leaves:* Usually alternate, occasionally whorled or in a basal rosette; straight-veined, stalkless; generally simple. (Some taxonomists feel that differences in the genera *Veratrum* and *Zigadenus* warrant their being put into their own family: Melanthiaceae).

Orchid Family

Orchidaceae. Herbs. *Flowers:* Irregular, with 3 outer sepals, 3 inner petals (2 are alike; the third forms a lip or large sac). Stamens and pistil joined together to form a column with anther at tip, stigma below. Inferior ovary, 1- to 3-celled. *Fruit:* 3-valved capsule. *Leaves:* Usually alternate, toothless, parallel-veined, simple.

Cattail Family

Typhaceae. Semi-aquatic perennial herbs. *Flowers:* Tiny, very dense terminal, cylindrical, brown, sausage-shaped spike of pistillate (female) flowers. Above are paler staminate (male) flowers on narrower spike. *Leaves:* Long, narrow, erect, stiff.

Gymnosperms

Gymnosperms evolved after ferns, but before true flowering plants (Angiosperms). *Gymnosperm* translates to "naked seed," as the seed is not produced in an ovary. The seed is usually produced on, or wedged between, woody, papery, or pulpy scales or bracts (think of a pine cone). Plant forms vary from woody shrubs to large trees. Leaves can be needles or scales.

Cypress Family

Cupressaceae. Evergreen trees and shrubs.
Fruit: Cones. Roundish, dry, scaly; or fleshy and berrylike.
Leaves: Opposite or whorled, short, scalelike.

Joint-Fir Family

Ephedraceae. Shrubs. *Flowers:* Conelike, primitive.
Fruit: 1 to 3 hard seeds. *Leaves:* Reduced to scales.

Pine Family

Pinaceae. Represented in Arizona by evergreen trees.
Flowers: Male flowers in naked catkinlike clusters; female flowers consist of naked ovules between bases of woody scales (at maturity, the cone). *Fruit:* Woody cone.
Leaves: Alternate or whorled; needlelike; borne mostly in bundles of 2, 3, or 5.

Ferns

Ferns include several families of nonflowering plants that reproduce by spores located in spore sacs, or *sori*, on undersides of fertile fronds. Each frond, or fern leaf, consists of a stem called a *rachis*—the section bearing leaflets. If the frond is compound, the leaflets are called *pinnae* (*pinna*, singular). When the leaf is divided again (*bipinnate*), the subleaflets are called *pinnules*. The lower part of the rachis (below the leafy section) is known as the *stipe* or *stalk*. The fifteen fern species in this book fit into the following five fern families.

Spleenwort Family

Aspleniaceae. Represented in Arizona by terrestrial perennials; creeping stems with scales. Sterile and fertile fronds are of similar size and shape. Frond blades are simple to 4-pinnate. Sori are borne along veins on the underside.

Bracken Fern Family

Dennstaedtiaceae. Represented in Arizona by mostly terrestrial perennials. Creeping stems usually with hair. Sterile and fertile fronds are of similar size and shape. Petioles not articulated. Frond blades are simple to compound. Sori are borne near or at the blade margin on the vein tip.

Shield Fern Family

Dryopteridaceae. Represented in Arizona by terrestrial perennials. Creeping stems usually growing on rocks. Sterile and fertile fronds are of similar size and shape. Petioles one-fifth to three-fourths the length of blades. Frond blades are linear to ovate; 1- to 2-pinnate. Sori are borne in 1 row between the midrib and margin on ultimate leaf segments.

Polypody Family

Polypodiaceae. Represented in Arizona by terrestrial perennials. Creeping stems in soil and occasionally on rocks. Short stems with scales. Sterile and fertile fronds are of similar size and shape. Frond blades are simple to pinnate. Sori are borne abaxially along veins.

Cloak Fern Family

Pteridaceae. Represented in Arizona by terrestrial perennials. Creeping stems bearing hair or scales. Sterile and fertile fronds are of similar size and shape. Petioles often bear persistent scales. Frond blades are 1- to 6-pinnate. Sori are borne abaxially along veins.

ANGIOSPERMS: WHITE TO CREAM FLOWERS

WHITE NEEDLE-FLOWER

Justicia longii (Siphonoglossa longiflora)
Acanthus Family (Acanthaceae)

Height: To 8".

Flowers: White, tubular; with 1 notched, smaller upper lip; 3-lobed, larger lower lip; brown-tipped stamens; flower to 2" long, ½" wide; clustered in leaf axils.

Leaves: Dark green, opposite, margins curled slightly upward; short-stalked, lance-shaped, to 2" long.

Blooms: April–October.

Elevation: 3,000 to 4,000'.

Habitat: Canyons and rocky slopes.

Comments: Flowers open in evening and close the following morning. Browsed by livestock and wild animals. Three species of *Justicia* in Arizona. Photograph taken in Molino Canyon, Santa Catalina Mountains, May 14.

BIGELOW BEARGRASS

Nolina bigelovii
Agave Family (Agavaceae)

Height: (Usually with a flower stalk) to 8'.

Flowers: White, tinged with green; to ⅛" long; very numerous. On upright flower stalk to 8' long; flowering upper half to two-thirds is branched, followed by thin, 3-lobed seed capsule; to ½" in diameter.

Leaves: Dark green, stiff and leathery; evergreen, rough-margined, narrow; to 4½' long, ¾" wide; in dense, basal cluster.

Blooms: June–July.

Elevation: 500 to 3,500'.

Comments: Erect, unbranched trunk. Occasionally poisonous to livestock. Four species of *Nolina* in Arizona. Photograph taken near Burro Creek and Wikieup, July 15. A very similar species, **Parry Nolina** (*Nolina parryi*), has sharply toothed leaf margins and slightly larger flowers and seed capsules.

BEARGRASS

Sacahuista
Nolina microcarpa
Agave Family (Agavaceae)

Height: Flower stalk to 8'.

Flowers: Creamy white, ⅛" wide, in dense plume-like, often bending or crooked, flowering cluster; to 3' long, followed by papery fruit capsule to ⅜" in diameter.

Leaves: Green, tough, grasslike; no marginal spines; loose fibers on margins at tips; to ½" wide, 4' long; in large, basal rosette.

Blooms: May–June.

Elevation: 3,000 to 6,500'.

Habitat: Rocky slopes and exposed areas on mountainsides.

Comments: Resembles a large, coarse grass. During drought, leaves are browsed by wildlife, but sheep and goats are occasionally poisoned. Mexicans use leaves for basketry. Native Americans used bud stalks for food. Four species of *Nolina* in Arizona. Photograph taken at Sedona, June 18.

ARIZONA YUCCA

Datil
Yucca baccata var. *brevifolia (Yucca arizonica)*
Agave Family (Agavaceae)

Height: Trunks to 8'. Another variety, **Banana Yucca** (*Yucca baccata* var. *baccata*) is nearly trunkless. Flower stalk to 5'.

Flowers: White, waxy, bell-shaped, with 6 yellow anthers and 6 petallike segments; to 3" long; on a stalk in long cluster; reddish buds, followed by large, fleshy, banana-like fruits, which grow to 5" long.

Leaves: Green and straight (variety *brevifolia*) or blue-green and slightly curved (variety *baccata*), broad, stiff, spine-tipped; white fibers on margins; to 3' long, 2" wide; in a basal rosette.

Blooms: April–July.

Elevation: 3,000 to 8,000' (variety *brevifolia* tending toward the lower end of the elevational range).

Habitat: Dry plains and slopes.

Comments: Has short flower stem scarcely taller than leaves. Nine species of *Yucca* in Arizona. Photograph taken near Willcox, April 22.

JOSHUA TREE

Yucca brevifolia
Agave Family (Agavaceae)

Height: To 30'.

Trunk: To 3' in diameter. Brown or gray, corky, and rough, deeply furrowed. Young trunks covered with dead leaves.

Flowers: Greenish white, waxy, bell-shaped; to 2½" long; in tight clusters on stalks to 1½" long at ends of branches; flowers do not open fully; followed by egg-shaped, greenish fruit about 4" long.

Leaves: Dark green, long and narrow, with pointed tip and toothed margins; to 14" long; clustered in dense rosettes at ends of branches.

Blooms: March–April.

Elevation: 2,000 to 3,500'.

Habitat: Rocky plains and hillsides.

Comments: Symbol of the Mojave Desert and the largest yucca, the Joshua tree was named by Mormon pioneers who likened its grotesque shape to the biblical Joshua lifting his arms in prayer. An evergreen, this narrow-leaved yucca can live between 100 and 300 years. It does not bloom every year, as flowering is governed by temperature and rainfall. Birds, woodrats, and a small species of night lizard make their homes in the Joshua tree. A healthy Joshua tree needs periods of low temperature to go into dormancy. Wood is used for splints and veneering. Nine species of *Yucca* in Arizona. The Joshua Tree Parkway northwest of Wickenburg is a good example of a Joshua "forest." Photographs taken on Joshua Tree Parkway, May 29 (tree) and March 11 (flowers).

SOAPTREE YUCCA

Palmilla
Yucca elata
Agave Family (Agavaceae)

Height: Trunks to over 20′ with 5′ flower stalks.

Flowers: Creamy white, bell-shaped; to 2″ long; in dense cluster on upright branch, followed by light brown, 3-celled, cylindrical seed capsule.

Leaves: Yellowish green, long and narrow, evergreen; with threadlike margins and a sharp spine at terminal end; to 2½′ long.

Blooms: May–July.

Elevation: 1,500 to 6,000′.

Habitat: Mesas, desert washes, sandy plains, and grasslands.

Comments: This yucca forms clumps. Arrangement of leaves channels moisture to plant's center. Native Americans use leaves for basket weaving. Flowers and buds were used as food. Roots, known as *amole*, used as substitute for soap. Nine species of *Yucca* in Arizona. Photograph taken in Sedona area, June 18.

SIERRA MADRE YUCCA

Mountain Yucca
Yucca madrensis (Yucca schottii)
Agave Family (Agavaceae)

Height: Flower stalk to 18′.

Flowers: White, waxy, bell-shaped, with 6 broad, pointed sepals; to 1½″ long; in short-stalked upright cluster, followed by green, fleshy, banana-like fruit growing to 5″ long, to 2″ in diameter; falling before winter.

Leaves: Bluish green; edges are reddish without teeth or threads; lance-shaped, flat, flexible, and leathery; sharp-pointed; to 2½′ long, 2″ wide.

Blooms: April–August.

Elevation: 4,000 to 7,000′.

Habitat: Hillsides and canyons.

Comments: Evergreen. Grows new asparagus-like flower stalk yearly. Native Americans ate buds, flowers, and young flower stalks. Leaf fibers made into mats, baskets, cloth, rope, and sandals. Prepared roots used as soap substitute. Nine species of *Yucca* in Arizona. Photograph taken in Madera Canyon, April 28.

NORTHERN WATER-PLANTAIN

Mud-Plantain
Alisma triviale
Water-Plantain Family (Alismataceae)

Height: To 3′.

Flowers: White, 3 rounded, notched petals; 3 green sepals showing between petals; 6 green-tipped stamens; flower to ⅜″ wide, in large, loose cluster.

Leaves: Dark green, leathery, broadly elliptical; basal, to 9″ long with stem, to 4″ long without stem; to 2″ wide.

Blooms: June–August.

Elevation: 4,000 to 8,000′.

Habitat: Shallow water and muddy areas.

Comments: Many-branched. Branches and branchlets arranged in whorls around stem. Three species of *Alisma* in Arizona. Photograph taken at Woodland Lake near Lakeside, July 6.

SMOOTH SUMAC

Scarlet Sumac
Rhus glabra
Cashew Family (Anacardiaceae)

Height: To 20′.

Trunk: To 4″ in diameter.

Bark: Brown; smooth or scaly.

Flowers: Whitish, 5 petals; to ⅛″ wide; in dense, spikelike terminal cluster to 8″ long; sexes usually on separate plants; followed by dark red, round fruit covered with short, sticky, red hairs; ⅛″ in diameter, in upright, terminal cluster when mature.

Leaves: Shiny green above, creamy beneath; turning bright red in fall; pinnately compound; to 12″ long; up to 31 lance-shaped, toothed leaflets, each to 4″ long.

Blooms: June–August.

Elevation: 5,000 to 7,000′.

Habitat: Roadsides and rich soil in ponderosa pine forests.

Comments: Birds feed on fruits; fruits and twigs browsed by deer. If chewed, fruit quenches thirst; a lemonadelike drink is made from fruit. Eight species of *Rhus* in Arizona. Photograph taken in Oak Creek Canyon, June 18.

LITTLE-LEAF SUMAC
Desert Sumac
Rhus microphylla
Cashew Family (Anacardiaceae)

Height: To 6'.

Flowers: Whitish, 5-petaled, with pinkish center; to 1/16" wide; in dense, roundish cluster to ¼" wide; followed by egg-shaped, sticky, hairy, reddish fruit, to ¼" long, in cluster.

Leaves: Green, pinnate, with 5 to 9 leaflets; airy, winged leaflet stem; to 1¾" long (in shady locations) but closer to ¾" in sunny locations, in clusters along stems.

Blooms: March–May.

Elevation: 3,500 to 6,000'.

Habitat: Dry slopes and mesas.

Comments: Sprawling, many-branched shrub with spine-tipped branches. Eight species of Rhus in Arizona. Photograph taken in Portal area on April 22.

SUGAR SUMAC
Chaparral Sumac
Rhus ovata
Cashew Family (Anacardiaceae)

Height: Shrub, or small tree to 15'.

Trunk: To 6" in diameter.

Bark: Grayish brown, shaggy, rough, scaly.

Flowers: Pinkish buds turning to cream color; 5 rounded petals; to ¼" wide; in crowded, terminal cluster to 2" long, followed by reddish, hairy fruits to ⅛" in diameter.

Leaves: Light green, shiny, thick, leatherlike; pinkish leaf stalk; oval, short-pointed at tip, rounded at base, curved upward at midvein; to 3¼" long.

Blooms: February–March.

Elevation: 3,000 to 5,000'.

Habitat: Mountain slopes in chaparral and in desert canyons.

Comments: Broadleaf evergreen. Fruit used by Native Americans as a sweetener. Eight species of Rhus in Arizona. Photograph taken in Superstition Mountains, March 25.

MEARNS SUMAC

New Mexican Evergreen Sumac
Rhus virens var. *choriophylla (Rhus choriophylla)*
Cashew Family (Anacardiaceae)

Height: To 7'.

Flowers: White, oval; to ⅛" long, 1⁄16" wide; in cluster to 2" long, 2" wide; followed by reddish to brown, hairy fruit, to ¼" long.

Leaves: Dark green, shiny, evergreen, leathery; to 4" long; pinnately compound with 3 to 5 oval leaflets, reddish purple stems, leaflets to 2½" long, 1¼" wide.

Blooms: July–September.

Elevation: 4,000 to 6,000' in southeastern Arizona.

Habitat: Rocky slopes and canyons

Comments: Eight species of *Rhus* in Arizona. Photograph taken at Chiricahua National Monument, April 25.

WATER PARSNIP

Berula erecta
Carrot Family (Apiaceae)

Height: To 3'.

Flowers: White, tiny, with 5 rounded, twisted, notched petals; 5 long pink-tipped stamens; in compound umbel to 1¾" wide with bracts around umbel and flower clusters; followed by nearly round fruit, to 1⁄16" long, in umbel.

Leaves: Light green, divided pinnately into segments; toothed or lobed, to 18" long; leaflets opposite, to 1½" long.

Blooms: June–August.

Elevation: 4,000 to 7,000'.

Habitat: Streamsides, in streams, and in other wet places.

Comments: Perennial. One species of *Berula* in Arizona. Photograph taken in vicinity of Lakeside, August 9.

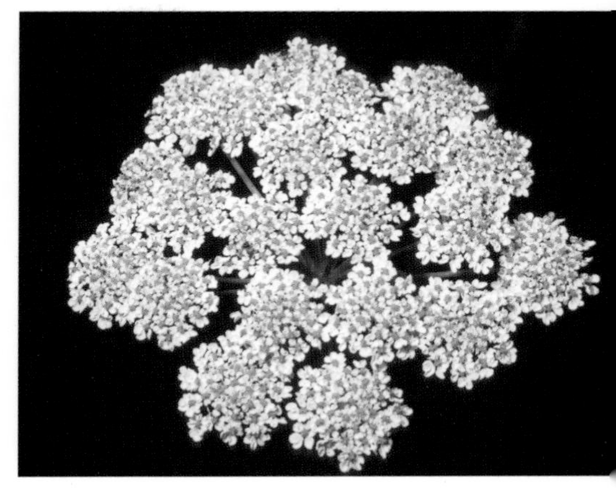

WATER HEMLOCK

Cicuta douglasii
Carrot Family (Apiaceae)

Height: To 7′.

Flowers: White, minute, in loose, flat-topped, terminal cluster to 5″ wide.

Leaves: Dark green, twice to 3 times pinnate, to 14″ long; lance-shaped, sharply toothed leaflets to 4″ long.

Blooms: July–September.

Elevation: 6,000 to 9,000′.

Habitat: Marshes, edges of streams, and low, wet areas.

Comments: Perennial herb. Roots and young growth are very poisonous to warm-blooded animals if ingested. Two species of *Cicuta* in Arizona. Photograph taken at Nelson Reservoir, August 3.

HEMLOCK-PARSLEY

Conioselinum scopulorum
Carrot Family (Apiaceae)

Height: To 4′.

Flowers: Greenish white, with 5 wavy petals, green pistil; to ⅛″ wide; in flat cluster or umbel to 4″ wide; followed by flattened fruits with winged ribs, to ¼″ long.

Leaves: Dark green, large, triangular-shaped; pinnately divided and cleft, clasping stem at base; to 9″ long.

Blooms: August–September.

Elevation: 6,000 to 9,500′.

Habitat: Moist spruce-fir forests.

Comments: Perennial herb. Attractive to flies. Two species of *Conioselinum* in Arizona. Photograph taken in vicinity of Hannagan Meadow, August 6.

POISON HEMLOCK

Conium maculatum
Carrot Family (Apiaceae)

Height: To 10', but usually less.

Flowers: White, 5-petaled; to ⅛" wide; in small cluster or umbel to ½" wide, grouped in compound umbel to 4" wide.

Leaves: Dark green, fernlike, very finely divided; triangular-shaped, to 2' long, 2' wide at widest part.

Blooms: May–August.

Elevation: 4,000 to 7,500'.

Habitat: Moist ground near streams, waste areas, and roadsides.

Comments: Biennial herb; many-branched. A native of Eurasia; now naturalized in U.S. Grayish green stems are hollow, grooved, and spotted or blotched with purple. All parts of plant contain very poisonous juices; an extract of this plant is what killed Socrates. Some children have used this plant's hollow stems as whistles. Mouthing these "instruments" can be fatal. One species of *Conium* in Arizona. Photograph taken in vicinity of Patagonia, May 10.

AMERICAN CARROT

Daucus pusillus
Carrot Family (Apiaceae)

Height: To 28".

Flowers: Whitish, tiny, long, with lacy bracts below flower cluster; cluster to 2" wide.

Leaves: Dark green, fernlike, lacy-lobed; to 3" long.

Blooms: March–May.

Elevation: Below 4,000'.

Habitat: Disturbed soil and roadsides.

Comments: Annual. A relative of the cultivated carrot; Native Americans ate roots raw and cooked. Two species of *Daucus* in Arizona. Photograph taken at Patagonia Lake State Park, April 26.

OSHA

Porter's Licorice-Root
Ligusticum porteri var. *porteri*
Carrot Family (Apiaceae)

Height: To 4'.

Flowers: White or pinkish, with 5 notched petals; to ⅛" wide; bractless; in wide, flat umbel to 3" wide; secondary umbels or umbellets to ¾" wide; followed by oblong fruit with narrow wings on ribs.

Leaves: Dark green, triangular, alternate, and fern-like; pinnate, much-divided, toothed; with base of stalk sheathing stem; to 12" long.

Blooms: June–August.

Elevation: 6,500 to 11,500'.

Habitat: Moist areas in mountains and in coniferous forests.

Comments: Perennial; hollow-stemmed. A forage plant. Roots used medicinally to treat numerous ailments. One species of *Ligusticum* in Arizona. Photograph taken in mountains above Greer, July 8.

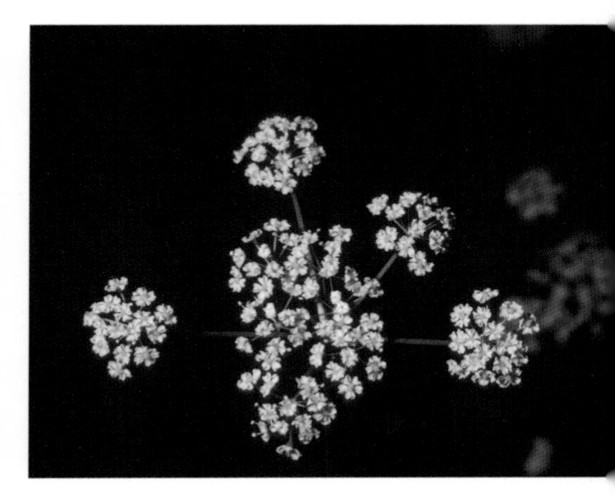

HOG FENNEL

Fendler's Cowbane
Oxypolis fendleri
Carrot Family (Apiaceae)

Height: To 2'.

Flowers: White, 5-petaled; to 1⁄16" wide; in loose, flat-topped umbel to 2" wide; followed by oblong to oval fruit with broad, thin, lateral wings, to ¼" long.

Leaves: Dark green, smooth, alternate, pinnate; to 5" long; bases of leaf stalks are expanded and sheath stem; up to 9 coarsely toothed, elliptical leaflets, each to 1½" long.

Blooms: July.

Elevation: 9,500 to 10,000'.

Habitat: Stream banks in partial shade.

Comments: Perennial herb. Unbranched, smooth stems. One species of *Oxypolis* in Arizona. Photograph taken in Mount Baldy Wilderness, July 8.

PARISH'S YAMPAH

Perideridia parishii
Carrot Family (Apiaceae)

Height: To 3'.

Flowers: White and lacy, with 5 tiny petals; to ⅛" wide; in small cluster to ½" wide; entire cluster or umbel to 2" wide.

Leaves: Dark green, only 1 or 2 per stem, to 3½" long; 1 to 3 leaflets with margins curved upward.

Blooms: July–September.

Elevation: 6,500 to 8,000'.

Habitat: Mountain meadows and moist pine forests.

Comments: Perennial. Its fleshy roots are edible, and were an important food source for Native Americans and pioneers. Raw roots have carrotlike flavor and can be ground into flour. Two species of *Perideridia* in Arizona. Photograph taken in vicinity of Woods Canyon Lake, August 2.

PALMER'S BLUESTAR

Amsonia palmeri (Amsonia hirtella var. *pogonosepala)*
Dogbane Family (Apocynaceae)

Height: To 3'.

Flowers: White to pale bluish gray, tubular, and starlike; 5 slightly twisted lobes; to ½" wide, ½" long; in terminal, branched cluster; followed by long, slender, cylindrical seed pod to 4" long; splitting into 2 sections when mature.

Leaves: Dark green, smooth to hairy, lance-shaped, with prominent midvein; to 3" long.

Blooms: March–April.

Elevation: 1,500 to 5,000'.

Habitat: Canyons and along streams.

Comments: Perennial herb. Stems contain a milky juice. Six species of *Amsonia* in Arizona. Photograph taken in Superstition Mountains, March 15.

CALIFORNIA FAN PALM
Desert Fan Palm
Washingtonia filifera
Palm Family (Arecaceae)

Height: To 60′ (to 30′ in Kofa Mountains).

Trunk: To 2′ in diameter.

Bark: Grayish brown, checkered, rough.

Flowers: White, fragrant, ⅜″ long; in branched, drooping clusters to 12′ long, followed by black, ½″-long oval fruits.

Leaves: Light green, leathery, fan-shaped, to 6′ long; 3 to 6′ broad with outer part consisting of narrowly folded segments, the edges of which have threadlike fibers. Leafstalks: thick, to 3″ wide, with hooked spines along edges.

Blooms: May–June.

Elevation: 2,500′.

Habitat: Canyons of desert mountains (only Kofa and Hieroglyphic Mountains in Arizona).

Comments: Largest native palm in this country. Dead brown leaves hang on tree, forming skirt covering trunk. Provides roosting sites for birds and bats. Fruits were eaten by Native Americans, who ground seeds for meal. One species of *Washingtonia* in Arizona. Photograph taken of an introduced specimen at Hassayampa Preserve, Wickenburg, February 27.

PINE-NEEDLE MILKWEED
Asclepias linaria
Milkweed Family (Asclepiadaceae)

Height: To 5′.

Flowers: Whitish, with pinkish buds; 5 untied petals; to ¼″ wide; with sort-horned hoods; flowers in terminal cluster, to 1″ wide; followed by smooth, shiny pod to 2″ long.

Leaves: Light green, linear, alternate, and soft; to 1½″ long; crowded all along stems.

Blooms: March–November.

Elevation: 1,500 to 6,000′.

Habitat: Mesas and dry, rocky slopes.

Comments: Twenty-nine species of *Asclepias* in Arizona. Photograph taken in vicinity of Tucson, November 10.

POISON MILKWEED

Horsetail Milkweed
Asclepias subverticillata
Milkweed Family (Asclepiadaceae)

Height: To 4'.

Flowers: White to cream-colored; to ½" wide; in round cluster to 1¼" wide; followed by a smooth, tapering seed pod to 4" long.

Leaves: Green, linear, to 5" long; in whorls at stem joints; tiny leaves in axils.

Blooms: May–September.

Elevation: 2,500 to 8,000'.

Habitat: Roadsides, sandy or rocky flats, and slopes.

Comments: Perennial herb. Each seed has a tuft of silky hairs. Stems have milky juice. Very poisonous to livestock. Queen and monarch butterfly caterpillars feed on the foliage of many milkweed species; ingested poisons are active in both caterpillars and adult butterflies, causing natural predators to avoid them. The foliage is fatal to other insects if eaten in large doses. Twenty-nine species of *Asclepias* in Arizona. Photograph taken near Show Low, July 22.

CLIMBING MILKWEED

Funastrum cynanchoides (Sarcostemma cynanchoides)
Milkweed Family (Asclepiadaceae)

Height: Vine with stems to 10' long.

Flowers: White and starlike, with 5 petals, 5 sepals; to ½" wide, in cluster to 4" wide; followed by smooth, plump, brownish pod to ⅝" wide, 4" long; containing seeds with silky hairs attached.

Leaves: Dark green, arrow-shaped, to 2½" long.

Blooms: May–September.

Elevation: 1,500 to 4,500'.

Habitat: Along streams, in washes, and on dry plains.

Comments: Perennial. Climbs on trees and shrubs. Has milky juice. Three species of *Funastrum* in Arizona. Photograph taken at Dead Horse Ranch State Park, September 9.

WESTERN YARROW

Milfoil
Achillea millefolium var. *occidentalis (Achillea millefolium* var. *lanulosa)*
Sunflower Family (Asteraceae)

Height: To 40".

Flowers: White (sometimes pinkish), to ⅛" wide; in flat-topped cluster of ray and disk flowers.

Leaves: Light green, finely dissected, fernlike; to 8" long at base.

Blooms: June–September.

Elevation: 5,500 to 11,500'.

Habitat: Fields, roadsides, clearings in pine forests, and waste ground.

Comments: Perennial herb. Native Americans and Spaniards used plant medicinally. Zuni use plant before fire ceremonies. Named for Achilles, said to have discovered healing powers of yarrow. One species of *Achillea* in Arizona. Photograph taken near Willow Springs Lake, July 21.

FRAGRANT SNAKEROOT

Ageratina herbacea (Eupatorium herbaceum)
Sunflower Family (Asteraceae)

Height: To 2'.

Flowers: White, rayless, with 5 starlike, pointed lobes; long white stamens; flower to 1⁄16" wide, flower head to ⅜" wide; in terminal cluster.

Leaves: Light green, heart-shaped, toothed, prominently veined; to 3" long (including stem), to 1¾" wide.

Blooms: June–October.

Elevation: 5,000 to 9,000'.

Habitat: Clearings in pine forests.

Comments: Six species of *Ageratina* in Arizona. Photograph taken in vicinity of Woods Canyon Lake, September 14.

CHEESEBUSH

Burrobush
Ambrosia salsola (Hymenoclea salsola)
Sunflower Family (Asteraceae)

Height: To 4'.

Flowers: Silvery white, tufted, in leaf axils of upper leaves; also terminal; to ⅜" wide; followed by fruit with silvery white wings.

Leaves: Dark green, very slender; lower leaves have 3 or more threadlike divisions; to 3" long.

Blooms: March–April.

Elevation: Below 4,000'.

Habitat: Arroyos, sandy washes, and rocky slopes.

Comments: Feathery branches. Has cheesy odor when foliage is crushed. Pollen can cause hay fever. Fourteen species of *Ambrosia* in Arizona. Photograph taken near Salome, March 28.

WESTERN PEARLY EVERLASTING

Anaphalis margaritacea
Sunflower Family (Asteraceae)

Height: To 3'.

Flowers: Minute, pearly white bracts surrounding yellowish center of disk flowers; flower head to ⅜" wide; in sprawling, terminal cluster.

Leaves: Shiny, dark green above, woolly white beneath; alternate, linear to linear-oblong, clasping stem, prominent midvein; to 5" long.

Blooms: July–October.

Elevation: 4,500 to 8,500'.

Habitat: Roadsides, open woodlands and forests, along streams, and in canyons.

Comments: Perennial herb; has woolly stems. Flowers last a long time, and dry well for floral arrangements. One species of *Anaphalis* in Arizona. Photograph taken along West Fork of Oak Creek Canyon, October 1.

SMALL-LEAF PUSSYTOES

Cat's Foot

Antennaria parvifolia

Sunflower Family (Asteraceae)

Height: To 6".

Flowers: Dirty white, tubular, minute; small heads in loose cluster; cluster to 1" wide.

Leaves: Gray, woolly, in basal rosette; spatula-shaped; about ¾" long, narrower on flower stalk.

Blooms: May–August.

Elevation: 5,000 to 12,000'.

Habitat: Open, sandy areas in coniferous forests.

Comments: Perennial herb. Native Americans use plant medicinally and for ceremonies. Five species of *Antennaria* in Arizona. Photograph taken near Greer, June 20.

KAIBAB PUSSYTOES

Antennaria rosulata

Sunflower Family (Asteraceae)

Height: To 1".

Flowers: Grayish white with brown stamens; minute; in cluster to ¼" wide. Flower heads single or up to 3, surrounded by leaves.

Leaves: Gray, woolly, spoon-shaped; to ½" long, in basal rosette.

Blooms: May–July.

Elevation: 5,500 to 11,000'.

Habitat: Ponderosa pine clearings and meadows.

Comments: Perennial herb. Five species of *Antennaria* in Arizona. Photograph taken near Ashurst Lake, June 1.

SAND SAGE
Artemisia filifolia
Sunflower Family (Asteraceae)

Height: To 5′.

Flowers: White to yellow, tiny, in terminal spikes to 8″ long; followed by tiny seeds.

Leaves: Grayish green, threadlike, covered with silvery hairs; occurring all along stems.

Blooms: August–November.

Elevation: 4,000 to 6,500′.

Habitat: Loose, sandy soil.

Comments: Many-branched shrub. A valuable browse plant. Used medicinally by Native Americans and pioneers. Fourteen species of *Artemisia* in Arizona. Photograph taken at Wupatki National Monument, September 8.

RAGWEED SAGEBRUSH
Bursage Mugwort
Artemisia franserioides
Sunflower Family (Asteraceae)

Height: To 2′ with flower stalk.

Flowers: Tiny, white to 3/16″ wide; in a long, slightly drooping, terminal cluster, to 6″ long.

Leaves: Light green above, grayish beneath, with fine hairs; bipinnate, alternate, very fragrant when rubbed; to 6″ long.

Blooms: August–September.

Elevation: 8,000 to 10,000′.

Habitat: Clearings in coniferous forests.

Comments: Leaves are very soft and velvety. Fourteen species of *Artemisia* in Arizona. Photograph taken at Lee Valley Reservoir in mountains above Greer, August 7.

YERBA-DE-PASMO

Baccharis pteronioides
Sunflower Family (Asteraceae)

Height: Shrub to 3′ tall, 3′ wide.

Flowers: Cream-colored, rayless, flower head to ¼″ wide; ¼″ long; heads on short, leafy branches arranged like a raceme.

Leaves: Dark green, sandpapery, sticky, lobed; to ⅝″ long; clustered along many-branched stems.

Blooms: April–September.

Elevation: 3,500 to 6,000′.

Habitat: Slopes and plains.

Comments: Spanish name, *yerba-de-pasmo*, means "chill weed." Infusions from leaves were supposedly used for chills. Browsed by livestock. Ten species of *Baccharis* in Arizona. Photograph taken near Prescott, May 27.

SEEP WILLOW

Batamote
Baccharis salicifolia

Sunflower Family (Asteraceae)

Height: To 12′.

Flowers: Creamy white, rayless, male and female flowers on separate plants; in clusters at ends of branches; followed (on female flowers) by seeds with silky tails.

Leaves: Dark green, shiny, waxy, sticky; lance-shaped, toothed; to 6″ long, ½″ wide.

Blooms: March–December.

Elevation: 2,000 to 5,500′.

Habitat: Along washes, streams, and seepage channels.

Comments: Not a true willow, but has willowlike growth. Leaves have a distinctive odor. Controls river and stream erosion by forming dense stands along watercourses. Acts as a "nurse" plant by protecting willow and cottonwood seedlings. Parts of plant used medicinally. Ten species of *Baccharis* in Arizona. Photograph taken at Granite Reef Dam area, March 1.

DESERT BROOM
Baccharis sarothroides
Sunflower Family (Asteraceae)

Height: Shrub to 10'.

Flowers: Whitish, rayless, small; male and female on separate plants (photo shows a male plant); to ⅜" long; in terminal clusters. Female plants develop white, silky, airborne seeds.

Leaves: Bright green, smooth, sticky; to 1½" long, ⅛" wide.

Blooms: September–February.

Elevation: 1,000 to 5,500'.

Habitat: Sandy washes, hillsides, along streams, and bottomlands.

Comments: Shrub with fast-growing, stiff stems. Important in erosion control. Certain Native Americans chewed stems to ease toothaches. Branches used as brooms by pioneers and Native Americans. Ten species of *Baccharis* in Arizona. Photograph taken at Stewart Dam area, October 18.

BRICKELLBUSH
Brickellia floribunda
Sunflower Family (Asteraceae)

Height: To 4½'.

Flowers: White to cream-colored, rayless; flower head to ⅜" wide, ½" long.

Leaves: Light green, heart-shaped; to 4" long (including stalk), 2" wide.

Blooms: September–October.

Elevation: 3,000 to 5,500'.

Habitat: Rich soil in canyons.

Comments: Entire plant very glandular, sticky, and has odd odor. More than 2 dozen species of *Brickellia* in Arizona. Photograph taken in vicinity of Christopher Creek, September 14.

LARGE-FLOWERED BRICKELLBUSH

Brickellia grandiflora (Eupatorium grandiflorum)
Sunflower Family (Asteraceae)

Height: To 3'.

Flowers: Whitish, upright to nodding, rayless; to ⅜" long; 20 to 40 flowers in a head, hanging at tips of stalks in various-sized clusters.

Leaves: Dark green, triangular, with toothed, prominent veins; to 4" long, smaller on upper stem.

Blooms: August–October.

Elevation: 5,000 to 9,000'.

Habitat: Rocky slopes and coniferous forests.

Comments: More than two dozen species of *Brickellia* in Arizona. Photograph taken at Sunset Crater National Monument, September 6.

WHITE TACKSTEM

Calycoseris wrightii
Sunflower Family (Asteraceae)

Height: To 12".

Flowers: White to ivory, with pinkish streaks or dots on back of petals; to 1½" wide.

Leaves: Grayish green, narrow; with narrow lobes near base of plant, sparser and smaller lobes toward ends of stems; to 3" long. Tiny stalked green glands on sepals, leaves and stems.

Blooms: March–May.

Elevation: 500 to 4,000'.

Habitat: Sandy soil of mesas, plains, and hillsides.

Comments: Annual. Named "tackstem" for its tack-shaped glands on stems. Two species of *Calycoseris* in Arizona. Photograph taken in Kofa Mountains, March 29.

FREMONT'S PINCUSHION
Desert Pincushion
Chaenactis fremontii
Sunflower Family (Asteraceae)

Height: To 16".

Flowers: White or pinkish, rayless; disk flowers forming pincushionlike flower head; to 1" wide.

Leaves: Green, divided into linear lobes; to 3" long.

Blooms: March–June.

Elevation: 1,000 to 3,500'.

Habitat: Plains and mesas.

Comments: Seven species of *Chaenactis* in Arizona. Photograph taken near Salome, March 28.

ESTEVE'S PINCUSHION
Chaenactis stevioides
Sunflower Family (Asteraceae)

Height: To 1'.

Flowers: White and rayless; disk flowers form pincushion with raylike outer disks, flower head to ¾" wide, ⅝" long.

Leaves: Grayish green, woolly, divided twice into many short, thick, very narrow segments; to 2" long.

Blooms: March–May.

Elevation: 1,000 to 6,500'.

Habitat: Dry mesas and plains.

Comments: Seven species of *Chaenactis* in Arizona. Photograph taken in vicinity of Kitt Peak, April 18.

BABY ASTER

Rose Heath
Chaetopappa ericoides (Leucelene ericoides)
Sunflower Family (Asteraceae)

Height: To 6".

Flowers: White rays, orange disks; to ⅝" wide; on tips of branches; fading to pale pink.

Leaves: Grayish green, finely haired, linear; to ¼" long, tightly adhering to the entire stem.

Blooms: March–September.

Elevation: 3,500 to 7,000'.

Habitat: Dry slopes and mesas.

Comments: One species of *Chaetopappa* in Arizona. Photograph taken at Portal, April 23.

CHAPARRAL FLEABANE

Erigeron oreophilus
Sunflower Family (Asteraceae)

Height: To 16".

Flowers: White, very narrow rays, yellow disk flowers; flower head to 1" wide.

Leaves: Grayish green, very hairy on both surfaces; alternate, 5- to 7-lobed, to 1" long.

Blooms: May–October.

Elevation: 4,500 to 9,500'.

Habitat: Oak woodlands and clearings in ponderosa forests.

Comments: Herb, with very hairy stem. More than two dozen species of *Erigeron* in Arizona. Photograph taken in vicinity of Willow Springs Lake, August 19.

WOOLLY DAISY

White Easterbonnets
Eriophyllum lanosum
Sunflower Family (Asteraceae)

Height: To 1½".

Flowers: White, woolly, with yellow centers; to ¼" wide; at ends of woolly stems.

Leaves: Grayish white, woolly, linear; to ¼" long.

Blooms: February–May.

Elevation: 1,000 to 3,000'.

Habitat: Dry, gravelly slopes and mesas.

Comments: Annual. Five species of *Eriophyllum* in Arizona. Photograph taken near Salome, March 28.

GUARDIOLA

Guardiola platyphylla
Sunflower Family (Asteraceae)

Height: To 4'.

Flowers: White, with 1 to 5 rays, 3 to 8 white disk flowers; flower to ½" wide; in loose, terminal clusters.

Leaves: Dark green, opposite, leathery; heart-shaped to oval; sharp-toothed, very short-stalked, prominent network; to 2" long.

Blooms: February–September.

Elevation: 3,000 to 5,000'.

Habitat: Rocky slopes and canyons.

Comments: Branching perennial. Mature stems are gray; immature stems, reddish. One species of *Guardiola* in Arizona. Photograph taken in vicinity of Kitt Peak, April 18.

TIDYTIPS
Layia glandulosa
Sunflower Family (Asteraceae)

Height: To 18".

Flowers: White ray flowers tipped with 3 equal teeth; yellow disk flowers; terminal; to 1½" wide.

Leaves: Green, narrow, with sticky hairs; basal leaves with 1 to 5 pairs of short lobes; to 3" long; stem leaves narrow and elliptical.

Blooms: February–April.

Elevation: Up to 5,000'.

Habitat: Desert washes, dry slopes, and mesas.

Comments: Annual. One species of *Layia* in Arizona. Photograph taken at Usery Mountain Recreation Area, March 1.

OXEYE DAISY
Field Daisy
Leucanthemum vulgare (Chrysanthemum leucanthemum)
Sunflower Family (Asteraceae)

Height: To 2'.

Flowers: White rays, yellow disks; to 2" wide; growing singly, on long stem.

Leaves: Dark green, coarsely toothed or pinnately lobed; basal, to 6" long; growing along stem to 3" long.

Blooms: June–October.

Elevation: Throughout Arizona.

Habitat: Roadsides and fields.

Comments: Introduced from Europe; now naturalized. Two species of *Leucanthemum* in Arizona. Photograph taken at Hannagan Meadow area, June 30.

BLACKFOOT DAISY

Melampodium leucanthum
Sunflower Family (Asteraceae)

Height: To 20".

Flowers: White ray flowers, 8 to 10, with purple veins; yellow disk flowers; to 1½" wide; numerous flower heads on rounded, shrublike plant.

Leaves: Ash gray, narrow, opposite; to 2" long.

Blooms: March–December.

Elevation: 2,000 to 5,000'.

Habitat: Dry, rocky slopes, desert grassland, and oak woodlands.

Comments: Perennial herb. Three species of *Melampodium* in Arizona. Photograph taken in Superstition Mountains, March 15.

MOHAVE DESERT STAR

Monoptilon bellioides
Sunflower Family (Asteraceae)

Height: To 2" high with tussocks to 10" wide.

Flowers: White to pinkish rays; yellow disk; to ¾" wide.

Leaves: Grayish green, narrow, stiffly haired; to 1" long.

Blooms: February–April.

Elevation: 200 to 3,500'.

Habitat: Sandy or rocky slopes, mesas, and desert flats.

Comments: Winter annual herb; in good years, often producing large patches of white on desert floor. Two species of *Monoptilon* in Arizona. Photograph taken at Organ Pipe Cactus National Monument, March 30.

GRAY'S FEVERFEW

Mariola
Parthenium confertum
Sunflower Family (Asteraceae)

Height: To 2'.

Flowers: White and buttonlike, with 5 tiny, cuplike ray flowers around outer margin; to ¼" wide; in clusters along upper stems.

Leaves: Grayish green, hairy, alternate; pinnately cleft with blunt, rounded lobes; to 3" long at base, smaller up along stem.

Blooms: April–October.

Elevation: 2,500 to 6,000'.

Habitat: Dry plains and mesas.

Comments: Has hairy stem. Sap contains rubber. This species common in southeastern Arizona. Two species of *Parthenium* in Arizona. Photograph taken near Portal, April 22.

EMORY'S ROCK DAISY

Perityle emoryi
Sunflower Family (Asteraceae)

Height: To 2'.

Flowers: White rays, yellow disks, to ½" wide.

Leaves: Dark green, finely haired, brittle, succulent; broadly triangular, with deeply toothed margins; to 1½" wide.

Blooms: February–May, and possibly to October.

Elevation: To 3,000'.

Habitat: Rocky desert slopes, cliffs, and washes.

Comments: Twelve species of *Perityle* in Arizona. Photograph taken at Cattail Cove State Park, February 25.

ARIZONA CUDWEED

Pseudognaphalium arizonicum (Gnaphalium arizonicum)
Sunflower Family (Asteraceae)

Height: To 2'.

Flowers: Whitish gray, rayless, tipped with yellowish brown; in slender heads; to ¼" long; to 47 per head, in cluster to ½" wide.

Leaves: Grayish, wooly, margins curled under; linear, alternate, pointing upward on stem; to 1" long.

Blooms: August–October.

Elevation: 5,000 to 7,500'.

Habitat: Pine forests.

Comments: Herb. Has woolly stems. Dried flower stalks last a long time. Ten species of *Pseudognaphalium* in Arizona. Photograph taken in Woods Canyon Lake area, September 14.

DESERT-CHICORY

Rafinesequia neomexicana
Sunflower Family (Asteraceae)

Height: To 20".

Flowers: White, ray flowers only; maroon stripes underneath; flower heads to 1½" wide.

Leaves: Dark green, narrowly lobed, to 6" long at base, smaller on upper stem.

Blooms: Mid-February–May.

Elevation: 200 to 3,000'.

Habitat: From deserts to mesas.

Comments: A weak-stemmed annual that usually grows among shrubs for support. Flower is similar to that of tackstem. Two species of *Rafinsequia* in Arizona. Photograph taken at Alamo Lake, February 26.

LEMMON'S CANDYLEAF

Stevia lemmoni
Sunflower Family (Asteraceae)

Height: To 3'.

Flowers: White, tubular, 5-lobed, with long stamens; flower to ⅛" wide, ½" long; in clusters to 2" wide.

Leaves: Grayish green, very hairy, rough; linear to elliptical, opposite, to 1½" long; in clusters along brownish stems.

Blooms: February–May.

Elevation: 2,500 to 5,500'.

Habitat: Rocky slopes and canyons.

Comments: Many-branched, mound-shaped shrub. Five species of *Stevia* in Arizona. Photograph taken in vicinity of Kitt Peak, April 18.

WHITE PRAIRIE ASTER

Symphyotrichum falcatum var. *commutatum (Aster commutatus)*
Sunflower Family (Asteraceae)

Height: To 2'.

Flowers: White rays, yellow disks; to ½" wide; numerous flower heads on branches.

Leaves: Grayish green, hairy, alternate, narrow; to ¼" long, growing all along stems.

Blooms: August–October.

Elevation: 5,000 to 8,000'.

Habitat: Roadsides and clearings in pine forests.

Comments: Perennial herb, with hairy stems. This species absorbs selenium from soil and may be toxic to livestock. More than two dozen species of *Symphyotrichum* in Arizona. Photograph taken at Upper Lake Mary, September 2.

STEMLESS DAISY

Townsendia exscapa
Sunflower Family (Asteraceae)

Height: To 2".

Flowers: Very narrow, white to pale pink rays; pinkish on undersides; yellow disk flowers; flower head to 2" wide; stemless; occurring singly, or in cluster nestled among rosette of leaves.

Leaves: Dark green, narrow; linear to spatula-shaped, hairy; to 2" long, in basal rosette.

Blooms: March–August.

Elevation: 4,500 to 7,000'.

Habitat: Mesas, hillsides, and clearings in ponderosa forests and oak woodlands.

Comments: Perennial. Seven species of *Townsendia* in Arizona. Photograph taken in Sharp Creek area northeast of Christopher Creek, April 22. (Much variation in flowers in this location, from all white disk flowers to pinkish, spoon-shaped rays on other specimens.)

TOWER DAISY

Townsendia formosa
Sunflower Family (Asteraceae)

Height: To 20".

Flowers: White, with pointed rays above, purplish tinge beneath; yellow disk flowers; daisylike, solitary; to 2½" wide; on tall, unbranched stem.

Leaves: Green, spatula-shaped, to 1¾" long; in basal rosette. Small, linear leaves along stem.

Blooms: June–September.

Elevation: 7,000 to 9,500'.

Habitat: Wet meadows and hillsides in moist coniferous forests.

Comments: Seven species of *Townsendia* in Arizona. Photograph taken south of Alpine, August 2.

DESERT ZINNIA

Zinnia acerosa
Sunflower Family (Asteraceae)

Height: To 10″.

Flowers: 4 or 6 white or light yellow ray flowers; yellow disk flowers, semi-drooping, to 1″ wide; on dense-growing, rounded clumps.

Leaves: Grayish green, very narrow, stiff; to 1″ long.

Blooms: March–October.

Elevation: 2,000 to 5,000′.

Habitat: Dry mesas and slopes.

Comments: Perennial. Three species of *Zinnia* in Arizona. Photograph taken near Why, March 30.

SOUTHERN CATALPA

Cigar-Tree
Catalpa bignonioides
Bignonia Family (Bignoniaceae)

Height: To 40′.

Trunk: To 2′ in diameter.

Bark: Brownish gray, scaly.

Flowers: White outside; white inside with 2 yellow stripes and spots and stripes of purplish brown; bell-shaped; 5 rounded, fringed lobes; to 1½″ long, 1½″ wide; in branched clusters to 9″ long; followed by dark brown, very narrow, cylindrical seed capsule to 14″ long.

Leaves: Dull green above, paler green and hairy beneath; heart-shaped, opposite, pointed at tip; to 10″ long, 7″ wide.

Blooms: June–July.

Elevation: Not available. Photograph taken at approximately 5,000′.

Habitat: Clearings and roadsides.

Comments: Fast-growing; has soft wood. Introduced to the state. One species of *Catalpa* in Arizona. Photograph taken in Oak Creek Canyon, Sedona, June 8.

NARROW-LEAVED POPCORN FLOWER

Cryptantha angustifolia
Forget-me-not Family (Boraginaceae)

Height: To 10".

Flowers: White, with 5 united petals; to ⅛" wide; in coiled cluster.

Leaves: Grayish, hairy, narrow; to 1½" long.

Blooms: February–June.

Elevation: Below 4,000'.

Habitat: Creosote bush desert in dry, sandy, or gravelly soil in western and southern Arizona.

Comments: A bristly plant. Thirty-nine species of *Cryptantha* in Arizona. Photograph taken north of Yuma, March 29.

BRISTLY HIDDENFLOWER

Cryptantha setosissima
Forget-me-not Family (Boraginaceae)

Height: To 3'.

Flowers: White, with 5 united petals, light yellow center; to ¼" wide; in bristly haired, coiled flower cluster.

Leaves: Grayish green, bristly haired, lance-shaped to linear-lobed; to 5" long.

Blooms: May–September.

Elevation: 6,000 to 8,500'.

Habitat: Pine belt.

Comments: Perennial, with bristly haired stems. Thirty-nine species of *Cryptantha* in Arizona. Photograph taken at North Rim of Grand Canyon National Park, June 25.

SWEET-SCENTED HELIOTROPE

Phlox Heliotrope
Heliotropium convolvulaceum
Forget-me-not Family (Boraginaceae)

Height: To 1′.

Flowers: White, fragrant, broadly funnel-shaped, with yellow "eye" in center; 5 lines of hairs beneath flower; to ¾" wide; in terminal clusters, or singly in leaf axils, or on stem between 2 leaves.

Leaves: Green, very hairy; short-stalked, oval (broadest below middle); to 1½" long.

Blooms: March–October.

Elevation: 4,500 to 6,000′.

Habitat: Roadsides and other dry, sandy areas.

Comments: Low, spreading herb with rigid hairs lying flat on stems and leaves. Five species of *Heliotropium* in Arizona. Photograph taken near St. Johns, August 4.

SALT HELIOTROPE

Heliotropium curassavicum
Forget-me-not Family (Boraginaceae)

Height: To 16".

Flowers: White to pinkish white, with yellowish center; funnel-shaped; 5 rounded lobes; to 3/16" wide; in paired, coiled flower cluster.

Leaves: Bluish green, thick, wavy, and fleshy; smooth, spatula-shaped, covered with bluish wax; to 1½" long.

Blooms: Most of the year.

Elevation: Below 6,000′.

Habitat: Moist, saline soil and dried ponds.

Comments: Pima Indians used powdered root to treat wounds. Quail feed on fruits. Five species of *Heliotropium* in Arizona. Photograph taken at Lyman Lake, June 28.

HEARTLEAVED BITTERCRESS

Cardamine cordifolia
Mustard Family (Brassicaceae)

Height: To 32".

Flowers: White, with 4 notched petals; to ⅝" wide, ¾" long; in cluster to 1½" wide; followed by long-stalked, upright, slender, slightly flattened seed pod, to 1½" long.

Leaves: Light green, heart-shaped, and shiny, with scalloped margins with hardened, rounded teeth; to 4" long.

Blooms: July–August.

Elevation: 9,000 to 11,000'.

Habitat: Stream banks, mountain streams, and wet alpine meadows.

Comments: Perennial. Three species of *Cardamine* in Arizona. Photograph taken in Mount Baldy Wilderness, August 13.

SPECTACLE POD

Dimorphocarpa wislizeni (Dithyrea wislizeni)
Mustard Family (Brassicaceae)

Height: To 2'.

Flowers: White, with 4 petals; to ½" long; in dense raceme; followed by a flat green, ½"-wide double pod, 2 round lobes resembling spectacles.

Leaves: Grayish, pinnately lobed; to 6" long (shorter on stem).

Blooms: February–October.

Elevation: 1,000 to 6,000'.

Habitat: Open areas of shady soil in deserts and grasslands.

Comments: Two species of *Dimorphocarpa* in Arizona. Photograph taken near Gila Bend, March 29.

DRYOPETALON

Dryopetalon runcinatum

Mustard Family (Brassicaceae)

Height: To 25"

Flowers: Bright white, with 4 petals; pinnately cleft into 4 to 9 lobes; to 3⁄8" wide; in terminal cluster, followed by long, very narrow, erect seed pod to 2½" long.

Leaves: Dark green, pinnately divided or lobed; to 6" long at base, shorter on upper stems.

Blooms: February–mid-May.

Elevation: 2,000 to 7,000'.

Habitat: Moist rock crevices in canyons.

Comments: Annual. One or more stems. One species of *Dryopetalon* in Arizona. Photograph taken in north-facing rock crevice in Chiricahua National Monument, May 7.

WESTERN PEPPERGRASS

Lepidium montanum

Mustard Family (Brassicaceae)

Height: To 2'.

Flowers: White and minute, with 4 petals; to 1⁄8" wide; in short, dense raceme; on slender branch, followed by oval pod with tiny notch at tip.

Leaves: Green, pinnately lobed at base; to 3" long; narrow on stem, to ½" long.

Blooms: April–September.

Elevation: 3,000 to 7,500'.

Habitat: Roadsides, fields, and other open areas.

Comments: Many-branched and shrublike; often grows in loose mounds. Sixteen species of *Lepidium* in Arizona. Photograph taken at Heber, August 4.

THURBER'S PEPPERGRASS
Lepidium thurberi
Mustard Family (Brassicaceae)

Height: To 2'.

Flowers: Pure white, with 4 petals, 6 stamens; ⅛" wide; in long, dense, terminal cluster; followed by elliptical or roundish pod with small notch.

Leaves: Light green, slightly haired or nearly hairless, pinnately divided into narrow lobes; to 2½" long.

Blooms: February–November.

Elevation: Below 5,000'.

Habitat: Roadsides and fields.

Comments: A conspicuous white roadside flower. Sixteen species of *Lepidium* in Arizona. Photograph taken near Portal, April 22.

WHITE WATERCRESS
Nasturtium officinale (Rorippa nasturtium-aquaticum)
Mustard Family (Brassicaceae)

Height: To 16", but usually grows horizontally.

Flowers: White, with 4 rounded petals; yellowish at base; to ¼" wide; in rounded, terminal cluster; followed by slender, curved, upward-pointing, smooth, shiny pod to ½" long, on ½"-long stem.

Leaves: Dark green, succulent; pinnately compound, alternate; to 6" long; 3 to 11 smooth, wavy leaflets, each to ½" long (terminal leaflet is the longest).

Blooms: April–August.

Elevation: 1,500 to 8,500'.

Habitat: Cool water of ponds, brooks, springs, and along mountain streams.

Comments: Perennial herb; native of Europe, now naturalized in North America. Floats or lies in water or mud. Succulent, reddish stems which root at nodes. Leaves have a peppery taste. One species of *Nasturtium* in Arizona. Photograph taken near Greer, July 7.

WHITE BLADDERPOD

Physaria purpurea (Lesquerella purpurea)
Mustard Family (Brassicaceae)

Height: To 20".

Flowers: White, streaked with purple, and fading to purplish; to ⅜" wide; in loose, open cluster; followed by globular fruit to ¼" long on ½"-long stem.

Leaves: Silvery green, hairy, narrowing toward base; slightly toothed; to 3" long; graduating upward on stem.

Blooms: January–May.

Elevation: 1,500 to 5,000'.

Habitat: Along washes, desert flats, and in the shade of bushes on mesas.

Comments: Eleven species of *Physaria* in Arizona. Photograph taken in Superstition Mountains, February 4.

SILVER BELLS

Twist Flower
Streptanthus carinatus ssp. *arizonicus (Streptanthus arizonicus)*
Mustard Family (Brassicaceae)

Height: To 3½' tall.

Flowers: White to cream, goblet-shaped, on short stems along erect branches; 4 petals; to ½" long; followed by slender, flat, seed pod to 3" long.

Leaves: Grayish green, rubbery, elongated, and triangular-shaped; base lobes projecting beyond stem; to 8" long.

Blooms: January–April.

Elevation: 1,500 to 4,500'.

Habitat: Desert washes or flats to open juniper-pinyon woodlands.

Comments: Two species of *Streptanthus* in Arizona. Photograph taken in Tucson area, March 31.

© MATT LAVIN, FLICKR.COM

WILD CANDYTUFT

Thlaspi montanum
Mustard Family (Brassicaceae)

Height: To 1', but usually much less.

Flowers: White, 4-petaled, with yellow-tipped stamens; to ⅜" long, ⅜" wide; in terminal cluster to 1¼" wide.

Leaves: Green, arrow-shaped, succulent, alternate, clasping stem; to ½" long; basal leaves are oval-shaped, toothed; to ¾" long.

Blooms: February–August.

Elevation: 4,000 to 12,000'.

Habitat: Mainly coniferous forests.

Comments: Perennial. Unbranched stem. Two species of *Thlaspi* in Arizona. Photograph taken in vicinity of Willow Springs Lake, April 22.

ELEPHANT TREE

Torote
Bursera microphylla
Torch Wood Family (Burseraceae)

Height: To 20'.

Trunk: To 1' in diameter.

Bark: Whitish to gray, papery, peeling.

Flowers: Whitish, less than ¼" long; followed by red, 3-angled, ¼"-long, very aromatic fruit.

Leaves: Green, pinnately compound, aromatic; to 2" long; with 10 to 30 narrow leaflets each to ¼" long.

Blooms: July.

Elevation: 1,000 to 2,500'.

Habitat: Rocky slopes of arid desert mountains.

Comments: Very stout tree whose tapering branches resemble an elephant's trunk and legs. Branches are very sensitive to frost, but roots will produce new growth. Leaves and stems produce copal, a resin once used as incense. Two species of *Bursera* in Arizona. Photograph taken at Organ Pipe Cactus National Monument, October 23.

HOP

Humulus lupulus var. *neomexicana (Humulus americanus)*
Hemp Family (Cannabaceae)

Height: Long, twining vine that climbs over rocks and up into trees.

Flowers: Cream-colored, small, to ⅛" wide, in loose clusters in leaf axils; followed by drooping cluster of overlapping bracts, to 1¼" long, to 1" wide.

Leaves: Dark green above, lighter with prickly veins beneath; 3- to 7-lobed, very rough, sharply toothed; to 10" long, 7" wide.

Blooms: July–August.

Elevation: 5,500 to 9,500'.

Habitat: Coniferous forests, rocky slopes, and stream banks.

Comments: Perennial. Stems are rough; main stem is candy cane–striped. One species of *Humulus* in Arizona. Photograph taken at Luna Lake, August 5.

UTAH HONEYSUCKLE

Lonicera utahensis
Honeysuckle Family (Caprifoliaceae)

Height: Shrub to 5'.

Flowers: Creamy white, trumpet-shaped, long-stamened, 5-lobed; to ¾" long; paired, on thin, ½"-long flower stem; followed by twin, orangish yellow to red, oval berries.

Leaves: Pale green above, lighter green and powdery beneath; broadly elliptical; to 1½" long.

Blooms: June–July.

Elevation: 8,000 to 11,000'.

Habitat: Openings in coniferous forests.

Comments: Berries are poisonous to humans but eaten by birds and other wildlife. Twelve species of *Lonicera* in Arizona. Photograph taken in Greer area, July 3.

BLUEBERRY ELDER

Blue Elderberry
Sambucus nigra ssp. *cerulean (Sambucus glauca)*
Honeysuckle Family (Caprifoliaceae)

Height: Shrub, or small tree to 20′.

Trunk: To 1′ in diameter.

Bark: Brown or gray, furrowed.

Flowers: Creamy white; to ¼″ wide; in dense, flat-topped cluster to 8″ wide; followed by loose cluster of dark blue, ¼″-diameter berries covered with a powdery coating.

Leaves: Dark green, pinnately compound, evenly toothed, to 8″ long; with 5 to 9 lance-shaped leaflets, to 4″ long.

Blooms: July–August.

Elevation: 6,500 to 9,500′.

Habitat: Openings in moist coniferous forests and meadows.

Comments: Forms clumps. Foliage browsed by livestock and deer. Berries attractive to birds; edible and used for wine, jelly, and pies. Bark used for fever medication. Two species of *Sambucus* in Arizona. Photograph taken south of Alpine, July 23.

MEXICAN ELDER

Common Elderberry
Sambucus nigra ssp. *canadensis (Sambucus mexicana)*
Honeysuckle Family (Caprifoliaceae)

Height: To 30′.

Trunk: To 18″ in diameter.

Bark: Light brown to gray with long, narrow, scaly ridges.

Flowers: Creamy white, fragrant; to ¼″ wide; in flat-topped cluster to 8″ wide; followed by dark blue to blackish fruits (with a whitish powdery coating). Fruits are sweet, juicy, edible, ¼″ in diameter, and grow in a cluster.

Leaves: Green above, paler green beneath, pinnately compound, to 14″ long; with 3 or 5 leaflets, elliptical or oval, finely saw-toothed, thick, leathery, to 3″ long, 1½″ wide.

Blooms: March–June.

Elevation: 1,000 to 4,000′.

Habitat: Along streams and rivers in woodlands, deserts, and desert grasslands.

Comments: Evergreen, but deciduous during long droughts. One of the largest of the native elders. Grows at lower elevations than any elder in Arizona. Fruits eaten by birds; used in pies and jellies; Native Americans dried fruits for future uses. Two species of *Sambucus* in Arizona. Photograph taken at Patagonia, April 27.

REDBERRIED ELDER

Red Elderberry
Sambucus racemosa var. *racemosa (Sambucus microbotrys)*
Honeysuckle Family (Caprifoliaceae)

Height: To 5′.

Flowers: White to cream-colored, fragrant, 5-petaled; to ½″ wide; in pyramidal cluster; followed by cluster of shiny, bright red, roundish berries to ¼″ in diameter.

Leaves: Dark green, pinnate, to 10″ long; 5 to 7 leaflets, each folded upward slightly lengthwise, toothed.

Blooms: June–July.

Elevation: 7,500 to 10,000′.

Habitat: Moist forests.

Comments: Berries eaten by birds. Foliage browsed by deer and livestock. Reportedly parts of this plant and its berries are poisonous if mouthed or eaten raw. Two species of *Sambucus* in Arizona. Photograph taken at Lee Valley Reservoir on July.

FENDLER'S SANDWORT

Arenaria fendleri (Eremogone fendleri)
Pink Family (Caryophyllaceae)

Height: To 10″.

Flowers: White, starlike, light green filaments; 5 petals; bright pink anthers; to ⅜″ wide; in open, branched cluster.

Leaves: Dark green, threadlike, sharply pointed; to 2½″ long.

Blooms: April–September.

Elevation: 4,000 to 12,000′.

Habitat: Clearings in ponderosa pine and mixed conifer forests.

Comments: Perennial herb. Three species of *Arenaria* in Arizona. Photograph taken at Upper Lake Mary, September 6.

SANDWORT

Arenaria lanuginosa ssp. *saxosa*
Pink Family (Caryophyllaceae)

Height: To 5".

Flowers: White and starlike, with pink anthers; to ½" wide.

Leaves: Dark green, opposite, elliptical to lance-shaped; to ½" long.

Blooms: May–September.

Elevation: 7,000 to 12,000'.

Habitat: Coniferous forests.

Comments: Grows in compact mounds. Three species of *Arenaria* in Arizona. Photograph taken near Greer, June 20.

MOUSE-EAR CHICKWEED

Cerastium fontanum ssp. *vulgare (Cerastium vulgatum)*
Pink Family (Caryopyhllaceae)

Height: To 16".

Flowers: White, greenish in center, with 5 deeply notched, 2-lobed petals, 5 yellow stamens; to ½" wide; in small, loose cluster at top of stalk.

Leaves: Yellowish-green, very hairy, opposite, oval to oblong; stalkless; to ¾" long.

Blooms: May–September.

Elevation: 2,500 to 8,000'.

Habitat: Fields and roadsides.

Comments: Perennial herb; a common weed introduced from Europe. Sticky hairs on stem. Boiled leaves can be eaten. Nine species of *Cerastium* in Arizona. Photograph taken in Greer area, July 4.

STARWORT
Chickweed
Pseudostellaria jamesiana (Stellaria jamesiana)
Pink Family (Caryophyllaceae)

Height: To 1'.

Flowers: White; 5 triangular petals, each with a V-shaped notch; to ½" wide; in loose clusters.

Leaves: Dark green, opposite, hairy, lance-shaped; deep center vein; clasping stem; to 4" long.

Blooms: April–July.

Elevation: 7,000 to 8,500'.

Habitat: Moist coniferous forests and mountain meadows.

Comments: Sticky, weak stems. One species of *Pseudostellaria* in Arizona. Photograph taken at North Rim of Grand Canyon National Park, June 25.

CHICKWEED
Starwort
Stellaria longipes ssp. *longipes*
Pink Family (Caryophyllaceae)

Height: To 16".

Flowers: White, 5 deeply cleft, pointed petals longer than the 5 sepals; black-tipped stamens; flower to ½" wide; 1 to 3 on long, erect stalk in leaf axil.

Leaves: Dark green, opposite, shiny, ascending, and stemless; linear to lance-shaped; sharp-pointed at tip to ¾" long.

Blooms: May–August.

Elevation: 8,500 to 10,000'.

Habitat: Wet meadows and moist spruce-fir forests.

Comments: Perennial herb. Six species of *Stellaria* in Arizona. Photograph taken in mountains above Greer, July 8.

SANDPAPER BUSH

Mortonia scabrella (Mortonia sempervirens ssp. *scabrella)*
Bittersweet Family (Celastraceae)

Height: To 4'.

Flowers: White and small, with 5 petals to ¼" wide; in narrow cluster to 3" long.

Leaves: Yellowish green, lighter green on margins; alternate; pointing upward on stems; elliptical, curved slightly inward; rough; to ⅜" wide, ½" long; crowded along stems in spiral arrangement, progressively smaller toward tips of branches.

Blooms: March–September.

Elevation: 3,000 to 5,500'.

Habitat: Mesas and dry plains.

Comments: Many stiff, erect stems. One species of *Mortonia* in Arizona. Photograph taken in vicinity of Tucson, November 12.

© MAX LICHER

RUSSIAN THISTLE

Tumbleweed
Salsola tragus
Goosefoot Family (Chenopodiaceae)

Height: To 4'.

Flowers: Whitish, minute, without petals; growing at base of leaves on upper branches; followed by the drying and enlargement of the 5 flower parts that cover tiny fruit.

Leaves: Grayish green, fleshy on young plants; to 2" long; replaced by bractlike, small, awl-shaped leaves ending in spines.

Blooms: May–October.

Elevation: 150 to 7,000'.

Habitat: Roadsides, overgrazed range, and disturbed soil.

Comments: Not a true thistle. This annual, a native of Russia, was accidentally brought to South Dakota in the 1870s in a shipment of flax seed. Multiple-branching, the often reddish stems form a large, prickly, bushy ball. When dry, the plant breaks off at ground level, allowing winds to roll tumbleweed, scattering thousands of seeds as it moves along. Three species of *Salsola* in Arizona. Photograph taken at Wupatki National Monument, September 8.

WESTERN CLAMMYWEED

Polanisia dodecandra ssp. *trachysperma (Polanisia trachysperma)*

Cleome Family (Cleomaceae)

Height: To 3′.

Flowers: White to cream, with 4 petals and 6 to 20 long, pink to purple stamens of varying lengths; to ¾″ wide; in terminal clusters on many branches, followed by erect, cylindrical pod, to 3″ long. Flowers and fruits are present at the same time.

Leaves: Dark green, clammy-feeling, finely haired; 3 elliptical or broadly lance-shaped leaflets, each to 1½″ long.

Blooms: May–October.

Elevation: 1,000 to 6,500′.

Habitat: Sandy washes.

Comments: Annual. Foliage gives off objectionable odor, especially when handled. Stems are sticky and hairy. A favorite of bees and butterflies. One species of *Polanisia* in Arizona. Photograph taken at Saguaro Lake, May 20.

FIELD BINDWEED

Convolvulus arvensis

Morning Glory Family (Convolvulaceae)

Height: Trailing vine to 4′.

Flowers: White or pinkish, funnel-shaped; to 1″ wide; growing on one side of stalk; single; in leaf axil.

Leaves: Dark green, variable, arrow-shaped to triangular; growing on one side of stalk; to 2″ long.

Blooms: May–September.

Elevation: Throughout Arizona.

Habitat: Roadsides, fields, and lots.

Comments: Perennial herb, from Europe; now naturalized. A troublesome, deep-rooted weed. Source of a blood-clotting material. Two species of *Convolvulus* in Arizona. Photograph taken at Nelson Reservoir, August 3.

SILVER MORNING GLORY

Evolvulus sericeus
Morning Glory Family (Convolvulaceae)

Height: Prostrate, spreading; stems to 6" long.

Flowers: White or bluish (depending on variety); with funnel spreading into flattened disk; to ½" wide.

Leaves: Dark green, edged in silver above, gray beneath; tightly folded together; narrowly linear, pointed at both ends; to ½" long.

Blooms: May–September.

Elevation: 3,500 to 5,500'.

Habitat: Dry mesas and plains.

Comments: Sun-loving. Four species of *Evolvulus* in Arizona. Photograph taken at Lynx Creek Ruins, Prescott, May 27.

RED-OSIER DOGWOOD

Kinnikinnick
Cornus sericea ssp. *sericea (Cornus stolonifera)*
Dogwood Family (Cornaceae)

Height: Normally a shrub to 8'; in rare instances, tree-sized.

Bark: Gray to brown, smooth or furrowed.

Trunk: to 3" in diameter.

Flowers: Creamy white, with 4 petals; less than ¼" wide; in flat-topped cluster at tip of branch to 2½" wide; followed by bluish white, ¼" berrylike fruits.

Leaves: Dark green tinged with pink above; pale green or whitish beneath; oval to elliptical; to 5" long.

Blooms: May–July.

Elevation: 5,000 to 9,000'.

Habitat: Moist locations, along streams and in canyons, in ponderosa pine and Douglas fir forests.

Comments: Has reddish twigs and branches. Spreads by underground, prostrate stems, often forming very large clumps. Controls erosion on banks of streams. Stems are flexible and used for making baskets. One species of *Cornus* in Arizona. Photograph taken in vicinity of Mormon Lake, June 2.

RAGGED ROCK FLOWER

Crossosoma bigelovii
Crossosoma Family (Crossosomataceae)

Height: Straggly shrub to 6'.

Flowers: White, with 5 petals, numerous stamens; very fragrant; to 2" wide.

Leaves: Bluish green, alternate, thick, smooth; somewhat oval; to ¾" long, ⅜" wide.

Blooms: February–May.

Elevation: 1,500 to 4,000'.

Habitat: Dry, rocky slopes and canyons.

Comments: A rough-barked shrub. One species of *Crossosoma* in Arizona. Photograph taken in Superstition Mountains, February 4.

WILD CUCUMBER

Gila Manroot
Marah gilensis
Gourd Family (Cucurbitaceae)

Height: Long vine climbing over shrubs and small trees.

Flowers: White to cream, star-shaped, to ⅜" wide; followed by round, green, fleshy fruit with stout, smooth spines, to 2" in diameter.

Leaves: Dark green, pointy or rounded lobes, to 3" wide.

Blooms: March–April.

Elevation: Below 5,000'.

Habitat: Thickets along washes and streams.

Comments: Perennial. Very large tuberlike root. One species of *Marah* in Arizona. Photograph taken at Bartlett Dam area, March 24.

PRETTY DODDER

Cuscuta indecora
Dodder Family (Cuscutaceae)

Height: Twining, matted mass of yellowish stems ranging from several inches to several feet wide.

Flowers: Cream-colored, tiny, fleshy, and tubular; to ¼" long; in small clusters.

Leaves: None.

Blooms: July–August.

Elevation: Not available. Photograph taken at 3,000'.

Habitat: Roadsides, canyons, and slopes.

Comments: Annual. A matted mass of yellowish stems. Parasitic; rootless. Seeds germinate in soil. Seedlings break contact with ground and twine about host plant, using suckers to absorb water and nutrients. Often spreads viral plant diseases. This species found on wide variety of shrubs and trees. Sixteen species of *Cuscuta* in Arizona. Photograph taken on Apache Trail, March 23.

ARIZONA MADRONE

Arbutus arizonica
Heather Family (Ericaceae)

Height: To 40'.

Trunk: To 1½' in diameter.

Bark: Light gray, in squarish plates.

Flowers: White to pinkish, urn-shaped; to ¼" long; in loose, terminal cluster to 2½" long; followed by a cluster of orange-red, warty, berrylike fruit, ⅜" in diameter.

Leaves: Shiny, light green above, paler beneath; lance-shaped, leathery, with reddish leaf stems; to 3" long, 1" wide.

Blooms: April–September (usually June).

Elevation: 4,000 to 8,000'.

Habitat: Oak woodlands in mountains of southeastern Arizona.

Comments: Evergreen tree with compact, rounded crown. Related to manzanita, except reddish bark occurs only on smaller branches. One species of *Arbutus* in Arizona. Photograph taken at Cave Creek, Portal, April 22.

POINTLEAF MANZANITA
Arctostaphylos pungens
Heather Family (Ericaceae)

Height: To 6'.

Flowers: White to pink, nodding, and bell-shaped; to ¼" long, in clusters at tips of branches; followed by reddish brown, berrylike fruit, to ¼" in diameter (resembling a miniature apple).

Leaves: Green to bluish green, thick, leathery, elliptical; pointed at tip and base; to 1½" long.

Blooms: March–May.

Elevation: 4,000 to 8,000'.

Habitat: Chaparral and dry hillsides and in ponderosa pine belt.

Comments: Evergreen, with smooth, red bark and crooked branches. Often grows in dense thickets, preventing erosion. Rapidly reseeds or grows from root sprouts in burned areas. Leaves twist on stalks to a vertical position to prevent excess evaporation during drought periods. Infrequently browsed, though berries are eaten by rodents, bears, and birds. Flowers attract hummingbirds. Native Americans use berries for food and for making a beverage. Jelly is made from unripened fruits. *Manzanita* is Spanish for "little apple." Four species of *Arctostaphylos* in Arizona. Photograph in flower taken north of Superior, April 20; in fruit, September 2. **Pringle Manzanita** (*Arctostaphylos pringlei* ssp. *pringlei*) is a tall shrub with rounded leaves, and is frequently found in chaparral with pointleaf manzanita. It blooms from April to June. The **Greenleaf Manzanita** (*Arctostaphylos patula*), found at the North Rim of the Grand Canyon, is a low shrub to 3' tall, with bright green, nearly oval leaves.

WOOD NYMPH
Moneses uniflora
Heather Family (Ericaceae)

Height: Flower stalk to 5".

Flowers: White or very pale pink, waxy; 5 rounded, spreading petals with crinkly margins; 10 golden stamens with swollen bases; thick, green stigma with 5 pointed lobes; flower to ¾" wide, solitary and downward-facing on curve tip of stem.

Leaves: Dark green, roundish, finely toothed, thick; in basal cluster; to ¾" long, ⅝" wide.

Blooms: July–August.

Elevation: 9,500 to 11,500'.

Habitat: Moist, cool spruce-fir forests.

Comments: Perennial herb. Evergreen. One species of *Moneses* in Arizona. Photograph taken in mountains above Greer, July 8.

SIDE-BELLS PYROLA

Orthila secunda (Pyrola secunda)

Heather Family (Ericaceae)

Height: Flower stalk to 6".

Flowers: White to greenish white, bell-shaped; to ¼" long, arranged on only one side of flower stem.

Leaves: Dark green, shiny, basal; oval to elliptical; fine toothed; to 2½" long.

Blooms: July–August.

Elevation: 7,000 to 9,500'.

Habitat: Moist coniferous forests.

Comments: Herbaceous perennial; an evergreen. One species of *Orthila* in Arizona. Photograph taken near Mexican Hay Lake, July 2.

SHORTLEAF WINTERGREEN

Pyrola chlorantha (Pyrola virens)

Heather Family (Ericaceae)

Height: Flower stalk to 12" long.

Flowers: White to greenish white, bell-shaped, drooping; 5-petaled, with large style extending below petals; to ½" wide; occurring along leafless flower stalk.

Leaves: Dark green, roundish, basal; leaf blade to 1½" wide, 1½" long; to 3" long, including stem.

Blooms: July–August.

Elevation: 6,500 to 10,000'.

Habitat: Rich soil of coniferous forests.

Comments: Perennial herb; evergreen. Native Americans use plant medicinally and to make paint for ceremonials. Five species of *Pyrola* in Arizona. Photograph taken near Willow Springs Lake, July 6.

WHITEVEIN WINTERGREEN

Pyrola picta
Heather Family (Ericaceae)

Height: Flower stalk to 8".

Flowers: Greenish white or cream-colored, globe-shaped, waxy, nodding, 5-petaled; style turned to one side, flower to ¼" long, to ½" wide, hanging in terminal raceme on pinkish stem.

Leaves: Dark green with white or pinkish white veins above and pinkish below; shiny, oval to elliptical, with reddish stems; to 3" long, 1½" wide; in basal rosette.

Blooms: July–August.

Elevation: 8,000 to 9,500'.

Habitat: Coniferous forests.

Comments: Perennial herb. Five species of *Pyrola* in Arizona. Photograph taken in mountains above Greer, August 9.

RATTLESNAKE WEED

White Margin Spurge
Euphorbia albomarginata (Chamaesyce albomarginata)
Spurge Family (Euphorbiaceae)

Height: Creeper, ½" high, with stems to 10" long.

Flowers: Tiny, white, flowerlike cups, to ⅛" wide; lacking sepals and petals; maroon pad at base of each cup, containing many simple flowers.

Leaves: Green, round or oblong; smooth, often edged with white; to ⅜" long.

Blooms: February–October.

Elevation: 1,000 to 6,000'.

Habitat: Open areas in grasslands, disturbed areas, roadsides, pinyon-juniper woodlands, and clearings in ponderosa forests.

Comments: Perennial herb. New plants started when roots form at stem joints. Also reproduces by seed. Its milky sap may irritate skin on contact. At one time people believed plant to be an important snakebite remedy. More than 3 dozen species of *Euphorbia* in Arizona. Photograph taken at Pine, September 2.

WHITE-BALL ACACIA

Fern Acacia
Acacia angustissima (Acaciella angustissima)
Pea Family (Fabaceae)

Height: To 3'.

Flowers: White, at times tinged with pink, with numerous stamens crowded into a ball-shaped head, to ½" in diameter; in clusters on stems in leaf axils and in elongated, terminal cluster; followed by brown, flattened pod (at maturity); to 3" long.

Leaves: Green to bluish green, bipinnate; first leaflets to 14 pairs, secondary leaflets to 33 pairs; linear-oblong, leaflet to ⅛" long, 1⁄16" wide.

Blooms: May–September.

Elevation: 3,000 to 6,500'.

Habitat: Roadsides and dry slopes commonly in chaparral areas.

Comments: No prickles or spines on branches; stems are deeply grooved and very hairy. Roots are perennial, but plant dies back to ground after hard frost. Browsed by horses and cattle; flowers attract butterflies, bees, and other insects. Inhibits soil erosion. There are numerous varieties of this species. Six species of *Acacia* in Arizona. Photograph taken near Superior, September 2.

STINKING MILKVETCH

Astragalus praelongus
Pea Family (Fabaceae)

Height: Sprawling to 4'.

Flowers: Cream-colored, pealike, and drooping; to ⅞" long; in raceme to 3" long; followed by whitish, inflated, warty pod to ½" wide, 1½" long.

Leaves: Grayish green, to 6" long; with 11 to 27 elliptical leaflets, each to ¾" long and slightly curved upward lengthwise.

Blooms: May–August.

Elevation: 3,000 to 6,500'.

Habitat: Sandy soil.

Comments: Coarse, malodorous plant; toxic to sheep. Its largest stems are reddish. The species of *Astragalus* , a very large genus, are difficult to identify. There are more than 6 dozen species of *Astragalus* in Arizona. Identifying them as either a milk vetch or a locoweed is in most cases as far as one can go. Photograph taken at Concho Lake, August 4.

FALSE MESQUITE

Calliandra humilis
Pea Family (Fabaceae)

Height: To 2" tall, sprawling to 8".

Flowers: White, numerous, with long, conspicuous stamens; flower head to ⅝" in diameter; followed by narrow, flat pod with thick, riblike margins. Reddish buds.

Leaves: Grayish green edged in red; hairy; leaf segments to ⅛" long, leaf to 2½" long. On some plants leaf segments close when touched and stay closed for several minutes.

Blooms: June–August.

Elevation: 4,000 to 9,000'.

Habitat: Dry soil in oak or pine woodlands.

Comments: Stems are reddish and very hairy. Two species of *Calliandra* in Arizona. Photograph taken in vicinity of Prescott, June 7.

SCRUFFY PRAIRIE CLOVER

White Prairie Clover
Dalea albiflora
Pea Family (Fabaceae)

Height: To 2'.

Flowers: Whitish, tiny, and pealike; in dense, elongated, terminal spike to 1½" long, ½" wide.

Leaves: Grayish green, pinnate, with linear lobes; covered with short, whitish hairs; to 1½" long.

Blooms: April–October.

Elevation: 3,500 to 7,500'.

Habitat: Clearings in ponderosa forests and roadsides.

Comments: Has hairy stems. Thirty species of *Dalea* in Arizona. Photograph taken near Camp Verde, September 30. This species resembles another species known as **White Prairie Clover** (*Dalea candida*); however, it has 10 stamens instead of the 5 found in *Dalea candida*.

WHITE PRAIRIE CLOVER

White Tassel-Flower
Dalea candida (Petalostemum occidentale)
Pea Family (Fabaceae)

Height: To 2½′.

Flowers: White, pealike; to ¼″ long; with yellow-tipped stamens; growing in axils of bracts on dense, terminal spike.

Leaves: Dark green, gland-dotted, hairless; pinnate, with 3 to 9 leaflets, very narrow to lance-shaped; leaf to 1″ long.

Blooms: May–September.

Elevation: 3,000 to 7,000′.

Habitat: Roadsides, plains, and mesas.

Comments: Perennial herb, with gland-dotted flower stems. Thirty species of *Dalea* in Arizona. Photograph taken in vicinity of Christopher Creek, September 14.

ARIZONA PEA

Nevada Pea
Lathyrus lanszwertii var. *leucanthus (Lathyrus arizonicus, L. leucanthus)*
Pea Family (Fabaceae)

Height: Sprawling or twining to 2′.

Flowers: White, pealike, with wide upper petal with pinkish veins; to ⅝″ wide, ¾″ long; 2 to 5 flowers in cluster arising from a leaf axil, followed by a flat seed pod.

Leaves: Dark green, thin, pinnate, with 2 to 10 broadly linear leaflets, each to 2″ long; tendrils small, bristlelike, and not prehensile. Stipules in leaf axils.

Blooms: May–October.

Elevation: 6,000 to 11,000′.

Habitat: Coniferous forests.

Comments: Not favored as browse by livestock, as are vetches. Seven species of *Lathyrus* in Arizona. Photograph taken on San Francisco Peaks, June 4.

FEATHER TREE

Lysiloma watsonii (Lysiloma microphylla var. *thornberi)*
Pea Family (Fabaceae)

Height: To 15'.

Trunk: To 5" in diameter.

Bark: Brownish gray, fissured and scaly.

Flowers: Creamy to white, in a dense ball; with numerous stamens; to ½" in diameter; followed by a large, dark brown, flat, broad pod to 9" long, 1" wide.

Leaves: Bright green, bipinnately compound, alternate; oval leaflets to ¼" long, ⅛" wide; leaf to 7" long, 4" wide.

Blooms: April–June.

Elevation: 2,800 to 4,000'.

Habitat: Rocky hillsides in upper desert.

Comments: A very localized species in southern Arizona. Dies back in hard winters. One species of *Lysiloma* in Arizona. Photograph taken at Desert Botanical Garden, Phoenix, April 30.

WHITE SWEETCLOVER

Melilotus albus
Pea Family (Fabaceae)

Height: To 6' or more.

Flowers: White and pealike; to ¼" long; in long, spikelike raceme of 30 to 80 flowers; spike to 8" long.

Leaves: Light green, pinnately divided into 3 lance-shaped, toothed leaflets, each to 1" long.

Blooms: July–October.

Elevation: 100 to 7,500'.

Habitat: Roadsides and fields.

Comments: A native of Eurasia; now naturalized in the U.S. Forage plant; enriches soil with nitrogen. Smells like newly mown hay. Excellent honey producer. Three species of *Melilotus* in Arizona. Photograph taken near Nutrioso, July 23.

WAIT-A-MINUTE BUSH

Cat Claw
Mimosa aculeaticarpa var. *biuncifera (Mimosa biuncifera)*
Pea Family (Fabaceae)

Height: To 8'.

Flowers: White to pinkish, with a ball-like head; to ⅝" in diameter; on long flower stalks in leaf axils; followed by a cluster of slightly curved, flat pods, brown to reddish brown on upper surface, greenish beneath; to 2" long, ⅛" wide; with a few marginal prickles.

Leaves: Dark green, bipinnate, with primary leaflets to 7 pairs, secondary leaflets to 13 pairs; haired; leaf to 2" long.

Blooms: May–August.

Elevation: 3,000 to 6,000'.

Habitat: Mesas, hillsides, desert grassland, and chaparral.

Comments: Nodes of stems armed with claw-like spines to ¼" long. Brown bark; honey plant; occasionally browsed by livestock. Forms thickets which prevent soil erosion and provide cover for wildlife. Four species of *Mimosa* in Arizona. Photograph taken near Lake Pleasant, May 7.

GRAHAM'S MIMOSA

Mimosa grahamii
Pea Family (Fabaceae)

Height: To 2'.

Flowers: Creamy white, ball-shaped; to ¾" wide; with purplish red tinged corolla, creamy white filaments tipped with pale yellow anthers; each ball on 1"-long stem in leaf axil; followed by pod to 1½" long, 5⁄16" wide with prickles on margins.

Leaves: Grayish green, very hairy, bipinnate; oblong leaflet to ¼" long; leaf to 6 ½" long; midrib beneath is lined with short, curved spines.

Blooms: April–August.

Elevation: 4,000 to 6,000'.

Habitat: Hillsides and dry slopes in canyons in southern Arizona.

Comments: An uncommon species. Stems have irregularly spaced spines to 1⁄16" long. Four species of *Mimosa* in Arizona. Photograph taken at Cave Creek Canyon near Portal, May 4.

WHITE CLOVER

Trifolium repens
Pea Family (Fabaceae)

Height: Creeper with flower stalks to 1′ high.

Flowers: White to pinkish, pealike; to ⅛″ wide, ⅜″ long; clustered in round flower head to ¾″ wide on leafless stem.

Leaves: Dark green, finely toothed, to 1″ wide; compound, with 3 oval, notched leaflets, each with a yellowish-green semicircle near the base.

Blooms: April–October.

Elevation: Throughout U.S.

Habitat: Meadows, lawns, fields, and roadsides.

Comments: Introduced from Europe. More than twenty species of *Trifolium* in Arizona. Photograph taken near Willow Springs Lake, September 14.

SWEET-CLOVER VETCH

Showy Vetch
Vicia pulchella
Pea Family (Fabaceae)

Height: Climbing or trailing to 3′.

Flowers: White and pealike; to ¼″ long, ⅛″ wide; in crowded, narrow raceme with flowers pointed downward, all facing in one direction. Raceme, to 5″ long, bearing up to 20 flowers, followed by a flattened pod to 1¼″ long.

Leaves: Dark bluish green, pinnate; to 5″ long; ending in a tendril; stipule at leaf base; up to 18 linear to oblong leaflets, each to ¾″ long.

Blooms: July–September.

Elevation: 6,000 to 8,500.

Habitat: Ponderosa pine forests.

Comments: Grooved, weak stems. Often a tangled mass. Four species of *Vicia* in Arizona. Photograph taken in vicinity of McNary, July 13.

TRUMPET GOOSEBERRY

Ribes leptanthum
Currant Family (Grossulariaceae)

Height: Shrub to 5′.

Flowers: White to cream-colored, trumpet-shaped; to ¼″ long; in cluster; followed by dark red to black berry, to 5⁄16″ in diameter.

Leaves: Light green, alternate, broad, rounded; 5-lobed, toothed; to 1″ wide.

Blooms: May–June.

Elevation: 6,000 to 9,500′.

Habitat: Along streams and mountain meadows.

Comments: Deciduous. Browsed by livestock and deer. Its tart berries are eaten by birds, and were used (both fresh and dried) by Native Americans. This genus of plant serves as an alternate host to white pine blister rust, which kills 5-needled pines (these pines are not common in Arizona). Ten species of *Ribes* in Arizona. Photograph taken near Greer, June 22.

CLIFF FENDLERBUSH

False Mockorange
Fendlera rupicola
Hydrangea Family (Hydrangeaceae)

Height: Straggling shrub to 6′.

Flowers: White and fragrant; 4 spoon-shaped petals, fringed, hairy, and faintly edged in pink; 8 stamens; to 2″ wide; pinkish buds, single or in small clusters, each followed by 4-chambered, woody, grayish green, acornlike capsule to ½″ long.

Leaves: Shiny, rough, dark green above, dull green beneath; 3 prominent sunken veins; thick, elongated, opposite; to 1¾″ long.

Blooms: March–June.

Elevation: 3,000 to 7,000′.

Habitat: Dry, rocky, and gravelly slopes.

Comments: New shoots have reddish stems. Browsed by bighorn sheep, deer, and goats. The Navajo Indians use parts of the bush for ceremonial food and to smoke. Three species of *Fendlera* in Arizona. Photograph taken at Cave Creek, Portal, April 23. Unlike the similar **Utah Serviceberry** (*Amelanchier utahensis*) (page 83), cliff fendlerbush has narrow, pointed leaves and 4-petaled flowers followed by 4-chambered, woody capsules. **Littleleaf Mockorange** (*Philadelphus microphyllus*) (page 59) also resembles cliff fendlerbush, but its 4 petals are not spoon-shaped, and it has many more than 8 stamens.

LITTLELEAF MOCKORANGE

Philadelphus microphyllus
Hydrangea Family (Hydrangeaceae)

Height: To 4'.

Flowers: White and fragrant, with 4 broad and rounded petals, numerous yellow stamens; to ½" wide.

Leaves: Dark green, shiny; faintly lined in lighter green above, paler green beneath; oval to elliptical; to ½" long.

Blooms: June–July.

Elevation: 5,000 to 9,000'.

Habitat: Rocky slopes and canyons.

Comments: Erect shrub with shreddy bark. To some, its flowers smell like orange blossoms. Browsed by bighorn sheep. Native Americans used stems for bows, arrows, and pipe stems. Eight species of *Philadelphus* in Arizona. Photograph taken at the North Rim of Grand Canyon National Park, June 25. A similar plant, **Cliff Fendlerbush** (*Fendlera rupicola*) (page 58) has spoon-shaped petals and only 8 stamens.

NARROWLEAF YERBA SANTA

Eriodictyon angustifolium var. *amplifolium*
Waterleaf Family (Hydrophyllaceae)

Height: To 6½'.

Flowers: White to pale lavender, funnel-shaped, 5-lobed, with dark-tipped stamens; to ⅜" wide; in loose, terminal clusters on upper branches.

Leaves: Sticky, leathery, aromatic, evergreen; dull, dark green above (shiny when immature); lighter green and white-woolly beneath, with a prominent midvein and network of veins; linear, often slightly toothed; margins rolled under; to 4" long.

Blooms: April–August.

Elevation: 2,000 to 7,000'.

Habitat: Dry hillsides, roadsides, and washes.

Comments: Woody at base. Browsed by mule deer. Infusion made from leaves is used medicinally to treat respiratory problems. One species of *Eriodictyon* in Arizona. Photograph taken northeast of Superior, May 2.

VARILEAF PHACELIA

Phacelia heterophylla (Phacelia magellanica)
Waterleaf Family (Hydrophyllaceae)

Height: To 4′.

Flowers: White to greenish yellow, with long stamens; 5-lobed; to ½″ wide: in coiled, terminal cluster.

Leaves: Grayish green, very hairy, oval on upper stem; pinnately divided into 3 to 5 sharp-pointed leaflets; to 4″ long on lower stem.

Blooms: May–October.

Elevation: 4,000 to 9,500′.

Habitat: Moist coniferous forests.

Comments: Pinkish purple, prickly, hairy stems. Forty-six species of *Phacelia* in Arizona. Photograph taken in Oak Creek Canyon, May 29.

ALKALI PHACELIA

Phacelia neglecta
Waterleaf Family (Hydrophyllaceae)

Height: To 4″.

Flowers: White, bowl-shaped, with 5 united petals; to ¼″ wide.

Leaves: Dark green, broadly oval, slightly scalloped; hairy, thick, succulent, brittle; to 1¼″ long.

Blooms: March–April.

Elevation: Below 1,500′.

Habitat: Stony desert soils frequently of volcanic origin.

Comments: Forty-six species of *Phacelia* in Arizona. Photograph taken north of Yuma, March 29.

HORSE-MINT
Yellow Mint
Agastache pallidiflora
Mint Family (Lamiaceae)

Height: To 3′.

Flowers: White to pale yellowish green to pinkish, depending on subspecies; tubular, 2-lipped, yellowish green bracts; to ½″ long; in thick, terminal cluster.

Leaves: Yellowish green above, paler green beneath; opposite, with round-toothed margins; triangular, deeply veined; to 1½″ long.

Blooms: July–October.

Elevation: 7,000 to 10,000′.

Habitat: Moist soil in coniferous forests and along mountain streams.

Comments: Perennial herb. Has square stems. Six species of *Agastache* in Arizona. Photograph taken at Luna Lake, August 5.

HOREHOUND
Marrubium vulgare
Mint Family (Lamiaceae)

Height: To 3′.

Flowers: White, tiny; to ¼″ long; in whorls in leaf axils.

Leaves: Grayish green, oval, veiny, crinkly surfaced above, white-woolly beneath; in pairs; to 1½″ long; on stem below each flower whorl.

Blooms: April–September.

Elevation: Throughout the state.

Habitat: Disturbed places, roadsides, fields, and pastures.

Comments: Perennial herb, and a weed. Stems are white, woolly, and 4-angled. Horehound used as flavoring in candy. Introduced from Europe, now naturalized in U.S. One species of *Marrubium* in Arizona. Photograph taken at Harshaw, April 27.

SPOTTED HORSEMINT
Monarda punctata
Mint Family (Lamiaceae)

Height: To 16".

Flowers: White, long, narrow, hairy; 2-lipped, with upper lip arched upward, lower lip curved downward and speckled with lavender; to 1" long; in circular clusters around stem; surrounded by 8 broad, purplish bracts covered with a white down.

Leaves: Grayish green, opposite, lance-shaped; folded upward from center vein; tinged with lavender at base; slightly toothed, curved downward from stem; to 2" long; in clusters surrounding stem.

Blooms: End of June–August.

Elevation: 5,000 to 7,000'.

Habitat: Dry, sandy soil.

Comments: Square stem. Four species of *Monarda* recorded for Arizona. This species is very rare in the state. Photograph taken 28 miles north of St. Johns, June 28.

FUNNEL LILY
Androstephium breviflorum
Lily Family (Liliaceae)

Height: Flower stalk to 1'.

Flowers: Whitish to light pink, with darker central stripe on petals; 6 petals, black anthers, petals and sepals partly joined; stamen filaments are partly united to form a tube in the center. Flower to ⅝" wide, in loose, terminal cluster of up to 12 flowers.

Leaves: Grayish green, grasslike, narrow; few, grooved on upper surface, basal; to 8" long.

Blooms: March–April.

Elevation: 2,000 to 7,000'.

Habitat: Dry, sandy soil of slopes and plains.

Comments: Leafless stem arises from a bulb. Lacks onion odor. One species of *Androstephium* in Arizona. Photograph taken south of Parker, March 7.

AJO LILY
Hesperocallis undulata
Lily Family (Liliaceae)

Height: To 4′.

Flowers: White, trumpet-shaped, fragrant; to 2½″ long, clustered flower stalk.

Leaves: Bluish green, narrow, wavy-margined; in basal rosette; to 20″ long.

Blooms: Mid-February–mid-April.

Elevation: Below 2,000′ in southwestern Arizona.

Habitat: Dunes and sand-gravel flats.

Comments: Perennial. Resembles Easter lily; called ajo lily because its big, edible bulb resembles garlic (*ajo* in Spanish). Bulbs sometimes grow as deep as 2′ below surface of soil. Hawk moths pollinate flowers. Some taxonomists put this genus in the Asparagus family (Asparagaceae). One species of *Hesperocallis* in Arizona. Photograph taken near Salome, March 28.

LARGE SOLOMON'S SEAL
Maiathemum racemosum ssp. *amplexicaule*
(Smilacina racemosa)
Lily Family (Liliaceae)

Height: To 3′.

Flowers: White, starlike, with 6 petallike segments; flower to ⅛″ long in branched, dense, terminal raceme on arching stem, followed by a cluster of reddish berries dotted with purple; each berry to ¼″ long.

Leaves: Dark green, broadly lance-shaped to oval; clasping stem at base; to 6″ long.

Blooms: May–July.

Elevation: 6,000 to 10,000′.

Habitat: Rich soil in coniferous forests.

Comments: Perennial herb. Two species of *Maiathemum* in Arizona. Photograph taken at Greer, June 17.

STAR SOLOMON'S SEAL

Maiathemum stellatum (Smilacina stellata)
Lily Family (Liliaceae)

Height: To 2'.

Flowers: White, starlike, with 6 petallike segments; to ¼" wide; loosely spaced on a terminal, zigzag raceme, followed by green berries with dark blue, vertical stripes, berries turning black with maturity.

Leaves: Dark green, lance-shaped; to 5" long.

Blooms: May–June.

Elevation: 7,500 to 10,000'.

Habitat: Moist, rich forests.

Comments: Perennial herb. Two species of *Maiathemum* in Arizona. Photograph taken near Greer, June 22.

FALSE HELLEBORE

Skunk Cabbage
Veratrum californicum var. *californicum*
Lily Family (Liliaceae)

Height: To 8'.

Flowers: Whitish to greenish, star-shaped; V-shaped green gland at base; to ¾" long, ½" wide; in branching, terminal cluster to 1' long.

Leaves: Yellowish green, oval, strongly veined; appear pleated; to 1' long.

Blooms: July–August.

Elevation: 7,500 to 9,500'.

Habitat: Wet meadows, around springs and bogs, and moist forests.

Comments: Perennial herb. Extremely poisonous to livestock as well as to bees and other insects. One species of *Veratrum* in Arizona. Photograph taken at Greer, July 20.

WHITE CAMAS

Mountain Death Camas
Zigadenus elegans
Lily Family (Liliaceae)

Height: To 3'.

Flowers: White to greenish white; 6 segments with green, heart-shaped gland at each base; to ¾" wide; in elongated, terminal cluster.

Leaves: Bluish green, onion-like; to ¼" wide, 12" long.

Blooms: July–September.

Elevation: 5,000 to 10,000'.

Habitat: Clearings in moist ponderosa pine and spruce-fir forests.

Comments: From a bulb similar to an onion. This species not as poisonous as **Death Camas** (*Z. virescens*) (at right). Three species of *Zigadenus* in Arizona. Photograph taken near Willow Springs Lake, September 15.

DEATH CAMAS

Zigadenus virescens
Lily Family (Liliaceae)

Height: To 2'.

Flowers: White to greenish, bell-shaped; to ¼" wide; widely spaced on long flower stalks that curve away from main stalk.

Leaves: Dark green, narrow, grasslike; to 12" long.

Blooms: July–September.

Elevation: 6,500 to 11,000'.

Habitat: Rich soil in moist coniferous forests.

Comments: From a bulb similar to an onion. All parts contain zygadenine, a heart depressant, and are poisonous. Three species of *Zigadenus* in Arizona. Photograph taken south of Alpine, August 2.

WHITE-BRACTED STICK LEAF

Sand Blazing Star
Mentzelia involucrata
Stick Leaf Family (Loasaceae)

Height: To 1′.

Flowers: White to pale cream, streaked inside with faint orange lines; translucent, erect, with 5 petals; to 1½″ wide, 1¼″ long.

Leaves: Grayish green, lance-shaped, with irregular teeth on margins; very rough due to stiff hairs; to 4½″ long.

Blooms: February–May.

Elevation: Below 3,000′.

Habitat: Desert washes, slopes, and flat areas in dry, sandy soil.

Comments: Annual. Very rough, sandpapery leaves stick to fabric, making them difficult to remove. Over twenty species of *Mentzelia* in Arizona. Photograph taken north of Yuma, March 29.

DESERT COTTON

Algodoncillo
Gossypium thurberi
Mallow Family (Malvaceae)

Height: To 7′.

Flowers: White to very pale pink, with 5 rounded, crinkly petals, each with a lavender spot near base; cup-shaped, with large filament tube in center; to 1″ long, 1½″ wide; followed by dark brown capsule, to ½″ in diameter.

Leaves: Dark green above, paler green beneath; palmate, 3- to 5-lobed; to 7″ wide, 7″ long.

Blooms: August–October.

Elevation: 2,500 to 5,000′.

Habitat: Rocky slopes, washes, and canyons.

Comments: Shrubby. An alternate host for the boll weevil. Bees and wasps are attracted to flower nectar; other insects feed on nectar produced by leaves and flower stalk. One species of *Gossypium* in Arizona. Photograph taken north of Superior, September 2.

WOODY BOTTLE-WASHER
Camissonia boothii ssp. *condensata*
Evening Primrose Family (Onagraceae)

Height: To 20".

Flowers: White, fading to pink, with 4 petals; long stamens curved inward toward center; to ½" wide on numerous long stems in center of plant; followed by splitting capsules on a woody core when mature and dry.

Leaves: Dark green, with reddish purple blotches and spots; mainly basal; lance-shaped to elliptical; to 4" long.

Blooms: February–May.

Elevation: Below 2,500'.

Habitat: Open desert.

Comments: Annual. Around two dozen species of *Camissonia* in Arizona. Photograph taken at Golden Shores, February 25.

PRAIRIE EVENING PRIMROSE
Oenothera albicaulis
Evening Primrose Family (Onagraceae)

Height: To 16".

Flowers: White (pink when aged), with 4 petals; to 1½" wide.

Leaves: Grayish green, to 2" long. Basal leaves are spoon-shaped; stem leaves are cleft into narrow lobes.

Blooms: March–August.

Elevation: 2,500 to 7,500'.

Habitat: Roadsides and dry, grassy, or sandy disturbed areas.

Comments: Twenty-one species of *Oenothera* in Arizona. Photograph taken near Nutrioso, August 3.

TUFTED EVENING PRIMROSE
Oenothera caespitosa
Evening Primrose Family (Onagraceae)

Height: To 4".

Flowers: White, turning to pink with age; 4 heart-shaped petals; long, yellow stamens and stigma; slightly fragrant; to 4" wide; opening in late afternoon, fading the following morning. Flower held above ground by long, thin calyx tube.

Leaves: Grayish green (some are tinged reddish), finely haired, narrow, with toothed margins; to 7" long, 1" wide; in basal rosette.

Blooms: April–September.

Elevation: 3,000 to 7,500'.

Habitat: Roadsides, ponderosa forest clearings, and dry, rocky slopes.

Comments: Twenty-one species of *Oenothera* in Arizona. Photograph taken northeast of Superior, April 20.

CUTLEAF EVENING PRIMROSE
Oenothera coronopifolia
Evening Primrose Family (Onagraceae)

Height: to 18".

Flowers: White, with 4 petals; to 1" wide.

Leaves: Grayish green, linear and linear-lobed, toothed; to ¾" long.

Blooms: June–August.

Elevation: 3,000 to 8,000'.

Habitat: Dry plains and sandy soil.

Comments: Perennial herb. Twenty-one species of *Oenothera* in Arizona. Photograph taken at Lee's Ferry, June 23.

DUNE PRIMROSE
Birdcage Evening Primrose
Oenothera deltoides
Evening Primrose Family (Onagraceae)

Height: To 1½'.

Flowers: White, turning pink; yellow toward center; 4 petals; saucerlike; to 3½" wide.

Leaves: Pale green, hairy, sometimes grooved or cleft; to 4½" long.

Blooms: February–May, depending on variety.

Elevation: Generally below 2,500', depending on variety.

Habitat: Sandy deserts and other open areas.

Comments: Sweet-scented and bushlike. There are numerous varieties of this species. The outer stems on some varieties curl upward and inward when they die, forming a cagelike structure. Twenty-one species of *Oenothera* in Arizona. Photograph taken near Tacna, March 29.

GIANT RATTLESNAKE PLANTAIN
Rattlesnake Orchid
Goodyera oblongifolia
Orchid Family (Orchidaceae)

Height: To 18".

Flowers: White to pinkish, hairy; to ⅝" long, ¾" wide; upper sepal and petals united in hood over lip; on long, densely flowered, hairy, naked stem; flowering section of stem to 4" long.

Leaves: Dark green, with mottled, white central line; oblong; to 4" long; in basal rosette.

Blooms: July–September.

Elevation: 8,000 to 9,500'.

Habitat: Rich, moist coniferous forests.

Comments: The giant rattlesnake plantain is named for its mottled leaves, which resemble rattlesnake skin. Two species of *Goodyera* in Arizona. Photograph taken at Greer, August 10.

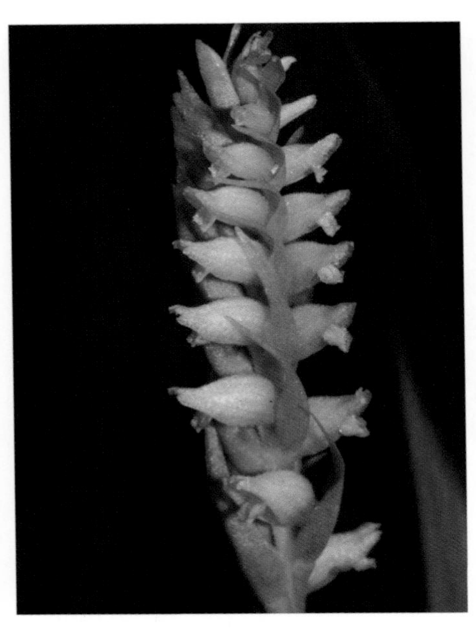

DWARF RATTLESNAKE PLANTAIN

Dwarf Rattlesnake Orchid
Goodyera repens
Orchid Family (Orchidaceae)

Height: Flower stem to 1′.

Flowers: White to greenish, occasionally pink-tinged; upper sepal and petals united in hood over lip; to ⅛″ long; along one side of slender flower spike.

Leaves: Dark green, shiny, fleshy, mostly basal; faint network, broadly oval; to 1½″ long.

Blooms: July–August.

Elevation: 9,000 to 10,000′.

Habitat: Moist mountain slopes in coniferous forests.

Comments: Two species of *Goodyera* in Arizona. Photograph taken in mountains above Greer, August 8.

HOODED LADIES' TRESSES

Pearl Twist
Spiranthes romanzoffiana
Orchid Family (Orchidaceae)

Height: Flower stem to 16″.

Flowers: White, 3 sepals, 3 petals; 2 upper petals form a hood; lip petal curved downward and pinched in on each margin near tip; flower to ½″ long, ⅛″ wide; flowers densely arranged in 3-spiraled rows on a slender spike.

Leaves: Light green, shiny, smooth, lanced-shaped; succulent, basal; to 8″ long, ½″ wide.

Blooms: August–September.

Elevation: 8,500 to 9,500′.

Habitat: Wet meadows, bogs, streamsides.

Comments: Often several stems in a clump. Three species of *Spiranthes* in Arizona. Photograph taken in Mount Baldy Wilderness, August 13.

BLUESTEM PRICKLEPOPPY
Southwestern Prickly Poppy
Argemone pleiacantha
Poppy Family (Papaveraceae)

Height: To 3'.

Flowers: White, tissue-paper thin, with 4 to 6 crinkled petals, numerous bright orange stamens; to 5" wide; followed by oblong, prickly seed pod to 1½" long.

Leaves: Bluish green, deeply lobed; clasping stem; prickles on veins and margins; to 8" long.

Blooms: April–November.

Elevation: 1,400 to 8,000'.

Habitat: Fields, roadsides, mesas, and washes.

Comments: Perennial herb; with very spiny stems, acrid yellow sap. All parts are poisonous; however, mourning doves feed on seeds. Large expanses of this species on rangeland indicate overgrazing. Nine species of *Argemone* in Arizona. Photograph taken at Theodore Roosevelt Lake, April 29.

PURSH PLANTAIN
Indian Wheat
Plantago patagonica (Plantago purshii)
Plantain Family (Plantaginaceae)

Height: Flower stem to 8".

Flowers: Buff-colored with brownish tinge toward center; 4 narrow, pointed petals; to ⅛" wide; flowers spaced on woolly bracted spike to 1¼" long.

Leaves: Grayish green, very hairy, narrow, and basal; to 4" long.

Blooms: February–July.

Elevation: 1,000 to 7,000'.

Habitat: Dry slopes and mesas.

Comments: Very hairy plant. Browsed by livestock. Twenty-one species of *Plantago* in Arizona. Photograph taken in Superstition Mountains, March 25.

NUTTALL'S LINANTHUS

Leptosiphon nuttallii (Linanthus nuttallii)
Phlox Family *(Polemoniaceae)*

Height: To 1′.

Flowers: White, yellow in center; with 5 broad, flared lobes on narrow tube; to ½″ wide; in terminal clusters on leafy stems.

Leaves: Green, rigid, narrowly linear; leaves appear whorled on stems; to ½″ long.

Blooms: July–November.

Elevation: 5,500 to 8,000′.

Habitat: Open ponderosa forests.

Comments: Perennial herb. Three species *Leptosiphon* in Arizona. Photograph taken near Willow Springs Lake, August 19.

DESERT SNOW

Linanthus demissus
Phlox Family (Polemoniaceae)

Height: To 4″.

Flowers: White, bell-shaped, with 5 flaring lobes, each with 2 reddish brown to purplish lines at base; to ½″ wide, ¼″ long; in few-flowered clusters on branches.

Leaves: Green, wiry, hairlike; to ⅜″ long; surrounding bases of flowers.

Blooms: March–May.

Elevation: Below 2,000′.

Habitat: Desert sands.

Comments: Annual. Many-branched. Four species of *Linanthus* in Arizona. Photograph taken in Kofa Mountains, March 29.

DESERT PHLOX

Vine Phlox
Phlox tenuifolia
Phlox Family (Polemoniaceae)

Height: To 3′.

Flowers: White, funnel-shaped, with yellow inner tube; to 1″ wide.

Leaves: Dark green, linear, to 2″ long.

Blooms: February–May, rare instances in the fall after summer rains.

Elevation: 1,500 to 5,000′.

Habitat: Along washes and on rocky slopes.

Comments: Perennial. In partial shade, stems are vinelike and supported by neighboring shrubs. In the open, plants form tufts. Fourteen species of *Phlox* in Arizona. Photograph taken in Superstition Mountains, February 4.

WHITE MILKWORT

Polygala alba
Milkwort Family (Polygalaceae)

Height: To 14″.

Flowers: White with greenish tinge at base; 3 petals; 5 sepals, 2 larger ones colored white like petals; to ⅛″ long; in dense, narrow spike to 3″ long; on top of slender stem.

Leaves: Dark green, linear, very narrow; in whorls at base, at intervals along upper stem; to 1¼″ long.

Blooms: May–September.

Elevation: 5,000 to 7,500′.

Habitat: Roadsides and fields.

Comments: Does not produce a milky sap. It was once believed that if cattle ate the plant their milk production increased. Fourteen species of *Polygala* in Arizona. Photograph taken at Heber, August 4.

WESTERN BISTORT

Smokeweed

Bistorta bistortoides (Polygonum bistortoides)

Buckwheat Family (Polygonaceae)

Height: To 2′.

Flowers: Snow white to pale pink with 5 petallike segments; to 3⁄16″ long; in dense, spikelike, terminal cluster to 2″ long, ¾″ wide; on long erect, nearly leafless stem.

Leaves: Dark green, with prominent, yellowish midvein; lance-shaped to elliptical, mainly basal; to 7″ long.

Blooms: June–September.

Elevation: 8,500 to 11,000′.

Habitat: Wet meadows and along mountain streams.

Comments: Stem is reddish near the ground. Roots are starchy and edible, raw or boiled. One species of *Bistorta* in Arizona. Photograph taken in Hannagan Meadow area, June 30.

ANTELOPE-SAGE

Eriogonum jamesii var. *jamesii*

Buckwheat Family (Polygonaceae)

Height: To 14″.

Flowers: White to cream-colored with long, whitish stamens; to ¼″ wide, ⅜″ long, in tight or loose cluster above whorl of small leaves.

Leaves: Grayish green above, very white and hairy beneath; lance-shaped; to 2″ long; in basal rosette.

Blooms: July–October.

Elevation: 5,000 to 9,000′.

Habitat: Rocky areas and cliffs in clearings in oak woodlands and pine forests.

Comments: Over fifty species of *Eriogonum* in Arizona. Photograph taken in Heber vicinity, August 4.

REDROOT BUCKWHEAT

Eriogonum racemosum
Buckwheat Family (Polygonaceae)

Height: Flowering stem to 1½'.

Flowers: White to pinkish, with dark pink ribs on back of petals; to ⅛" wide, ¼" long; in loose, spikelike clusters along upper half of gray, leafless flowering stems.

Leaves: Grayish green, on long leaf stem; basal cluster, erect, broadly oval, and feltlike; to 1¼" long, ¾" wide.

Blooms: June–October.

Elevation: 5,000 to 9,000'.

Habitat: Clearing in pine forests.

Comments: Over fifty species of *Eriogonum* in Arizona. Photograph taken in vicinity of Upper Lake Mary, September 2.

ROCKY MOUNTAIN SPRING BEAUTY

Claytonia lanceolata var. *rosea (Claytonia rosea)*
Purslane Family (Portulacaceae)

Height: To 4".

Flowers: White to pinkish, 5 petals, 5 pink-tipped stamens; to ⅝" wide; in loose raceme.

Leaves: Dark green above, pinkish beneath, narrow, and smooth; succulent, lance-shaped; 1 pair, opposite; to 1¾" long.

Blooms: February–May.

Elevation: 5,500 to 8,000'.

Habitat: Moist coniferous forests.

Comments: Perennial. Three species of *Claytonia* in Arizona. Photograph taken at Willow Springs Lake, April 22.

MINER'S LETTUCE
Indian Lettuce
Claytonia perfoliata (Montia perfoliata)
Purslane Family (Portulacaceae)

Height: To 14".

Flowers: White to pinkish; to ¼" wide, in tiny raceme rising above center of circular leaf.

Leaves: Dark green, fleshy, circular, flower-bearing leaf; to 2" wide. Other leaves lance-shaped on narrow stalks.

Blooms: February–May.

Elevation: 2,500 to 7,500'.

Habitat: Moist places in shade, along brooks, and around springs.

Comments: Annual. Used by pioneers and Native Americans as a potherb and for salads. Three species of *Claytonia* in Arizona. Photograph taken at Saguaro Lake area, February 6.

CHAMISSO'S MONTIA
Montia chamissoi
Purslane Family (Portulacaceae)

Height: Creeping stems to 6" long.

Flowers: White or pink, 5 petals, 5 pink-tipped stamens; to ½" wide; in loose cluster of up to 8 flowers.

Leaves: Dark green, lance-shaped; very succulent; opposite; to 1½" long.

Blooms: June–August.

Elevation: 6,000 to 9,500'.

Habitat: Springs and other wet areas in coniferous forests.

Comments: Perennial. Runner branches end in bulblets. Two species of *Montia* in Arizona. Photograph taken at Lee Valley Reservoir, August 11.

ROCK JASMINE

Pygmyflower
Androsace septentrionalis
Primrose Family (Primulaceae)

Height: To 10".

Flowers: White, funnel-shaped, starlike, with 5 roundish lobes; to ⅛" wide; on leafless, erect stems.

Leaves: Dark green to reddish green, lance-shaped, irregularly toothed; in basal rosette; to 1¼" long.

Blooms: April–September.

Elevation: 7,000 to 12,000'.

Habitat: Meadows and clearings in coniferous forests.

Comments: Many-stemmed. Two species of *Androsace* in Arizona. Photograph taken in Greer area, July 4.

COMMON BANEBERRY

Snakeberry
Actaea rubra
Buttercup Family (Ranunculaceae)

Height: To 3'.

Flowers: White, with 4 to 10 small petals, many stamens; to ⅛" long; in elongated cluster to 5" long; followed by cluster of ½" purplish red berries.

Leaves: Green, few but very large; pinnately divided into oval, sharply toothed leaflets, each to 3½" long, with up to 2 dozen leaflets per triangular leaf.

Blooms: May–July.

Elevation: 7,000 to 10,000'.

Habitat: Moist mountain forests and stream banks.

Comments: Perennial herb with poisonous berries. One species of *Actaea* in Arizona. Photograph taken at Greer, July 21.

DESERT ANEMONE
Desert Windflower
Anemone tuberosa
Buttercup Family (Ranunculaceae)

Height: To 16".

Flowers: White to pinkish purple, petalless, with petallike structures that are actually sepals; to 1½" wide; at ends of erect stems.

Leaves: Green, divided several times into sections; basal and in whorl midway on stem; to 4" long, 2" wide.

Blooms: February–April.

Elevation: 2,500 to 5,000'.

Habitat: Rocky desert slopes.

Comments: Perennial herb. Two species of *Anemone* in Arizona. Photograph taken in Superstition Mountains, February 4.

MARSH MARIGOLD
Cowslip
Caltha leptosepala
Buttercup Family (Ranunculaceae)

Height: To 8".

Flowers: White, with 5 to 12 petallike sepals, not true petals; a center of numerous yellow stamens; to 1½" wide.

Leaves: Dark green, shiny; heart-shaped, minutely scalloped; basal; to 3" long.

Blooms: May–September.

Elevation: 7,500 to 11,000'.

Habitat: Wet, boggy meadows and along streams.

Comments: Perennial. One species of *Caltha* in Arizona. Photograph taken at Hannagan Meadow, June 24.

ROCK CLEMATIS

Alpine Clematis
Clematis columbiana var. *columbiana (Clematis pseudoalpina)*
Buttercup Family (Ranunculaceae)

Height: Trailing or climbing woody vine to 5′.

Flowers: White to violet or purple; 4 drooping, petallike sepals on leafless stem; to 2″ long, followed by seeds in cluster of long silky plumes.

Leaves: Dull green, compound, divided twice into threes; toothed; to 2″ long.

Blooms: June–July.

Elevation: 7,000 to 9,000′.

Habitat: Rich soil in moist coniferous forests.

Comments: Plumes aid in distribution of seeds by the wind. Six species of *Clematis* in Arizona. Photograph taken at Greer, June 17.

TEXAS VIRGIN'S BOWER

Barbas De Chivato
Clematis drummondii
Buttercup Family (Ranunculaceae)

Height: Woody vine to 20′ or more.

Flowers: White, petalless, with 4 white, petallike sepals, and numerous white stamens; to 1″ wide; followed by fluffy, white plume.

Leaves: Grayish green, downy-haired, thin; pinnately compound, cleft or lobed; to 2″ long.

Blooms: March–September.

Elevation: Below 4,000′.

Habitat: Open ground.

Comments: Six species of *Clematis* in Arizona. Photograph taken at Patagonia Lake State Park, April 26.

WHITE VIRGIN'S BOWER

Clematis ligusticifolia
Buttercup Family (Ranunculaceae)

Height: Woody vine to 20' or more.

Flowers: White and petalless, with 4 white, petal-like sepals, numerous white stamens; to 1" wide; followed by fluffy, white plume on female plant.

Leaves: Green, thin, smooth, pinnately compound; 5 to 7 toothed or 3-lobed leaflets, each to 3" long.

Blooms: May–September.

Elevation: 3,000 to 8,000'.

Habitat: Along streams and in other moist places.

Comments: Perennial vine; clings by twisting leaf stems that form tendrils. Weak stems climb over shrubs, up into trees, and over rocks. Chewed by Native Americans and pioneers as sore throat treatment. Infusion of leaves used for treating sores on horses. Leaves and stems taste like pepper; certain species of *Clematis*, however, are poisonous. Very fragrant; a favorite of bees. Six species of *Clematis* in Arizona; this species hybridizes with **Texas Virgin's Bower** (*Clematis drummondii*) (page 79) and varies greatly in leaflet size, shape, and dentation. Photograph taken in vicinity of Nutrioso, August 18.

WATER BUTTERCUP

Water Crowfoot
Ranunculus aquatilis var. *diffusus*
Buttercup Family (Ranunculaceae)

Height: Aquatic flower; projects about 1" above surface of water.

Flowers: White; 5 petals, yellow at base; to ¾" wide.

Leaves: Submerged, cut into numerous forked, hairlike segments.

Blooms: May–August.

Elevation: 4,500 to 9,000'.

Habitat: Ponds and slow streams.

Comments: Aquatic perennial; forms dense beds in ponds. Food for ducks and geese. Over twenty species of *Ranunculus* in Arizona. Photograph taken at Nelson Reservoir, August 3.

FENDLER'S CEANOTHUS
Buckbrush
Ceanothus fendleri
Buckthorn Family (Rhamnaceae)

Height: Spiny shrub to 6', but usually 3'.

Flowers: White to pinkish, with 5 petals, 5 stamens; to 3⁄16" wide, in terminal, pyramid-shaped cluster to 1½" long; followed by reddish brown, dry, 3-lobed fruit, to 3⁄16" in diameter.

Leaves: Grayish green, somewhat thick; velvety surface, hairy beneath, prominent veins, elliptical, alternate; to 1" long.

Blooms: April–October.

Elevation: 5,000 to 10,000'.

Habitat: Pinyon-juniper woodlands and pine forests.

Comments: Forms thickets. Has gray, felty stems and straight, slender spines up to 1" long. Mature twigs are reddish brown. Browsed by deer and livestock. Foliage and stems are eaten by porcupines and rabbits. Native Americans used leaves for tea and fruits for food and medicines. Seven species of *Ceanothus* in Arizona. Photograph taken in vicinity of Portal, April 22. This is the only species of *Ceanothus* with spines.

GREGG'S CEANOTHUS
Desert Ceanothus
Ceanothus greggii (Ceanothus pauciflorus)
Buckthorn Family (Rhamnaceae)

Height: To 8', but usually less.

Flowers: Whitish to pinkish, with 5 spoon-shaped, hooded petals; to 3⁄8" wide; in crowded cluster on branches; followed by cluster of fruit; each capsule to 3⁄16" in diameter.

Leaves: Shiny green above, grayish and felty beneath with a visible network; opposite, thick, leathery; to 1" long near base of shrub, in clusters along upper branches.

Blooms: March–May.

Elevation: 3,000 to 7,000'.

Habitat: Oak woodlands.

Comments: Light gray, felt-covered bark on old branches. Young wood is pinkish and felt-covered. Browsed by deer; seeds eaten by birds and small mammals. Seven species of *Ceanothus* in Arizona. Photograph taken northeast of Superior on April 3. Branches of this species are not spine-tipped.

DEERBRUSH

White Lilac
Ceanothus integerrimus
Buckthorn Family (Rhamnaceae)

Height: Shrub to 8'.

Flowers: White, occasionally pink or bluish, fragrant, 5-petaled; to 1/16" wide; in a spikelike flower cluster to 6" long; followed by cluster of dry, crested capsules.

Leaves: Dark green, elliptical, with 3 prominent main veins; to 1½" long.

Blooms: May–October.

Elevation: 3,500 to 7,000'.

Habitat: Open woodland and chaparral.

Comments: Browsed by deer; wildlife feed on bark and seeds. Native Americans made a soapy mixture from the bark; root bark was used medicinally. Blooms are a source of honey. Seven species of *Ceanothus* in Arizona. Photograph taken in a canyon north of Superior, April 20.

WRIGHT'S MOCK BUCKTHORN

Sageretia wrightii
Buckthorn Family (Rhamnaceae)

Height: Straggling shrub to 8'.

Flowers: Cream-colored, starlike, with 5 petals; to ⅛" wide; followed by a small, fleshy drupe.

Leaves: Dark green, shiny, oval, pointed at tip; to 1¼" long.

Blooms: March–September.

Elevation: 1,500 to 5,000'.

Habitat: Along washes and in canyons among rocks.

Comments: Grayish, woolly stems. Somewhat spiny; spines to ¼" long. One species of *Sageretia* in Arizona. Photograph taken in vicinity of Saguaro Lake, August 26.

UTAH SERVICEBERRY

Shadberry
Amelanchier utahensis
Rose Family (Rosaceae)

Height: Shrub or small tree to 15′.

Bark: Smooth, gray.

Flowers: White, with 5 long, narrow petals; to ½″ wide; 3 to 6 in a cluster; followed by small, applelike fruits to ⅜″ in diameter, maturing to bluish purple.

Leaves: Dark green, nearly round to elliptical; toothed on margins from midleaf to tip; to 1½″ long, 1″ wide.

Blooms: April–May.

Elevation: 2,000 to 7,500′.

Habitat: Rocky slopes in pinyon-juniper woodlands and ponderosa pine forests.

Comments: Browsed by livestock and deer. Fruits are eaten by birds and rodents. Berries are used for wine, jelly, and jam. Two species of *Amelanchier* in Arizona. Photograph taken at Double Springs, in vicinity of Mormon Lake, June 1. Unlike the similar **Cliff Fendlerbush** (*Fendlera rupicola*) (page 58), Utah serviceberry has rounded, toothed leaves and 5-petaled flowers, followed by small, fleshy, applelike fruits.

FERNBUSH

Chamaebatiaria millefolium
Rose Family (Rosaceae)

Height: To 6′.

Flowers: White, sticky, with 5 crinkly petals, yellow stamens; to ⅜″ wide; in elongated cluster to 4″ long at end of branch; followed by small, dry pod.

Leaves: Grayish green, much-divided, fernlike; fragrant, scaly, sticky; to 1½″ long.

Blooms: July–November.

Elevation: 4,500 to 8,000′.

Habitat: Pinyon-juniper-sagebrush areas.

Comments: Evergreen. Bark is tinged red and shrubby. Browsed by deer, sheep, and goats. One species of *Chamaebatiaria* in Arizona. Photograph taken at Walnut Canyon National Monument, September 7.

CERRO HAWTHORN

Crataegus erythropoda
Rose Family (Rosaceae)

Height: Spiny shrub to 9', or small tree to 20'.

Trunk: To 4" in diameter.

Bark: Reddish brown or gray, and scaly.

Flowers: White, with 5 petals; to ¾" wide; in up to 2½" compact cluster of 5 to 10 flowers; followed by orange-red, ⅜" roundish fruits.

Leaves: Dark green and shiny above, paler beneath; broadly oval, coarsely toothed; prominently veined, often shallowly lobed; to 2½" long, 1¾" wide.

Blooms: April–May.

Elevation: 4,500 to 8,000'.

Habitat: Along streams and in moist canyons.

Comments: Has widely spreading branches, with reddish brown, zigzag twigs. Has numerous 2"-long, nearly straight, shiny, dark reddish spines on branches. Fruits are called "haws." Birds and other animals feed on fruits. Three species of *Crataegus* in Arizona. Photograph taken in Oak Creek Canyon, Sedona, September 9. The similar **River Hawthorn** (*Crataegus rivularis*) has fewer spines, which are blackish, curved, and only 1" long, and leaves that are double-toothed.

APACHE-PLUME

Fallugia paradoxa
Rose Family (Rosaceae)

Height: Shrub to 6'.

Flowers: Pure white, 5-petaled, roselike, and yellow-centered; to 2" wide; followed by seeds on long, white-to-pinkish long-lasting plumes.

Leaves: Grayish green, slightly downy, pinnately divided into 3 to 7 linear lobes; to ¾" long.

Blooms: April–October.

Elevation: 4,000 to 8,000'.

Habitat: Roadsides, dry washes, dry hillsides, and chaparral.

Comments: Evergreen shrub. An erosion deterrent along banks of washes. Native Americans used stems for arrowshafts. Good browse for sheep, cattle, goats, and deer. One species of *Fallugia* in Arizona. Photograph taken at Sunset Crater National Monument, September 7. Differs from the similar **Stansbury Cliff-Rose** (*Purshia stansburiana*) (page 203), by its numerous branches at base, more shrublike appearance, short-haired leaves, white flowers, and fuller plumes.

BRACTED STRAWBERRY
Woodland Strawberry
Fragaria vesca ssp. *bracteata (Fragaria bracteata)*
Rose Family (Rosaceae)

Height: Creeper, with flower stalks to 2".

Flowers: White with yellow center, 5 petals, to 1" wide; followed by ¾"-long, red, cone-shaped berry with seeds barely attached to berry's surface.

Leaves: Dark green, thin, compound, with 3 leaflets, each to ¾" long; toothed on margins.

Blooms: May–September.

Elevation: 7,000 to 9,500'.

Habitat: Coniferous forests.

Comments: Perennial herb. Hairs on stem spread out. Fruits are eaten by birds and mammals, including humans. Two species of *Fragaria* in Arizona. Photograph taken at Greer, June 18. This species recognizable by its dark green, toothed leaflets and by berries with seeds barely attached.

WILD STRAWBERRY
Fragaria virginiana ssp. *glauca (Fragaria ovalis)*
Rose Family (Rosaceae)

Height: Creeper, with flower stalk to 2".

Flowers: White with fuzzy, yellow center; 5 petals; to 1" wide; followed by ¾"-long, red, cone-shaped berry with seeds partly buried in flesh.

Leaves: Light green, slightly covered with whitish bloom; haired on margins; thick, compound, with 3 leaflets, each to ¾" long; toothed on margins mainly from center to tip.

Blooms: May–October.

Elevation: 7,000 to 11,000'.

Habitat: Coniferous forests.

Comments: Perennial herb, with somewhat flattened hairs on stem. Plant spreads by runners. Fruits relished by wildlife. Two species of *Fragaria* in Arizona. Photograph taken in mountains above Greer, July 3. This species identifiable by its partially toothed leaflets, which whiten slightly, and its strawberries with seeds partly buried in flesh.

MOUNTAIN SPRAY

Rock Spiraea
Holodiscus dumosus
Rose Family (Rosaceae)

Height: To 8′.

Flowers: Creamy white to pinkish, with yellow center; 3⁄16″ wide; in spikelike, terminal cluster to 2″ long; followed by small, dry, brown seed case.

Leaves: Green above, lighter green and velvety beneath; oval to wedge-shaped; sawtooth-notched on margins of broad outer end; to 1½″ long.

Blooms: June–September.

Elevation: 5,500 to 10,000′.

Habitat: Rocky slopes, ponderosa pine and spruce clearings, and often in lava flows.

Comments: Deciduous. Young twigs are hairy, older branches are gray and shreddy. Buds are pink. Dried blooms remain into fall and winter. Plant is browsed by deer. One species of *Holodiscus* in Arizona. Photograph taken at Sunset Crater National Monument, June 22.

© J BREW (BREWBOOKS), FLICKR.COM

ROCK MAT

Petrophyton caespitosum
Rose Family (Rosaceae)

Height: To 8″.

Flowers: White to light pink, with 5 petals, long stamens; to ⅛″ long; in dense, spikelike raceme.

Leaves: Grayish green and spatula-shaped, with silky hairs; to ¼″ long; basal rosettes form dense mats.

Blooms: June–October.

Elevation: 5,000 to 8,000′.

Habitat: Rock crevices and ledges.

Comments: A well-named plant; in Greek *petra* means "rock" and *phyton* means "plant." One species of *Petrophyton* in Arizona. Photograph taken at Walnut Canyon National Monument, September 7.

MOUNTAIN NINEBARK
Physocarpus monogynus
Rose Family (Rosaceae)

Height: Small shrub to 4′.

Flowers: White, with 5 petals, 5 sepals, and yellow stamens; to ⅝″ wide; in rounded, terminal cluster.

Leaves: Dull, dark green above, paler green and powdery beneath; palmately lobed, toothed, prominent network; to 3″ long.

Blooms: June–July.

Elevation: 8,000 to 9,500′.

Habitat: Pine and mixed conifer forests.

Comments: Old bark of the shrub continually shreds. Native Americans used boiled roots as poultice. One species of *Physocarpus* in Arizona. Photograph taken south of Alpine, June 30.

COMMON CHOKECHERRY
Prunus virginiana
Rose Family (Rosaceae)

Height: To 25′.

Trunk: To 8″ in diameter.

Bark: Shiny and reddish brown, darker and scaly as plant ages.

Flowers: White, with 5 rounded petals; to ½″ wide; in cylindrical cluster to 4″ long; followed by shiny, dark red or black, juicy, bitter cherry, to ⅜″ in diameter.

Leaves: Shiny green above, lighter green beneath (occasionally slightly hairy); elliptical, with finely, sharply saw-toothed margins; to 4″ long, 2″ wide.

Blooms: April–June.

Elevation: 4,500 to 8,000′.

Habitat: Roadsides, along streams, and in forests and woodland clearings.

Comments: Cherries are very bitter, with poisonous pits. Bears are fond of them. Cherries are used for making syrup, jelly, and wine. There are 2 varieties of this species *demissa* and *melanocarpa*. Seven species of *Prunus* in Arizona. Photograph taken at Chiricahua National Monument, April 25.

RED RASPBERRY
Rubus idaeus ssp. *strigosus (Rubus strigosus)*
Rose Family (Rosaceae)

Height: To 5′.

Flowers: White, with 5 petals, 5 prominent sepals which are longer than petals and have spiny undersides; to 1″ wide; followed by bright red fruits when mature.

Leaves: Dark green above, grayish green beneath; pinnate, 3 to 7 toothed leaflets; to 8″ long.

Blooms: June–July.

Elevation: 7,000 to 11,000′.

Habitat: Ponderosa pine and spruce forests.

Comments: Pinkish stems are prickly. Fruits eaten by birds and other wildlife. People eat them raw or use for making jams and jellies; leaves and twigs are used for tea. Six species of *Rubus* in Arizona. Photograph taken near Woods Canyon Lake, June 5.

NEW MEXICAN RASPBERRY
Thimbleberry
Rubus neomexicanus
Rose Family (Rosaceae)

Height: To 5′, rarely to 9′.

Flowers: White, with 5 rounded petals and bright yellow stamens; to 3″ wide; followed by red, juicy raspberry.

Leaves: Green, finely haired above, velvety beneath; toothed, 3- to 5-lobed; to 3″ wide.

Blooms: May–September.

Elevation: 5,000 to 9,000′.

Habitat: Moist slopes and woods.

Comments: Young shrubs have brownish bark with vertical stripes. Browsed by deer. Fruits eaten by humans and by birds and other wildlife. Six species of *Rubus* in Arizona. Photograph taken on Mount Graham, April 21.

THIMBLEBERRY
Rubus parviflorus var. *parvifolius*
Rose Family (Rosaceae)

Height: To 6′, usually closer to 3′.

Flowers: White, with 5 rounded petals, numerous yellowish tan stamens, and 5 sharply pointed sepals; to 2″ wide; in terminal cluster of 3 or more; followed by a very seedy, red, raspberrylike fruit.

Leaves: Dark green, 5-lobed, toothed, hairy above, deeply veined; to 8″ wide.

Blooms: July–September.

Elevation: 8,000 to 9,500′.

Habitat: On slopes in ponderosa pine and spruce-fir forests.

Comments: A thornless, sprawling shrub. Berries eaten by birds and other wildlife. Six species of *Rubus* in Arizona. Photograph taken in mountains above Greer, July 8.

ARIZONA MOUNTAIN-ASH
Serbo
Sorbus dumosa
Rose Family (Rosaceae)

Height: To 10′.

Bark: Young growth is pinkish and hairy; mature bark is smooth and gray.

Flowers: White, with 5 petals, numerous stamens; to ¼″ wide; in flat-topped cluster; followed by orange-red berry at maturity.

Leaves: Dark green, paler green beneath; pinnate, sharp-toothed; to 7″ long; 5 to 13 elliptical leaflets to 2″ long.

Blooms: June–July.

Elevation: 7,500 to 10,000′.

Habitat: Moist soil in coniferous forests.

Comments: Not a true ash. Berries eaten by birds and small mammals. One species of *Sorbus* in Arizona. Photograph taken at Black Canyon Lake area, June 4.

ARIZONA ROSEWOOD

Vauquelinia californica
Rose Family (Rosaceae)

Height: Shrub, or small tree to 25′.

Trunk: To 8″ in diameter.

Bark: Gray to reddish brown, thin; shaggy or divided into small, square scales.

Flowers: White, with 5 rounded petals; to ¼″ wide; in dense, flat-topped, terminal cluster to 3″ wide; followed by hard, woody seed capsules, each to ¼″ wide, splitting into 5 sections and remaining all winter.

Leaves: Yellowish green and smooth above, with yellowish, sunken midvein; white, finely haired beneath; narrow, lance-shaped, thick, and stiff; leathery, spiny saw-toothed; to 4″ long, ½″ wide.

Blooms: May–June.

Elevation: 2,500 to 5,000′.

Habitat: Canyons and mountains in upper desert and oak woodlands.

Comments: Evergreen; slow growing. Wood is hard and heavy. One species of *Vauquelinia* in Arizona. Photograph taken on Peralta Trail in Superstition Mountains, May 20.

BUTTONBUSH

Button-Willow
Cephalanthus occidentalis
Madder Family (Rubiaceae)

Height: Shrub, or small tree to 10′.

Trunk: To 4″ in diameter.

Bark: Brown or gray, ridged, scaly.

Flowers: White with very tiny, brownish dots; narrow, tubular, with very long stamens tipped with yellow; white buds with yellow tips; fragrant, 4-lobed; to ½″ long, ⅛″ wide; clustered in round ball, to 1½″ in diameter. Followed by a rough button of seeds; green turning to brown; to ¾″ in diameter.

Leaves: Shiny, dark green above, slightly paler beneath; finely toothed, with prominent midvein; broadly lance-shaped to elliptical; pointed; opposite or in whorls of 3; to 5″ long, 2½″ wide.

Blooms: June–September.

Elevation: 1,000 to 5,000′.

Habitat: Wet ground bordering lake sand streams.

Comments: Deciduous. Has green stems with warts; poisonous foliage. Bark is used medicinally. Waterfowl feed on seeds. Plant is a source for honey. One species of *Cephalanthus* in Arizona. Photograph taken at Saguaro Lake, August 26.

COMMON BEDSTRAW
Goosegrass
Galium aparine
Madder Family (Rubiaceae)

Height: Sprawling to 3'.

Flowers: White to greenish, with 4 petals; to ⅛" wide; in small clusters on stems rising from the leaf axils.

Leaves: Dark green, hairy, linear to lance-shaped; to 3" long; 6 to 8 in whorls around stem.

Blooms: March–May.

Elevation: 2,000 to 8,000'.

Habitat: Along streams, in canyons, and in woodlands.

Comments: Weak-stemmed annual supported by other plants. Barbed stems cleave to fabric and fur. This species can be found in Europe and is likely exotic to Arizona. Early settlers used plants as mattress stuffing. Seeds used as a coffee substitute. More than a dozen species of *Galium* in Arizona. Photograph taken in Superstition Mountains, March 26. This species has leaves in whorls of 6 to 8 and 4-angled, spiny stems.

NORTHERN BEDSTRAW
Galium boreale ssp. *septentrionale*
Madder Family (Rubiaceae)

Height: To 2'.

Flowers: White, with 4 petals; to ⅛" wide; in dense, round-tipped clusters.

Leaves: Dark green, narrow, nearly smooth; lance-shaped to linear, with 3 prominent veins; to 2" long; in whorls of 4 around stem, smaller side shoots arise from main stems.

Blooms: July–September.

Elevation: 6,000 to 9,500'.

Habitat: Rocky slopes and clearings in pine forests.

Comments: More than a dozen species of *Galium* in Arizona. Photograph taken in Greer area, July 4. This species has leaves in whorls of 4 and a nearly smooth stem and leaves.

DESERT BEDSTRAW

Galium stellatum var. *eremicum*
Madder Family (Rubiaceae)

Height: To 2'.

Flowers: Cream-colored, with dark yellow stamens; 4 petals joined at base; starlike; to ⅛" wide, in loose clusters.

Leaves: Dark green, lance-shaped, sharp-pointed, and rough; in whorls of 4 or 5 on lower stems; to ¼" long.

Blooms: January–May.

Elevation: Below 3,000'.

Habitat: Dry, rocky slopes.

Comments: Many-branched shrubby plant with reddish, square stems. More than a dozen species of *Galium* in Arizona. Photograph taken in Superstition Mountains, March 26.

NARROWLEAF HOPTREE

Common Hoptree
Ptelea trifoliata ssp. *angustifolia*
Rue Family (Rutaceae)

Height: Shrub, or small tree to 20'.

Trunk: To 8" in diameter.

Bark: Brownish gray, and smooth.

Flowers: White to yellowish green, fragrant, with 4 petals; to ½" wide, in cluster in leaf axils, followed by light brown, flat, roundish, winged, hoplike fruit, to ½" wide.

Leaves: Dark green, long-stalked; to 5½" long; compound, with 3 ovate, slightly wavy-toothed leaflets, to 3" long.

Blooms: May–June.

Elevation: 3,500 to 8,500'.

Habitat: Canyons, pinyon-juniper woodlands, and ponderosa pine forests.

Comments: Deciduous. Plant parts have disagreeable odor. Fruit is used as hops substitute in brewing. One species of *Ptelea* in Arizona, with many varieties. Photograph taken at Oak Creek Canyon north of Sedona, May 29.

BASTARD TOADFLAX

False Toadflax

Comandra umbellata ssp. *pallida (Comandra pallida)*

Sandalwood Family (Santalaceae)

Height: To 14".

Flowers: White to pinkish white, starlike, petalless; 5 to 6 pointed lobes surround greenish center, 5 greenish white stamens; to ⅜" wide; in terminal cluster.

Leaves: Green, narrow, lance-shaped; to 1½" long.

Blooms: April–August.

Elevation: 4,000 to 9,500'.

Habitat: Mountainsides, oak woodlands, and ponderosa pine forests.

Comments: A root parasite on many different plants. One species of *Comandra* in Arizona. Photograph taken on mountainside at Madera Canyon, April 29.

WESTERN SOAPBERRY

Jaboncillo

Sapindus saponaria var. *drummondii (Sapindus marginatus)*

Soapberry Family (Sapindaceae)

Height: To 50'; in Arizona, usually much shorter and often just a large-sized shrub.

Bark: Grayish brown and furrowed.

Flowers: White, with 5 round-tipped petals; to ¼" wide; in large, branched, terminal cluster to 9" long; followed by a ½"-diameter, smooth fruit with a yellowish, translucent flesh.

Leaves: Light green splashed with yellow; smooth above, hairy beneath; alternate, deciduous, pinnate; 7 to 19 lance-shaped, unequal-sided leaflets; terminal leaflet smaller; toothless; leaflets to 4" long, leaf to 12" long.

Blooms: May–August.

Elevation: 2,400 to 6,000'.

Habitat: Canyon slopes, along streams, desert grasslands, and oak woodlands.

Comments: Slow grower, with hairy stems. Fruits contain saponin, a poisonous substance, and have been used as a soap for washing clothes. One species of *Sapindus* in Arizona. Photograph taken in vicinity of Portal, May 5.

GUM BULLY

Sideroxylon lanuginosum ssp. *rigidum (Bumelia lanuginosa* var. *rigida)*
Sapote Family (Sapotaceae)

Height: Shrub to 15' in Arizona.

Trunk: To 8" in diameter.

Bark: Gray, rough.

Flowers: White, 5-petaled, bell-shaped; 5-lobed, fragrant; to ⅛" wide; clustered at leaf axils; followed by juicy, blackish, egg-shaped fruits to ⅜" long.

Leaves: Shiny, dark green above, dense gray, matted hairs beneath; alternate or in small clusters; leathery; elliptical or lance-shaped; rounded at tip and narrowing to base; tiny, pinkish stems; to 2" long, to ½" wide.

Blooms: June–July.

Elevation: 3,000 to 5,000'.

Habitat: Along washes and streams.

Comments: Forms dense thickets in some areas. Twigs are often tipped with straight spines. Gray spines at leaf clusters grow to ⅝" long. Gum from stem used as a chewing gum. Wood used for making tool handles and cabinets. One species of *Sideroxylon* in Arizona. Photograph taken at Catalina State Park, November 10.

YERBA MANSA

Anemopsis californica
Lizard Tail Family (Saururaceae)

Height: To 16".

Flowers: Very tiny, numerous and dense; on conical spike to 2" long; with 7 or 8 1"-long, white, petallike bracts surrounding base of cone.

Leaves: Grayish green, leathery, oblong; mostly basal; to 6" long; smaller leaves erect on flower stem.

Blooms: May–August.

Elevation: 2,000 to 5,500'.

Habitat: Moist, saline soil.

Comments: Perennial herb. Root used medicinally by Native Americans and Spaniards. One species of *Anemopsis* in Arizona. Photograph taken at Hassayampa River Preserve, Wickenburg, May 7.

NEW MEXICO ALUM-ROOT
Heuchera novomexicana
Saxifrage Family (Saxifragaceae)

Height: Flower stalk to 2′.

Flowers: White and small, in small cluster on weak, 2′-long flower stalk.

Leaves: Dark green above, lighter beneath; roundish and scalloped; hairy, basal; on 4″-long stalks; to 2″ wide, 2″ long.

Blooms: May–June.

Elevation: 8,000 to 9,000′.

Habitat: Mixed coniferous forests.

Comments: Hairy plant. Six species of *Heuchera* in Arizona. Photograph taken south of Hannagan Meadow, June 30.

REDFUZZ SAXIFRAGE
Saxifraga eriophora (Micranthes eriophora)
Saxifrage Family (Saxifragaceae)

Height: To 6″, but usually less.

Flowers: White, with 5 petals, reddish hairs on calyx lobes; to 3⁄16″ wide; in loose, terminal cluster.

Leaves: Dark green above, reddish hairs beneath; succulent, somewhat thick; oval to elliptical, with scalloped or toothed margins; to 1″ long, ½″ wide; in basal rosette.

Blooms: March–mid-May.

Elevation: 5,000 to 8,500′.

Habitat: Moist mountain slopes in coniferous forests, often very mossy areas.

Comments: Reddish hairs on stems. Six species of *Saxifraga* in Arizona. Photograph taken on Mount Graham, May 3.

DIAMONDLEAF SAXIFRAGE

Rockfoil

Saxifraga rhomboidea (Micranthes rhomboidea)

Saxifrage Family (Saxifragaceae)

Height: Flower stem to 1′.

Flowers: White, 5-petaled, greenish in center with light yellow anthers; to ¼″ wide; in terminal cluster to ¾″ wide on erect, sticky, hairy stem.

Leaves: Dark green to light green, basal, diamond-shaped; toothed or scalloped, succulent, sparsely haired on margins; to 2″ long, 1¼″ wide.

Blooms: April–July.

Elevation: 5,500 to 11,000′.

Habitat: Moist meadows.

Comments: Perennial. Six species of *Saxifraga* in Arizona. Photograph taken at Willow Springs Lake, April 22.

MAIDEN BLUE-EYED MARY

Blue Lips

Collinsia parviflora

Figwort Family (Scrophulariaceae)

Height: To 1′, but usually less.

Flowers: Two whitish, rounded, upper lobes flaring upward; 2 deep blue lower lobes; bent forward, with a folded, purplish lobe between them; tubular; to ⅛″ wide, ⅜″ long; in terminal, leafy cluster on stem.

Leaves: Dark green above, bright maroon beneath; lance-shaped; in whorls at flower clusters, opposite on lower stem, to 1½″ long.

Blooms: February–June.

Elevation: 4,000 to 8,000′.

Habitat: Areas moist in the spring.

Comments: Annual herb, with reddish stems. One species of *Collinsia* in Arizona. Photograph taken northeast of Superior, April 3.

BRITTLESTEM
Mabrya acerifolia (Maurandya acerifolia)
Figwort Family (Scrophulariaceae)

Height: Prostrate, to 10" long.

Flowers: White to greenish white, 5-lobed, tubular; to 1" long.

Leaves: Dark green, downy, sticky; heart-shaped to kidney-shaped; coarsely toothed; to 1" wide, more wide than long.

Blooms: March–May.

Elevation: About 2,000'.

Habitat: Shaded cliffs and rock ledges.

Comments: Mat-forming plant with brittle stems. Stems often hang down from moist rock ledges. One species of *Mabrya* in Arizona. Photograph taken in Superstition Mountains, April 6.

PARRY LOUSEWORT
Pedicularis parryi
Figwort Family (Scrophulariaceae)

Height: To 20".

Flowers: White to yellowish white; narrow; upper lip compressed sideways and arched; lower lip 3-lobed and bent downward; to 1" long; on tall spike to 8" long.

Leaves: Dark green with some red; fernlike and narrow; deeply lobed and toothed; to 5" long, ½" at widest; mainly basal.

Blooms: June–September.

Elevation: 7,500 to 12,000'.

Habitat: Moist mountain meadows and stream banks.

Comments: Perennial herb; partially root-parasitic. *Pediculus* means "louse" in Latin; in Roman times seeds were used to kill lice. Five species of *Pedicularis* in Arizona. Photograph taken in Greer area, July 3.

SMALL GROUNDCHERRY

Chamaesaracha coronopus
Nightshade Family (Solanaceae)

Height: To 10" tall, spreading on ground to 1½'.

Flowers: Whitish to yellowish green, flat, wheel-shaped with 5 spreading lobes; to ¾" wide; on stems in leaf axils; followed by yellow, berrylike seed pod, to ¼" in diameter.

Leaves: Dark green, rough, thick, alternate; covered with scaly down; narrowly oblong, occasionally shallowly lobed; to 4" long.

Blooms: April–September.

Elevation: 2,500 to 7,500'.

Habitat: Dry plains, mesas, roadsides, and disturbed ground.

Comments: Perennial herb; many-branched. Native Americans eat seed pods. Three species of *Chamaesaracha* in Arizona. Photograph taken at Dead Horse Ranch State Park, May 28.

SACRED DATURA

Thornapple
Datura wrightii (Datura meteloides)
Nightshade Family (Solanaceae)

Height: To 4'.

Flowers: White tinged with lavender; trumpet-shaped, united petals; to 6" long; followed by prickly seed capsule to 2" in diameter.

Leaves: Grayish green, oval to heart-shaped, to 6" long.

Blooms: April–November.

Elevation: 1,000 to 6,000'.

Habitat: Washes and roadsides from deserts to mesas.

Comments: Perennial herb. Flowers open in early evening and close following day when struck by the sun's rays. All parts of plant extremely poisonous if ingested. *Datura* was one of most important medicinal plants to early Native Americans. Four species of *Datura* in Arizona. Photograph taken north of Payson, September 2.

DESERT THORN

Desert Wolfberry
Lycium macrodon
Nightshade Family (Solanaceae)

Height: To 10′.

Flowers: Greenish white, tubular, with 5 pointed lobes; to ½″ long; followed by fruit constricted below middle; 2- to 4-seeded.

Leaves: Dark green, linear to spatula-shaped, narrow; to 1½″ long, ¼″ wide.

Blooms: February–May.

Elevation: 500 to 2,000′.

Habitat: Desert and plains.

Comments: Branches end in spines. Immature twigs are woolly. Eleven species of *Lycium* in Arizona. Photograph taken at Desert Botanical Garden, Phoenix, March 3. This species native to Pinal County and southern Arizona.

DESERT TOBACCO

Tabaquillo
Nicotiana obtusifolia (Nicotiana trigonophylla)
Nightshade Family (Solanaceae)

Height: To 3′.

Flowers: White to greenish white, trumpet-shaped; to ¾″ long; in loosely branched clusters.

Leaves: Dark green, sticky, oval to lance-shaped; to 6″ long; upper leaves stalkless, with 2 lobes clasping sticky stem.

Blooms: Throughout the year.

Elevation: Below 6,000′.

Habitat: Washes and other sandy areas.

Comments: Perennial herb. Leaves contain nicotine. Four species of *Nicotiana* in Arizona. Photograph taken at Usery Mountain Recreation Area, March 18.

WHITE NIGHTSHADE
Solanum americanum (Solanum nodiflorum)
Nightshade Family (Solanaceae)

Height: To 30".

Flowers: White, with 5 starlike, united petals with yellow beak of stamens; to ¾" wide; followed by a shiny, black, pea-sized berry.

Leaves: Light green, triangular, lobed; to 3½" long.

Blooms: March–December.

Elevation: Not available. Photograph taken at 2,000'.

Habitat: Wasteland and roadsides in Maricopa and Pinal counties and probably elsewhere.

Comments: Introduced from tropical America. Fifteen species of *Solanum* in Arizona. Photograph taken in Superstition Mountains, March 26.

WILD POTATO
Solanum jamesii
Nightshade Family (Solanaceae)

Height: To 1'.

Flowers: White, deeply 5-cleft, with 5 orange stamens; to ¾" wide, in loose cluster.

Leaves: Dark green, alternate, sparsely haired; to 4¾" long, pinnately compound; 5 to 9 leaflets.

Blooms: July–September.

Elevation: 5,500 to 8,500'.

Habitat: Coniferous forests, clearings, and wooded slopes.

Comments: Perennial herb. Has small tubers, and was once used as food by Native Americans. Fifteen species of *Solanum* in Arizona. Photograph taken in vicinity of Nutrioso, August 3.

DESERT HACKBERRY

Celtis pallida
Elm Family (Ulmaceae)

Height: To 20′.

Flowers: Whitish, tiny, to ¹⁄₁₆″ wide; single or in small clusters on new growth; followed by twinned or single green, egg-shaped berry, to ¼″ long, which ripens to yellowish or orange in fall.

Leaves: Dark green, elliptical to oval, alternate, toothed or untoothed; smooth when young, rough when older; to 1½″ long, ¾″ wide.

Blooms: Summer.

Elevation: 1,500 to 3,500′.

Habitat: Washes, canyons, and open desert.

Comments: Spiny, evergreen shrub. Spines on new branches to ⅝″ long; on older branches to 1″ long. Fruits provide food for wildlife; cover for many species of birds. Native Americans use fruits for food. Wood occasionally used for fence posts. Two species of *Celtis* in Arizona. Photograph taken at Usery Mountain Recreation Area, September 3.

TEXAS FROG FRUIT

Phyla nodiflora (Phyla incisa)
Vervain Family (Verbenaceae)

Height: Creeping stems to 2′ or more.

Flowers: White, 4-lobed; to 1/6″ wide; arranged in circle, alternating from white with 1 yellowish orange lobe to white with 1 lavender lobe; compact flower head to ½″ wide, on long stem in leaf axil.

Leaves: Dark green, hairy, rough, and granular; prominent midvein; 2-toothed on each margin near pointed tip; oblong to wedge-shaped; to 1″ long.

Blooms: April–November.

Elevation: 400 to 7,000′.

Habitat: Open ground along lakesides, riverbanks, and in damp woodlands.

Comments: Plant roots at nodes; has hairy stems. Three species of *Phyla* in Arizona. Photograph taken at Lyman Lake State Park, June 28.

ANGIOSPERMS: YELLOW FLOWERS

BIGTOOTH MAPLE

Acer grandidentatum
Maple Family (Aceraceae)

Height: To 40′.

Trunk: To 8″ in diameter.

Bark: Light brown to gray; smooth or scaly.

Flowers: Yellow, 3/16″ long; in drooping cluster, followed by greenish, U-shaped, paired, winged seeds or "keys" to 1¼″ long. Male and female flowers on same tree.

Leaves: Shiny, dark green above, lighter beneath; turning red and yellow in fall; red stems, 3 to 5 lobes, not toothed; often appearing on short stems along length of trunk; to 4″ wide.

Blooms: April.

Elevation: 4,700 to 7,000′.

Habitat: Moist canyons, alongside streams, and in ponderosa pine forests.

Comments: Has spreading, rounded crown. Browsed by livestock and deer. Used for fuel, and can be tapped for syrup in late winter. Three species of *Acer* in Arizona. Photograph taken at Cave Creek, Portal, April 23. Toothless leaf margins identify this maple.

BOXELDER

Acer negundo
Maple Family (Aceraceae)

Height: To 50′.

Trunk: To 2½′ in diameter.

Bark: Gray to brown, becoming deeply furrowed with age.

Flowers: Yellowish green, 3/16″ long, female flowers in hanging, terminal cluster; male flowers in flat-topped cluster; on separate trees; each followed by paired, clustered, long, V-shaped, winged seed cases ("keys") to 1½″ long.

Leaves: Bright green above, lighter green and hairy beneath; thick, pinnately compound; to 6″ long; 3 to 7 toothed leaflets to 4″ long.

Blooms: April–May.

Elevation: 3,500 to 8,000′.

Habitat: Along streams, ponds, and lakes in oak woodlands and ponderosa pine forests.

Comments: A rapid grower, but short-lived. Three species of *Acer* in Arizona. Photograph taken at Oak Creek Canyon, September 9.

GOLDEN-FLOWERED AGAVE

Agave chrysantha
Agave Family (Agavaceae)

Height: Flower stalk to 20′.

Flowers: Golden yellow with no purplish tinge, in dense clusters of up to 300 flowers, on tall flower stalk.

Leaves: Bluish green, evergreen, thick; prickles on margins to ⅜″ long; crowded in basal cluster to 40″ wide, 32″ high.

Blooms: June–August.

Elevation: 3,000 to 6,000′.

Habitat: Foothills, mountains, and canyons.

Comments: After several years, plant produces flower stalk and then dies. Native Americans roasted emerging flower stalk for food. Twelve species of *Agave* in Arizona. Photograph taken at Superior, April 12, shows huge asparagus-like flower stalk. Flowering clusters similar to **Parry's Agave** (*Agave parryi* ssp. *parryi*) (at right), but lack purplish red tinge on buds.

PARRY'S AGAVE

Mescal
Agave parryi ssp. *parryi*
Agave Family (Agavaceae)

Height: Foliage to 20″, flowering stalk to 18′.

Flowers: Buds reddish orange, after opening turn yellow; 6 petallike parts to 2½″ long; facing skyward, in large, flattened, terminal clusters.

Leaves: Grayish green, spatula-shaped, concave on upper surface; hooked spines on margins; sharp, terminal spine; leaf to 20″ long, in large, basal rosette.

Blooms: June–August.

Elevation: 4,500 to 8,000′.

Habitat: Dry, rocky slopes.

Comments: After approximately 25 years plant sends up flowering stalk; after blooming, it dies. New plants already formed on root system take over. Pollinated by insects and hummingbirds. The juice of this species can be irritating to the skin. Native Americans use plant for food, fiber, soap, beverages, and medicines. Twelve species of agave in Arizona. Photograph taken near Payson, June 27.

SHIN DAGGER

Amole

Agave schotti var. *schottii*

Agave Family (Agavaceae)

Height: Unbranched flower stalk to 9'.

Flowers: Light yellow, waxy, and sweetly fragrant; 6 flower segments, long yellow stamens; flower to 2½" long, 1" wide; in crowded, narrow, elongated cluster to 40" long; followed by woody fruit capsule to ¾" long, ⅜" in diameter.

Leaves: Dark green to yellowish green; linear, frequently curved; margins have fine, curved fibers, spine-tipped, concave or flat on upper surface; without marginal spines; to 16" long, in crowded, basal rosette.

Blooms: May–October.

Elevation: 4,000 to 7,000'.

Habitat: Dry, exposed slopes.

Comments: Perennial. Resembles a yucca. Plants grow crowded together forming large nuts which cover good-sized areas. Twelve species of *Agave* in Arizona. Photograph taken at Molino Canyon, Mount Lemmon, May 13.

WOOLLY TIDESTROMIA

Tidestromia lanuginosa

Amaranth Family (Amaranthaceae)

Height: Prostrate; but occasionally up to 1½' high, to a mound of 5' wide.

Flowers: Yellow-green, petalless; 5 pointed sepals, 5 yellow-tipped stamens; flower to ⅛" wide, in clusters in leaf axils.

Leaves: Grayish green, very downy, in pairs; heart-shaped to oval to roundish, in clusters; to 2" long (including stem), 1" wide.

Blooms: June–October.

Elevation: 100 to 5,500'.

Habitat: Fields, dry plains, and roadsides.

Comments: Annual; very hairy pinkish stems. Host plant of beet leafhopper. Collects blowing sand. Three species of *Tidestromia* in Arizona. Photograph taken in Mesa, June 16.

SKUNK BUSH

Rhus aromatic var. *trilobata (Rhus trilobata)*
Cashew Family (Anacardiaceae)

Height: Shrub to 10'.

Flowers: Yellow, appearing before leaves; to ¾" wide; in dense cluster on spike; followed by cluster of sticky, bright, orange-red berries.

Leaves: Dark green, shiny; compound with 3 oval, coarsely toothed, lobed leaflets; to 1¼" long.

Blooms: March–June.

Elevation: 2,500 to 7,500'.

Habitat: Pinyon-juniper areas, canyons, mesas, and slopes.

Comments: Closely related to poison ivy. Leaves look like miniature poison ivy leaves and turn red in fall; emit strong odor when bruised. Berries not poisonous; small mammals and birds eat berries. Sheep, antelope, and deer browse on twigs and foliage. A lemonade-like drink is made from fruits. Native American women used stems in basket weaving. Eight species of *Rhus* in Arizona. Photograph taken near Nutrioso, August 3.

MOUNTAIN PARSLEY

Alpine False Spring Parsley
Pseudocymopterus montanus
Carrot Family (Apiaceae)

Height: To 2', but highly variable.

Flowers: Minute; color varies greatly from yellow to reddish purple; in flat-topped, terminal cluster of variable size.

Leaves: Dark green, pinnately compound, varies substantially in shape and size.

Blooms: May–October.

Elevation: 5,500 to 12,000'.

Habitat: Ponderosa pine forests, mixed conifer forests, and grasslands.

Comments: Perennial herb. One species of *Pseudocymopterus* in Arizona. Photographed in Willow Springs Lake area, August 19.

CORN-KERNEL MILKWEED

Broadleaf Milkweed
Asclepias latifolia
Milkweed Family (Asclepiadaceae)

Height: To 3'.

Flowers: Pale yellow and white, with 5 upward-pointing hoods above a pedestal; 5 downward-pointing petals below pedestal; very corn kernel–like in appearance; to ⅜" wide, ½" long, followed by a broad, tapered pod to 5" long.

Leaves: Green to bluish green, roundish, smooth, and leathery; succulentlike; with reddish midvein, whitish side veins; to 6" long, 6" wide.

Blooms: June–August.

Elevation: 3,000 to 7,000'.

Habitat: Roadsides, plains, and mesas.

Comments: Perennial herb. Produces a milky sap. Twenty-nine species of *Asclepias* in Arizona. Photograph taken north of Springerville, August 5.

DESERT MILKWEED

Asclepias subulata
Milkweed Family (Asclepiadaceae)

Height: To 4'.

Flowers: Yellowish to cream; to ½" wide, in cluster to 2" wide; followed by a smooth, tapered seed pod to 4" long.

Leaves: Mostly leafless on mature plants. New growth has 2"-long leaves, which soon drop.

Blooms: April–October.

Elevation: Below 3,000'.

Habitat: Dry mesas, slopes, flats, and sandy washes.

Comments: Perennial herb. Has numerous, erect, gray-green stems that produce a milky sap. Each seed has a tuft of silky hairs. Twenty-nine species of *Asclepias* in Arizona. Photograph taken at Usery Mountain Recreation Area, May 17.

BUTTERFLY WEED

Asclepias tuberosa

Milkweed Family (Asclepiadaceae)

Height: To 3'.

Flowers: Brilliant yellow to orange (page 214); to ½" wide, ½" long; 5 small sepals, 5 petals (bent back) and 5 hoods, in flat-topped, erect, terminal cluster to 3" wide; followed by a narrow, tapered pod to 5" long.

Leaves: Light green, narrowly arrow-shaped; to 4 ½" long.

Blooms: May–September.

Elevation: 4,000 to 8,000'.

Habitat: Dry, open grasslands and open areas in pine forests.

Comments: Perennial, bushy herb with hairy stems. Seeds have white, silky hairs. Unlike most milkweeds, sap of this species is not milky. Twenty-nine species of *Asclepias* in Arizona. Photograph taken at Oak Creek Canyon, June 18.

SAN FELIPE FOETID MARIGOLD

Adenophyllum porophylloides (Dyssodia porophylloides)

Sunflower Family (Asteraceae)

Height: To 2'.

Flowers: Yellowish orange, to 3/16" long; bracts have dark glands; large flower head to ½" wide, 1" long.

Leaves: Dark green, opposite and alternate; pinnately 3- or 5-parted into narrow lobes; to 1" long, 7/8" wide.

Blooms: March–October.

Elevation: 500 to 3,000'.

Height: Mesas, washes, and rocky slopes.

Comments: A straggly bush. When crushed, plant gives off disagreeable odor. Four species of *Adenophyllum* in Arizona. Photograph taken at Kings Canyon, Tucson, April 17.

PALE AGOSERIS

Mountain Dandelion
Agoseris glauca var. *glauca*
Sunflower Family (Asteraceae)

Height: Flower stem to 18".

Flowers: Yellow; all ray flowers, center ones shorter; to 1½" wide; single; terminal on erect, leafless stem; followed by round seed heads with soft, white bristles attached.

Leaves: Dark green, basal, producing milky sap when crushed; shape varies from narrow to broadly lance-shaped; toothed or not, or deeply pinnately divided; to 14" long.

Blooms: May–October.

Elevation: 6,500 to 10,000'.

Habitat: Mountain meadows and coniferous forests.

Comments: Perennial herb. Three species of *Agoseris* in Arizona. Photograph taken at Lake Mary, June 1.

WILD CHRYSANTHEMUM

Ragleaf
Amauriopsis dissecta (Bahia dissecta)
Sunflower Family (Asteraceae)

Height: To 3'.

Flowers: Yellow rays, darker yellow disks; to ¾" wide; numerous; in open, branched cluster.

Leaves: Green, pinnate, divided 2 or 3 times; basal leaves deeply lobed, to 3¼" long.

Blooms: August–October.

Elevation: 5,000 to 9,000'.

Habitat: Roadsides, pinyon-juniper woodlands, and clearings in ponderosa pine forests.

Comments: Biennial or short-lived perennial. One species of *Amauriopsis* in Arizona. Photograph taken in vicinity of Mormon Lake, September 3.

MEADOW ARNICA

Leafy Arnica
Arnica chamissonis
Sunflower Family (Asteraceae)

Height: To 2½′.

Flowers: Yellow rays notched at tips, with brownish yellow to orangish disks; to 1½″ wide; single or up to 3 terminal flower heads. Bracts on flower head have tufts of white hairs inside tips.

Leaves: Grayish green, felty-haired, opposite; clasping stem; 5 to 12 pairs; toothed, oblong to lance-shaped; to 4½″ long.

Blooms: July–August.

Elevation: 8,000 to 9,500′.

Habitat: Moist mountain meadows and mountain lakesides.

Comments: Single-stemmed, sticky-haired plant. Thirteen species of *Arnica* in Arizona. Photograph taken at Lee Valley Reservoir in mountains above Greer, August 7. A similar species, **Heartleaf Arnica** (*Arnica cordifolia*), has pointed ray flowers, heart-shaped leaves, and solitary flower heads.

BAHIA

Bahia absinthifolia
Sunflower Family (Asteraceae)

Height: To 16″.

Flowers: Bright yellow rays, thick, orange disk flowers; to 2″ wide.

Leaves: Grayish, hairy, 3-lobed, to 2½″ long.

Blooms: April–October.

Elevation: 2,500 to 5,500′.

Habitat: Mesas, plains, and slopes.

Comments: Three species of *Bahia* in Arizona. There are varieties and intermediates of this species which vary in leaf and flower characteristics. Photograph taken in Tucson area, April 18.

PARISH GOLDENEYE

Bahiopsis parishii (Viguiera deltoidea var. *parishii)*
Sunflower Family (Asteraceae)

Height: Shrub to 4'.

Flowers: Yellow rays with lighter yellow tips; darker yellow disks; to 1¼" wide; solitary, on long stalks at branch tips.

Leaves: Dark green, shiny, crinkled, mostly opposite; triangular-shaped, hairy, toothed; to 1½" long.

Blooms: February–June.

Elevation: 1,000 to 3,500'.

Habitat: Rocky slopes, canyons, and mesas.

Comments: Many-branched, with grayish bark. Two species of *Bahiopsis* in Arizona. Photograph taken in vicinity of Tortilla Flat, March 19.

DESERT MARIGOLD

Baileya multiradiata
Sunflower Family (Asteraceae)

Height: To 2'.

Flowers: Yellow, daisylike; to 2" wide; on long, nearly leafless stem.

Leaves: Grayish, woolly, well-divided, and lobed; to 3" long.

Blooms: March–October at various intervals; year-round under ideal conditions.

Elevation: To 5,000'.

Habitat: Roadsides, slopes, and sandy, gravelly areas.

Comments: Annual herb. On overgrazed land, sheep and goats are frequently poisoned by feeding on this marigold. With age, flower petals become bleached and tissue paper–like. Three species of *Baileya* in Arizona. Photograph taken at Usery Mountain Recreation Area, March 18.

WILLOW RAGWORT
Willow Groundsel
Barkleyanthus salicifolius (Senecio salignus)
Sunflower Family (Asteraceae)

Height: To 7'.

Flowers: Yellow, ray flowers vary in number from 2 to 8; flower head to ⅞" wide.

Leaves: Dark green, willowlike, narrow; tapered at both ends; to 4" long.

Blooms: February–May.

Elevation: 2,500 to 5,000'.

Habitat: Along streams and moist washes.

Comments: Sprawling bush. One species of *Barkleyanthus* in Arizona. Photograph taken at Harshaw, April 27.

CHUCKWALLA'S DELIGHT
Sweet Bush
Bebbia juncea var. *aspera*
Sunflower Family (Asteraceae)

Height: To 4', usually straggly.

Flowers: Yellow, rayless, with hairy bracts; tubular, sweet-smelling; flower head to ½" wide, ⅝" long; solitary or several; terminal on many branches.

Leaves: (When present) dark green, sparse, alternate; linear to lance-shaped; lobed, rough, hairy; to 2" long.

Blooms: Much of the year.

Elevation: Below 4,000'.

Habitat: Canyons, rocky slopes, sandy washes, and roadsides.

Comments: Bushy, slender-branched plant with hairy stems. Attracts butterflies. One species of *Bebbia* in Arizona. Photograph taken in vicinity of Granite Reef Dam, April 25.

LYRELEAF GREENEYES

Chocolate Flower
Berlandiera lyrata
Sunflower Family (Asteraceae)

Height: To 4'.

Flowers: Yellow rays (5 to 12) veined with reddish tint beneath; maroon disk; prominent, nearly flat, broad bracts; flower head to 1½" wide.

Leaves: Grayish green, velvetlike; deeply lobed or pinnately divided into segments, with largest at end; to 5" long.

Blooms: April–October.

Elevation: 4,000 to 5,000'.

Habitat: Roadsides and fields of southeastern Arizona.

Comments: Perennial herb. Gives off a faint smell of chocolate when rays are pulled from flower head. Native Americans used flower heads as food seasoning. One species of *Berlandiera* in Arizona. Photograph taken in vicinity of Sierra Vista, April 26.

COULTER'S BRICKELLBUSH

Brickellia coulteri
Sunflower Family (Asteraceae)

Height: To 3'.

Flowers: Yellowish green, rayless, with brownish to pinkish bracts; slender flower head to ⅝" long, ¼" wide; approximately 15 flowers in loose, terminal cluster.

Leaves: Light green, triangular-shaped, tapering to tip; toothed, opposite, to 1½" long.

Blooms: March–November.

Elevation: 2,000 to 4,000'.

Habitat: Canyons and dry, rocky slopes.

Comments: Forms a roundish shrub with brownish, hairy, brittle stems. Sticky plant. More than 2 dozen species of *Brickellia* in Arizona. Photograph taken at Kings Canyon, Tucson, April 17.

YELLOW TACKSTEM

Calycoseris parryi
Sunflower Family (Asteraceae)

Height: To 12".

Flowers: Butter yellow, with pinkish streaks or dots on back of petals; to 1½" wide.

Leaves: Grayish green, narrow; with narrow lobes near base of plant, sparser and smaller lobes toward ends of stems; to 3" long. Tiny stalked red glands on sepals, leaves and stems.

Blooms: March–May.

Elevation: 500 to 4,000'.

Habitat: Sandy soil of mesas, plains, and hillsides.

Comments: Annual. Named "tackstem" for its tack-shaped glands on stems. Two species of *Calycoseris* in Arizona. Photograph taken at Organ Pipe Cactus National Monument, March 30.

YELLOW STAR THISTLE

Centaurea solstitialis
Sunflower Family (Asteraceae)

Height: To 20".

Flowers: Bright yellow disk flowers, erect; to ½" wide; yellowish spines to ¾" long extending from green bracts surrounding flower head; followed by seeds attached to hairy fluff.

Leaves: Grayish green, finely haired, spineless; wavy, divided at base; to 2" long; linear and decreasing in size along stem; grasping and extending like wings along entire stem.

Blooms: May.

Elevation: Below 4,000'.

Habitat: Disturbed areas, roadsides, and fields.

Comments: Annual weed. Well-branched. Native of Europe; now naturalized in parts of U.S. Over a dozen species of *Centaurea* in Arizona. Photograph taken in vicinity of Granite Reef Dam, May 14.

PARRY'S THISTLE
Cirsium parryi
Sunflower Family (Asteraceae)

Height: To 40".

Flowers: Yellowish green, very narrow, hairy; in flower head with outer bracts fringed with bristles and a spine; flower head to 1¼" wide; occurring singly or several in a cluster.

Leaves: Dark green above, grayish green beneath, hairy, spiny-margined; to 10" long.

Blooms: July–September.

Elevation: 7,500 to 9,500'.

Habitat: Clearings in coniferous forests and mountain meadows.

Comments: Has hairy stems. Around two dozen species of *Cirsium* in Arizona. Photograph taken in mountain clearing above Greer, August 8.

CALLIOPSIS
Golden Tickseed
Coreopsis tinctoria
Sunflower Family (Asteraceae)

Height: To 3'.

Flowers: Yellow with reddish brown at base; 5 to 7 rays; to 1¼" wide.

Leaves: Dark green; pinnate, with 2 to 3 pairs of linear lobes; to 4" long.

Blooms: June–September.

Elevation: Not available. Photograph taken at 7,000'.

Habitat: Moist fields, waste areas, and roadsides.

Comments: Annual. An escapee from cultivation. Five species of *Coreopsis* in Arizona. Photograph taken at Upper Lake Mary, September 2.

BRITTLEBUSH
Incienso
Encelia farinosa
Sunflower Family (Asteraceae)

Height: To 4'.

Flowers: Bright yellow, daisylike rays and disks; to 2" wide, in branched clusters; on tall, brittle stems forming a yellow canopy above foliage. In some areas of Arizona, a form of this species has flower heads with brownish-red disk flowers.

Leaves: Greenish gray to silvery gray and woolly; oval, oblong or triangular-shaped; to 4" long; leaves are smaller during dry conditions.

Blooms: November–May (in frost-free areas).

Elevation: Below 3,000'.

Habitat: Slopes, washes, and flats.

Comments: Shrubby perennial. A rounded bush. During drought, leaves turn brown and drop, and are replaced by tiny new leaves. Stems are easily broken; exude a resin or gum which was chewed by Native Americans and burned as incense (*incienso*) in early missions. Four species of *Encelia* in Arizona. Photograph taken at Usery Mountain Recreation Area, March 7.

RAYLESS ENCELIA
Green Brittlebush
Encelia frutescens
Sunflower Family (Asteraceae)

Height: To 3'.

Flowers: Yellowish orange, rayless, dome-shaped; flower head to ¾" wide; solitary; on hairy stem.

Leaves: Dark green, very shiny above, hairy on margins and beneath; wavy-margined, rough, oval to oblong; to ½" wide, 2" long.

Blooms: January–September.

Elevation: Below 4,000'.

Habitat: Mesas and rocky slopes.

Comments: Low, branching shrub often to 3' wide. Young stems are pinkish, older stems are whitish. Four species of *Encelia* in Arizona. Photograph taken near Salome, March 28.

ENGELMANN'S DAISY

Cutleaf Daisy
Engelmannia peristenia (Engelmannia pinnatifida)
Sunflower Family (Asteraceae)

Height: To 3'.

Flowers: Bright yellow, usually with 8 rays, slightly curled under at tips; yellow disks: to 1¼" wide; in loose, terminal cluster on upper stems.

Leaves: Dark green above, lighter green beneath; alternate, very rough and hairy, deeply pinnately lobed; clasping stem; basal leaves to 8" long, becoming shorter on upper stems.

Blooms: May–September.

Elevation: 4,000 to 6,500'.

Habitat: Dry prairies and hills.

Comments: Perennial herb. Hairy-stemmed. One species of *Engelmannia* in Arizona. Photograph taken in vicinity of Portal, May 4.

TURPENTINE BUSH

Ericameria laricifolia
Sunflower Family (Asteraceae)

Height: Bush to 40".

Flowers: Golden yellow, with up to 11 ray flowers, up to 13 disk flowers; to ⅜" long, ⅜" wide; in dense cluster; terminal on leafy branches.

Leaves: Light green to grayish green; linear, leathery, crowded, glandular-dotted; to ¾" long, 1⁄16" wide.

Blooms: August–December.

Elevation: 3,000 to 6,000'.

Habitat: Canyons, rocky slopes, and mesas.

Comments: Foliage smells like turpentine when crushed. Plant contains small amount of rubber. Attractive to bees and other insects. Over a dozen species of *Ericameria* in Arizona. Photograph taken east of Camp Verde, September 30.

RABBIT BRUSH
Chamisa
Ericameria nauseosa (Chrysothamnus nauseosus)
Sunflower Family (Asteraceae)

Height: To 5'.

Flowers: Yellow, feathery, slender, and rayless; to ¼" wide, ½" long; in terminal cluster; followed by seeds with white bristles attached.

Leaves: Grayish green, linear, very narrow; to 2½" long; at intervals along stem of soft, matted hairs.

Blooms: July–October.

Elevation: 3,000 to 9,000'.

Habitat: Dry plains, dry mountainsides, grassland, open woodlands, and roadsides.

Comments: Perennial shrub, with slender, flexible branches. Twigs are covered with feltlike hairs. Eaten by rabbits and browsed by deer, elk, and pronghorn. Provides shelter for birds and small mammals. Flowers attract insects, and yield yellow dye used by Navajo Indians. Inner bark is a source of green dye. The Hopi use plant for kiva fuels and wind breaks, and in arrow and wicker work. Latex obtained from plant is of no commercial value. Fourteen species of *Ericameria* in Arizona; this species has many varieties. Photograph taken at Upper Lake Mary, September 2.

PRINGLE'S WOOLLY SUNFLOWER
Eriophyllum pringlei
Sunflower Family (Asteraceae)

Height: To 2".

Flowers: Yellow, all disk flowers; to 1/16" wide, in ¼"-wide clusters.

Leaves: Grayish green, woolly haired, linear, 3-lobed at tips; to 3/16" long; surrounding flower cluster.

Blooms: March–May.

Elevation: 1,500 to 3,000'.

Habitat: Sandy desert flats and slopes.

Comments: Grows in tufts. Five species of *Eriophyllum* in Arizona. Photograph taken at Catalina State Park, April 2.

SLENDER BLANKETFLOWER
Reddome Blanketflower
Gaillardia pinnatifida
Sunflower Family (Asteraceae)

Height: To 20".

Flowers: Yellow ray flowers; reddish to greenish disk flowers in rounded dome; to 1¾" wide.

Leaves: Green, pinnately lobed, to 3" long.

Blooms: May–October.

Elevation: 3,500 to 7,000'.

Habitat: Fields, plains, mesas, and clearings in ponderosa pine forests.

Comments: Perennial herb. Hopi Indians use this species as a diuretic. Five species of *Gaillardia* in Arizona. Photograph taken at Portal, April 23.

DESERT SUNFLOWER
Desert Gold
Geraea canescens
Sunflower Family (Asteraceae)

Height: To 3'.

Flowers: Yellow rays, golden orange disks, to 2" wide; terminal on branches.

Leaves: Grayish green, very hairy, diamond-shaped, with toothed margins; to 3" long.

Blooms: January–June; abundant in April.

Elevation: Below 3,000'.

Habitat: Sandy desert roadsides and flats.

Comments: Annual. Fragrant; attractive to bees and hummingbird moths. Seeds are a food source for small rodents and birds. One species of *Geraea* in Arizona. Photograph taken north of Yuma, March 29.

CURLYTOP GUMWEED
Mountain Gumplant
Grindelia nuda var. *aphanactis (Grindelia aphanactis)*
Sunflower Family (Asteraceae)

Height: To 16".

Flowers: Yellow, rayless, ball-shaped flower head of rounded, linear, semi-hooked, sticky bracts; to 1¼" wide.

Leaves: Dark green, narrow, straplike, sticky, with toothed margins, to 2⅜" long.

Blooms: June–October.

Elevation: 5,000 to 7,000'.

Habitat: Roadsides, fields, and clearings in ponderosa forests.

Comments: Annual or biennial. Sticky-gummy plant. Nine species of *Grindelia* in Arizona. Photograph taken at Nelson Reservoir, August 3. Recognizable by its absence of ray flowers.

CURLYCUP GUMWEED
Rosinweed
Grindelia squarrosa
Sunflower Family (Asteraceae)

Height: To 3'.

Flowers: Yellow rays, darker yellow disk flowers; daisylike; tips of bracts surrounding flower head are rolled back and very sticky; to 1½" wide.

Leaves: Dark green, oblong, stemless, clasping stem at base, to 4" long.

Blooms: July–September.

Elevation: 4,000 to 7,500'.

Habitat: Dry, open fields, and waste places.

Comments: Perennial herb; sticky-gummy plant. Plant used medicinally. Invades overgrazed ranchland. Nine species of *Grindelia* in Arizona. Photograph taken near Show Low, August 5. Recognizable by presence of ray flowers.

BROOM SNAKEWEED

Turpentine Weed
Gutierrezia sarothrae
Sunflower Family (Asteraceae)

Height: To 2'.

Flowers: Yellow, tiny, 3 to 8 ray flowers to ⅛" long; 2 to 8 disk flowers; flower heads to ¼" long; in dense clusters at ends of branches.

Leaves: Dark green, very narrow, to ⅛" wide, 2½" long; at intervals along stem.

Blooms: July–December.

Elevation: 3,000 to 8,000'.

Habitat: Plains, pinyon-juniper woodlands, ponderosa forest clearings, and roadsides.

Comments: Perennial herb. Plant forms mound. Where abundant, it indicates overgrazed land. Poisonous to livestock; causes abortion of fetuses. Seeds eaten by birds. Once used medicinally on sheep for snakebite. Dried stems made into primitive brooms. Chewed leaves were once placed on ant, bee, or wasp stings to reduce swelling. Seven species of *Gutierrezia* in Arizona. Photograph taken at Canyon de Chelly National Monument, July 27. A similar species., **Threadleaf Snakeweed** (*Gutierrezia microcephala*), has fewer and more slender flower heads.

GUMHEAD

Gymnosperma glutinosum
Sunflower Family (Asteraceae)

Height: To 4'.

Flowers: Yellow, with inconspicuous rays; flower to ⅜" wide, ½" long; in clusters on stems off main stem; followed by minute pappus (hairs).

Leaves: Dark green, alternate, linear; rough, curled a bit, to 2" long; occurring all up stem; smaller leaves in clusters at base of leaf stem.

Blooms: March–December.

Elevation: 1,000 to 6,000'.

Habitat: Rocky canyons and slopes.

Comments: A slightly woody plant toward base; has light brown stem. Plant used medicinally in Mexico. One species of *Gymnosperma* in Arizona. Photograph taken at Organ Pipe Cactus National Monument, November 14.

ARIZONA SNEEZEWEED
Helenium arizonicum
Sunflower Family (Asteraceae)

Height: To 4'.

Flowers: Yellow, 3-lobed ray flowers; purplish brown, globular disk flowers; to 2" wide; occurring singly at tips of stems.

Leaves: Dark green, narrow, to 5" long; becoming smaller on upper stems.

Blooms: July–September.

Elevation: 7,000 to 8,000'.

Habitat: Roadsides and clearings in ponderosa forests.

Comments: Five species of *Helenium* in Arizona. Photograph taken at Willow Springs Lake area, July 21.

ASPEN SUNFLOWER
Five-Nerve Helianthella
Helianthella quinquenervis
Sunflower Family (Asteraceae)

Height: To 4'.

Flowers: Pale yellow rays, with greenish yellow disks; sunflowerlike, hairy bracts; to 3" wide; solitary or in a few-flowered cluster.

Leaves: Grayish green, hairy, with 5 prominent veins; broadly lance-shaped; to 10" long; mostly basal.

Blooms: July–October.

Elevation: 5,000 to 10,000'.

Habitat: Mountain meadows, slopes, and coniferous clearings, often near aspen groves.

Comments: Perennial herb. Wildlife graze on flower heads. Four species of *Helianthella* in Arizona. Photograph taken in Greer area, July 5. A similar species, **Parry's Dwarf Sunflower** (*Helianthella parryi*) has narrower leaves and smaller flowers.

COMMON SUNFLOWER

Mirasol
Helianthus annuus
Sunflower Family (Asteraceae)

Height: To 9′.

Flowers: Bright yellow ray flowers, maroon disk flowers; to 5″ wide.

Leaves: Dull green, stiff hairs; lower leaves are broadly triangular to heart-shaped; irregularly toothed; to 12″ long.

Blooms: May–October.

Elevation: 100 to 7,000′.

Habitat: Roadsides and fields.

Comments: Annual; state flower of Kansas. Frost-sensitive. Flowers are heliotropic (face the sun as it moves across sky). Seeds eaten by birds, rodents, and humans. Native Americans use seeds to make purple and black dye; yellow dye is made from the flowers. Eight species of *Helianthus* in Arizona. Photograph taken near Mexican Hay Lake, July 21.

© MAX LICHER

PRAIRIE SUNFLOWER

Helianthus petiolaris ssp. *fallax*
Sunflower Family (Asteraceae)

Height: To 4′.

Flowers: Yellow rays, reddish brown disk flowers; to 2″ wide.

Leaves: Green, oblong to lance-shaped, toothed; to 2″ long.

Blooms: April–October.

Elevation: 1,000 to 7,500′.

Habitat: Roadsides, fields, and cultivated land.

Comments: Annual. Has hairy stems. Eight species of *Helianthus* in Arizona. Photograph taken at Sunset Crater National Monument, September 7. Shorter plant, with more compact flower heads than **Common Sunflower** (*Helianthus annuus)* (at left).

ANNUAL GOLDENEYE

Annual Viguiera
Heliomeris longifolia var. *annua (Viguiera annua)*
Sunflower Family (Asteraceae)

Height: To 3'.

Flowers: Yellow, pointed rays, yellowish orange disks, flat at first, later cone-shaped; flower head to ⅞" wide; terminal, and numerous on branches.

Leaves: Dark green, very narrow, rough, opposite and alternate; with margins rolled under; to 2½" long, ⅛" wide.

Blooms: May–October.

Elevation: 2,500 to 7,000'.

Habitat: Roadsides, fields, and hillsides.

Comments: In fall, turns hillsides golden yellow, especially in Yavapai County. Forage for sheep. Two species of *Heliomeris* in Arizona. Photograph taken north of Payson, September 30. This species of *Heliomeris* recognizable by its numerous small flowers and very narrow leaves with margins rolled under.

GOLDENEYE

Heliomeris longifolia var. *longifolia (Viguiera longifolia)*
Sunflower Family (Asteraceae)

Height: To 3'.

Flowers: Yellow rays, darker yellow disks, large bracts; to 2" wide; at tips of slender stalks.

Leaves: Shiny, dark green, willowlike; lance-shaped to linear; to 2" long.

Blooms: July–October.

Elevation: 4,500 to 8,000'.

Habitat: Clearings in ponderosa pine forests.

Comments: Perennial herb. Many-branched with reddish stems. Two species of *Heliomeris* in Arizona. Photograph taken at Upper Lake Mary, September 2.

MANY-FLOWERED GOLDENEYE

Heliomeris multiflora (Viguiera multiflora)
Sunflower Family (Asteraceae)

Height: To 4'.

Flowers: Yellow ray flowers, darker yellow disk flowers tinged with brown; disk is flat but as flower fades it becomes cone-shaped. Green bracts beneath flower head are in 3 layers: lowest are long and turn down; middle ones are short and turn down; upper ones are short and turn up. Flower head to 2½" wide, terminal on stem.

Leaves: Green, lance-shaped, opposite on lower stem; to 2¼" long, ⅜" wide.

Blooms: May–October.

Elevation: 4,500 to 9,500'.

Habitat: Mountain meadows, slopes, and ponderosa pine forests.

Comments: Perennial herb, with branches. Two species of *Heliomeris* in Arizona. Photograph taken near Willow Springs Lake, September 13. This species of *Heliomeris* recognizable by its layers of bracts and larger flowers and leaves.

CAMPHORWEED

Telegraph Plant
Heterotheca subaxillaris (Heterotheca psammophila)
Sunflower Family (Asteraceae)

Height: To 5'.

Flowers: Bright yellow rays, orange disk flowers, to 1" wide; in loose clusters on branches on a single, erect stem resembling a telegraph pole; followed by seed heads like tiny, mature dandelions.

Leaves: Light green, wavy, egg-shaped; thick, toothed, with clasping bases; to 3½" log.

Blooms: March–November.

Elevation: 1,000 to 5,500'.

Habitat: Roadsides, pastures, vacant lots, and other disturbed places.

Comments: Crushed leaves smell like camphor. Ten species of *Heterotheca* in Arizona. Photograph taken at Oak Creek Canyon, September 9.

HAIRY GOLDEN ASTER

Rosinweed
Heterotheca villosa (Chrysopsis villosa)
Sunflower Family (Asteraceae)

Height: 30".

Flowers: Yellow rays varying in number, darker yellow disks; to 1" wide; terminal, in cluster.

Leaves: Gray to grayish green, woolly haired; alternate, oblong to spoon-shaped (widest toward tip); crinkled; to 1¼" long; occurring all along stems.

Blooms: May–October.

Elevation: 1,500 to 8,500'.

Habitat: Plains, mesas, and clearings in ponderosa pine forests.

Comments: Brownish, hairy stems. Ten species of *Heterotheca* in Arizona. There are six varieties of this species, differing in hairiness and other characteristics. Photograph taken at Black Canyon Lake, September 29.

FENDLER'S HAWKWEED

Hieracium fendleri
Sunflower Family (Asteraceae)

Height: To 20".

Flowers: Yellow, small, all rays; to ½" wide, ¾" long.

Leaves: Green above, lighter green beneath; very hairy, in basal rosette; widest between middle and tip; to 5" long.

Blooms: May–August.

Elevation: 6,000 to 9,500'.

Habitat: Pine forests.

Comments: Perennial herb. Stem is very hairy, with ⅛"-long hairs. Around twenty species of *Hieracium* in Arizona. Photograph taken in vicinity of Ashurst Lake, June 1.

FINELEAF WOOLLYWHITE

Yellow Cut-Leaf
Hymenopappus filifolius
Sunflower Family (Asteraceae)

Height: To 30".

Flowers: Bright yellow, petalless; disk flowers enlarged with long stamens and style projecting upward and outward; flower head to ⅞" wide; number of flower heads varies.

Leaves: Grayish green above, gray beneath; feltlike hairs; pinnately divided into threadlike segments; leaves 3" to 8" long, depending on variety; basal in some varieties; in others 1 to 7 leaves extend up stem.

Blooms: May–September.

Elevation: Varies depending on variety.

Habitat: Rocky slopes or sandy soil or clearings in ponderosa pine forests depending on variety.

Comments: Very variable species (several varieties). Photograph taken in Ashurst Lake Area, September 5. Five species of *Hymenopappus* in Arizona. The flowers of **Mexican Woollywhite** (*Hymenopappus mexicanus*) are very similar, but the basal, woolly leaves are single, lance-shaped to partly lobed, and up to 6" long, ½" wide. Plant is 3' tall with lower reddish stem.

BITTERWEED

Hymenoxys bigelovii
Sunflower Family (Asteraceae)

Height: To 16".

Flowers: Yellow, notched rays; orange disks to ¾" wide; flower head to 2" wide.

Leaves: Dark green, very narrow; with whitish midvein in grooved, upper surface; hairy beneath, very hairy at base; clasping stem; mainly basal, to 7" long; with a few shorter, alternate leaves on upper stem.

Blooms: April–July.

Elevation: 5,500 to 7,500'.

Habitat: Pine forests.

Comments: Flower resembles a tall, sparsely leaved gaillardia. Fourteen species of *Hymenoxys* in Arizona. Photograph taken in vicinity of Upper Lake Mary, June 2.

COOPER'S GOLDFLOWER

Hymenoxys cooperi
Sunflower Family (Asteraceae)

Height: To 3'.

Flowers: Bright yellow, straggly rays, orangish disks; to 1" wide, on stalks up to 4" long; numerous stalks of flowers on plant.

Leaves: Grayish green, woolly, pinnate, with linear lobes in some varieties, wider leaves in others; to 3" long.

Blooms: May–September.

Elevation: 2,000 to 8,000'.

Habitat: Dry, rocky areas.

Comments: Two varieties of this species. Woolly stems. Fourteen species of *Hymenoxys* in Arizona. Photograph taken at North Rim of Grand Canyon National Park, July 13.

WESTERN SNEEZEWEED

Owl-Claws
Hymenoxys hoopesii (Helenium hoopesii)
Sunflower Family (Asteraceae)

Height: To 3'.

Flowers: Yellow to orange-yellow, narrow, straggly, droopy ray flowers, each tipped with 3 teeth; flat flower head to 3" wide.

Leaves: Grayish green, long, narrow, and woolly; progressively smaller toward flower head; to 1' long.

Blooms: June–September.

Elevation: 7,000 to 11,000'.

Height: Mountain meadows and coniferous forests.

Comments: Has woolly stems. Causes illness in sheep and is poisonous to cattle if eaten. Some people are allergic to pollen. Root is used medicinally. Yellow dye made from flower heads. Fourteen species of *Hymenoxys* in Arizona. Photograph taken at Lee Valley Reservoir, July 2.

SOUTHERN JIMMYWEED

Isocoma pluriflora (Isocoma wrightii)
Sunflower Family *(Asteraceae)*

Height: To 3′ high, 3′ wide.

Flowers: Yellow, to ½″ wide, ½″ long; flower head to 15 flowers, in terminal clusters on stems.

Leaves: Grayish green, very rough, linear with shorter linear leaf cluster on stem at base; alternate; to 2¼″ long, all along stem.

Blooms: June–September.

Elevation: Below 5,000′.

Habitat: Roadsides, plains, and mesas.

Comments: A weed. Often takes over on overgrazed land. If large amounts are eaten by cattle, it causes "milk-sickness" or "trembles," a disease transmitted through milk to humans. Six species of *Isocoma* in Arizona. Photograph taken at Usery Mountain Recreation Area, August 20. A similar species, **Burroweed** (*Isocoma tenuisecta*) is common in the Tucson area. Its leaves are glandular and linear with prominent side lobes.

PRICKLY LETTUCE

Wild Lettuce
Lactuca serriola
Sunflower Family (Asteraceae)

Height: To 6′.

Flowers: Yellow, composed solely of ray flowers; to ⅓″ wide; on short stalks, on branched, upper flowering stems; followed by miniature, dandelion-like tufts.

Leaves: Bluish green, clasping stem; with prominent white midvein smooth above, prickly veins beneath; prickly on margins, cut into deep, irregular lobes; to 10″ long at base, graduating to smaller size on upper stems.

Blooms: May–October.

Elevation: 1,000 to 8,000′.

Habitat: Roadsides and disturbed soil.

Comments: A weed. Stems filled with milky juice; main stem branches where flowering occurs. Introduced from Europe; now naturalized. Seven species of *Lactuca* in Arizona. Photograph taken at Ashurst Lake, September 4.

GOLDFIELDS

Lasthenia gracilis (Lasthenia californica, Baeria chrysostoma ssp. *gracilis)*
Sunflower Family (Asteraceae)

Height: To 8".

Flowers: Yellow ray flowers notched at tips; darker yellow disk flowers; to 1" wide; terminal on slender stems.

Leaves: Light green, hairy on both surfaces; narrow, linear, opposite; to 1½" long.

Blooms: March–May.

Elevation: 1,500 to 4,500'.

Habitat: Deserts, mesas, and plains.

Comments: Annual herb. Often grows in dense patches, forming carpets of gold. Grazed by horses. Fragrant flowers attract a species of small fly. Four species of *Lasthenia* in Arizona. Photograph taken in Superstition Mountains, March 22.

FENDLER'S DANDELION

Desert Dandelion
Malacothrix fendleri
Sunflower Family (Asteraceae)

Height: To 6".

Flowers: Yellow, notched petals, pink-striped on back; to 1" wide.

Leaves: Grayish green, pinnate, triangularly lobed, to 2" long.

Blooms: March–June.

Elevation: 2,000 to 5,000'.

Habitat: Foothills, sandy plains, and mesas.

Comments: Annual herb. Eight species of *Malacothrix* in Arizona. Photograph taken at Portal, April 23.

DESERT DANDELION

Malacothrix glabrata (Malacothrix californica var. *glabrata)*
Sunflower Family (Asteraceae)

Height: To 16".

Flowers: Pale yellow rays, no disk flowers, centers are red until all petals expand; to 1¾" wide.

Leaves: Green, linear, pinnately lobed, mostly basal; to 5" long.

Blooms: March–June.

Elevation: Below 7,000'.

Habitat: Dry, sandy flats of low desert and mesas.

Comments: Annual herb. Eight species of *Malacothrix* in Arizona. Photograph taken near Golden Shores, March 9.

SENECIO

Packera franciscana (Senecio franciscanus)
Sunflower Family (Asteraceae)

Height: To 4", in tufts to 3" wide.

Flowers: Yellow rays, orange disks; flower head to ½" wide, single or a few at tips of stems.

Leaves: Grayish green, edged in reddish purple; oval to roundish, downy, crinkly, toothed; to 2" long.

Blooms: July–August on San Francisco Peaks.

Elevation: On San Francisco Peaks up to 12,000'.

Habitat: Tundra on San Francisco Peaks.

Comments: Stems are reddish. Found only in Arizona, this species is federally protected. Nineteen species of *Packera* in Arizona. Photograph taken at the Arboretum at Flagstaff, June 3.

AXHEAD BUTTERWEED

Packera multilobata (Senecio multilobatus)
Sunflower Family (Asteraceae)

Height: To 14".

Flowers: Yellow rays, orangish disk flowers, to 1" wide; numerous flower heads in wide, flat-topped cluster, with many clusters per plant.

Leaves: Dark green, divided into sharply toothed lobes or segments; to 4" long.

Blooms: May–August.

Elevation: 6,000 to 8,000'.

Habitat: Rocky slopes.

Comments: Several varieties of this species. Nineteen species of *Packera* in Arizona. Photograph taken at North Rim of Grand Canyon National Park, May 30.

NEW MEXICO BUTTERWEED

New Mexico Groundsel
Packera neomexicana var. *neomexicana (Senecio neomexicanus)*
Sunflower Family (Asteraceae)

Height: To 32".

Flowers: Yellow rays, darker yellow disk flowers; to ⅞" wide; in terminal cluster.

Leaves: Grayish green with woolly hairs; oval to lance-shaped; sharply saw-toothed, mostly basal; to 3" long; few and smaller leaves on stem.

Blooms: April–August.

Elevation: 3,000 to 9,000'.

Habitat: Pine forest and oak chaparral.

Comments: Perennial herb. Most abundant of *Packera* species in Arizona. Nineteen species of *Packera* in Arizona. Photograph taken at Chiricahua National Monument, April 24.

OAK CREEK RAGWORT

Packera quercetorum (Senecio quercetorum)
Sunflower Family (Asteraceae)

Height: To 4'.

Flowers: Yellow rays, orange disks; to 1¼" wide.

Leaves: Dark green, pinnately lobed, toothed; end lobe to 3" long, side lobes shorter; leaf to 8" long; upper leaves smaller and clasping stem.

Blooms: March–May.

Elevation: 3,500 to 6,000'.

Habitat: Oak woodland areas.

Comments: Reddish to purplish hollow stems. Nineteen species of *Packera* in Arizona. Photograph taken northeast of Superior, April 20.

FOETID-MARIGOLD

Lemonweed
Pectis angustifolia var. *angustifolia*
Sunflower Family (Asteraceae)

Height: To 6".

Flowers: Bright yellow, 8 to 10 rays; to ¼" wide; clustered at end of branches.

Leaves: Dark green, smooth, linear, with 3 to 5 pairs of linear lobes; to 1½" long.

Blooms: August–November.

Elevation: 700 to 7,000'.

Habitat: Dry, sandy, or gravelly mesas.

Comments: Hopi Indians extracted a dye from plant, and also consumed the plant raw or dried. Nine species of *Pectis* in Arizona. Photograph taken at Painted Rocks State Park, November 12.

TAILLEAF PERICOME

Yerba De Chivato
Pericome caudata (Pericome glandulosa)
Sunflower Family (Asteraceae)

Height: To 5'.

Flowers: Yellow to orange-yellow, rayless, to ½" wide; in branched clusters to 2" wide.

Leaves: Dark green, triangular, tip tapers to long slender tail; limp, drooping; to 5" long, 2½" wide at widest part.

Blooms: July–October.

Elevation: 6,000 to 9,000'.

Habitat: Roadsides and slopes in pine forests.

Comments: Perennial herb; named by Spaniards *yerba de chivato* ("herb of the he-goat") because of its goatlike smell. Plant used for treating ailments. Bushes often grow to 6' wide, forming mound of yellow flowers. One species of *Pericome* in Arizona. Photograph taken at Mormon Lake, September 3.

LEMMON'S ROCK DAISY

Perityle lemmonii (Laphania lemmoni)
Sunflower Family (Asteraceae)

Height: To 1', but generally prostrate.

Flowers: Yellow, rayless; to ½" long, to ¼" wide.

Leaves: Dark green edged in brown; woolly haired, opposite, maplelike; cleft into narrow lobes; to ⅜" long, ¼" wide.

Blooms: May–October.

Elevation: 3,000 to 7,000'.

Habitat: Crevices of boulders and cliffs.

Comments: Brittle, hairy stems. Twelve species of *Perityle* in Arizona. Photograph taken at Molino Canyon, Santa Catalina Mountains, May 13.

DESERT FIR

Pigmy Cedar
Peucephyllum schottii
Sunflower Family (Asteraceae)

Height: Shrub to 4½".

Flowers: Yellow, rayless, ½" wide; occurring at tips of branches.

Leaves: Green, hairlike, stiff, and dense; to ¾" long.

Blooms: Mid-February–June.

Elevation: Below 5,000'.

Habitat: Dry, rocky slopes, and along washes.

Comments: Perennial; many-branched shrub. One species of *Peucephyllum* in Arizona. Photograph taken at Cattail Cove State Park, February 23.

PAPER FLOWER

Cooper's Paperflower
Psilostrophe cooperi
Sunflower Family (Asteraceae)

Height: To 2'.

Flowers: Bright yellow, with varying number of broad rays, notched into 3 lobes; small disk flowers; to 1" wide; terminal on branches; turning papery with age; remaining on plant for weeks.

Leaves: Grayish green, woolly, linear; to 2½" long.

Blooms: Most of the year.

Elevation: 2,000 to 5,000'.

Habitat: Plains, mesas, and along washes.

Comments: Forms a bushy mound with tangled branches. Three species of *Psilostrophe* in Arizona. Photograph taken in Superstition Mountains, March 26.

PAPER DAISY
Psilostrophe tagetina
Sunflower Family (Asteraceae)

Height: To 18".

Flowers: Bright yellow, with 3 to 5 petals; 3-lobed; to ¾" wide, in clusters at ends of branches; becoming straw-colored and papery with age.

Leaves: Grayish green, very woolly, oblong to lance-shaped; twisted, growing along stems, to 2½" long.

Blooms: May–October.

Elevation: 4,000 to 7,500'.

Habitat: Plains, mesas, and pine forest clearings.

Comments: A many-branched, aromatic, rounded plant. Poisonous to sheep. Three species of *Psilostrophe* in Arizona. Photograph taken near Nutrioso, August 18.

CURLY-HEAD GOLDENWEED
Pyrrocoma crocea var. *genuflexa*
Sunflower Family (Asteraceae)

Height: To 16".

Flowers: Saffron-colored rays, yellowish disks; to 2" wide; usually single, but occasionally up to 3 per stem.

Leaves: Dark green above, with prominent midvein beneath; alternate, clasping stem; basal; linear leaves to 6" long; smaller, arrow-shaped leaves on stem, to 1½" long.

Blooms: July–October.

Elevation: 6,000 to 9,500'.

Habitat: Clearings in coniferous forests and mountain meadows.

Comments: Reddish, hairy stem. Four species of *Pyrrocoma* in Arizona. Photographed in vicinity of Willow Springs Lake, September 14.

MEXICAN HAT

Prairie Coneflower
Ratibida columnifera (Ratibida columnaris)
Sunflower Family (Asteraceae)

Height: To 3'.

Flowers: Drooping rays, yellow with reddish brown or all reddish brown), to 1½" long; disks are purplish brown and tubular, covering a cone-shaped column to 1½" long; terminal flower head to 3" wide.

Leaves: Green, narrow, pinnately cleft into 5, 7, or 9 narrow segments; to 6" long.

Blooms: June–October.

Elevation: 5,000 to 8,500'.

Habitat: Roadsides, fields, and open clearings in pine forests.

Comments: Perennial herb. Two species of *Ratibida* in Arizona. Flower photograph taken at Greer, July 21.

CUTLEAF CONEFLOWER

Brown-Eyed Susan
Rudbeckia laciniata var. *ampla*
Sunflower Family (Asteraceae)

Height: To 7'.

Flowers: Yellow ray flowers arching downward to 2" long; tiny, greenish yellow disk flowers forming cone. Flower head to 5" wide.

Leaves: Dark green, pinnate, leaves with 3 to 7 deeply toothed lobes; to 8" long.

Blooms: July–September.

Elevation: 5,000 to 8,500'.

Habitat: Rich soil in meadows, along mountain streams, and in moist canyons.

Comments: Perennial herb. Poisonous to livestock. Four species of *Rudbeckia* in Arizona. Photograph taken south of Alpine, August 2.

SANDPAPER MULES EARS

Scabrethia scabra (Wyethia scabra)
Sunflower Family (Asteraceae)

Height: To 30".

Flowers: Yellow rays, yellowish brown disks; long, spiny bracts on flower head turned downward; to 3" wide.

Leaves: Shiny, dark green, sandpapery, with whitish midvein; finely toothed, narrowly lance-shaped, mostly basal; to 20" long, shorter on stem.

Blooms: June–October.

Elevation: 5,000 to 6,000'.

Habitat: Dry slopes and mesas, often in very sandy conditions.

Comments: Perennial herb. One species of *Scabrethia* in Arizona. Photograph taken south of Kayenta, June 27.

NODDING GROUNDSEL

Bigelow Groundsel
Senecio bigelovii var. *bigelovii*
Sunflower Family (Compositae)

Height: To 3'.

Flowers: Yellow, all disk flowers; nodding flower head to ½" wide, ⅝" long, each on separate stalk; flower heads on elongated cluster; followed by seedlike fruits with fine, white hairs attached.

Leaves: Dark green, leathery, lance-shaped, toothed; clasping stem, alternate; to 7" long at base, progressively sorter toward flower heads. Also in Arizona is **Hall's Ragwort** (*Senecio bigelovii* var. *hallii*), with tomentose leaves.

Blooms: July–September.

Elevation: 7,000 to 11,000'.

Habitat: Moist soil of roadsides, mountain meadows, and clearings in coniferous forests.

Comments: Twenty-four species of *Senecio* in Arizona. Photograph taken in mountain meadow above Greer, August 10.

GROUNDSEL

Senecio eremophilus var. *macdougalii (Senecio macdougalii)*

Sunflower Family (Asteraceae)

Height: To 3'.

Flowers: 5 to 8 yellow rays, darker yellow disks; stamens curling upward; long, slender bracts tipped with black; flower head to ⅝" wide, ⅜" long.

Leaves: Dark green, lance-shaped, slightly hairy; lobes deeply cleft almost to midvein; leaf to 5" long toward base; alternate and graduating in size upward on stem.

Blooms: July–October.

Elevation: 6,500 to 10,500'.

Habitat: Coniferous forests and clearings in aspen groves.

Comments: Recognizable as a *Senecio* by its few ray flowers and as this particular species by its black-tipped bracts and its leaf shape. Twenty-four species of *Senecio* in Arizona. Photograph taken in mountains above Greer, August 13.

SAND WASH GROUNDSEL

Comb Butterweed

Senecio flaccidus var. *monoensis (Senecio douglasii* var. *douglasii)*

Sunflower Family (Asteraceae)

Height: To 5'.

Flowers: Yellow rays, orange disks, to 1½" wide.

Leaves: Yellowish green, pinnate with linear lobes curling upward; comblike; to 2½" long.

Blooms: Most of the year.

Elevation: 2,500 to 7,500'.

Habitat: Washes, dry slopes, mesas, and plains.

Comments: Twenty-four species of *Senecio* in Arizona. Photograph taken in Tucson area, March 3. The comblike leaves help to identify this species.

THREADLEAF GROUNDSEL

Felty Groundsel
Senecio flaccidus var. *flaccidus (Senecio douglasii* var. *longilobus)*
Sunflower Family (Asteraceae)

Height: To 4'.

Flowers: 8 to 13 yellow rays, yellowish orange disks; woolly bracts side by side; floppy petals; to 1⅛" wide; in clusters.

Leaves: Grayish green, very woolly, linear; divided into very narrow lobes; to 4" long.

Blooms: Throughout most of year at lower elevations; May–November elsewhere.

Elevation: 2,500 to 7,500'.

Habitat: Sandy washes, plains, and mesas.

Comments: Many-branched, woolly stemmed shrub. Once used medicinally by Native Americans. Very poisonous if eaten by cattle or horses. Twenty-four species of *Senecio* in Arizona. Photograph taken near Portal, May 5. Unlike **Broom Ragwort** (*Senecio spartioides* var. *multicapitatus*) (page 142), this species has woolly bracts and stems and divided, woolly leaves.

LEMMON'S BUTTERWEED

Senecio lemmonii
Sunflower Family (Asteraceae)

Height: To 3'.

Flowers: Yellow rays, orange disks, to 1⅛" wide.

Leaves: Dark green, shiny, alternate; lance-shaped, clasping stem; sunken midvein, toothed, to 5" long.

Blooms: February–May.

Elevation: 1,500 to 3,500'.

Habitat: Along washes and on rocky slopes.

Comments: Somewhat shrubby. Has reddish stems. Twenty-four species of *Senecio* in Arizona. Photograph taken at King's Canyon, Tucson, April 17.

BROOM RAGWORT

Senecio spartioides var. *multicapitatus (Senecio multicapitatus)*
Sunflower Family (Asteraceae)

Height: To 4'.

Flowers: Yellow rays, yellowish orange disks; narrow, floppy petals; to 1" wide; in loose clusters.

Leaves: Dark green, smooth, very narrow; threadlike segments; to 4" long, occurring all along stem.

Blooms: May–November.

Elevation: 5,000 to 7,000'.

Habitat: Clearings in pine forests, mesas, and plains.

Comments: Many-stemmed. Twenty-four species of *Senecio* in Arizona. Photograph taken at Upper Lake Mary, September 6. Unlike **Threadleaf Groundsel** (*Senecio flaccidus* var. *flaccidus*), (page 141) this species has smooth bracts and stems, and smooth, undivided leaves.

BROOM GROUNDSEL

Grass-Leaved Ragwort
Senecio spartioides
Sunflower Family (Asteraceae)

Height: To 2'.

Flowers: 8 yellow, slightly drooping rays, each with 2 notches; orange disks; flower head to ¾" wide.

Leaves: Dark green, alternate, very narrow; some with narrow lobes; to 3" long, along length of stem.

Blooms: July–October.

Elevation: 6,500 to 9,000'.

Habitat: Clearings in pine forests.

Comments: A bushlike plant with numerous stems. Twenty-four species of *Senecio* in Arizona. Photograph taken in vicinity of Nelson Reservoir, August 16.

WOOTON'S BUTTERWEED

Senecio wootonii
Sunflower Family (Asteraceae)

Height: To 2'.

Flowers: Yellow, 8 to 10 irregularly shaped rays; darker yellow disks; to ⅝" wide; in loose, terminal clusters.

Leaves: Grayish green, leathery, smooth; toothed or untoothed; spatula- to spoon-shaped with long, tapering, flat stalk; basal, to 10" long.

Blooms: May–September.

Elevation: 6,000 to 9,500'.

Habitat: Coniferous forests.

Comments: Has leafless stem. Twenty-four species of *Senecio* in Arizona. Photograph taken at Carnero Lake near Greer, July 11.

PHOTO LICENSED BY SHUTTERSTOCK

TALL GOLDENROD

Solidago altissima ssp. *altissima*
Sunflower Family (Asteraceae)

Height: To 5'.

Flowers: Bright yellow, heads to ¼" wide, ⅛" long; forming a pyramidal, terminal cluster with flowers mostly on one side.

Leaves: Green and lance-shaped; to 6" long, gradually smaller on upper stem.

Blooms: August–October.

Elevation: 2,500 to 8,500'.

Habitat: Roadsides and clearings.

Comments: Perennial herb. A variable species. Eleven species of *Solidago* in Arizona. Photograph taken at Willow Springs Lake, August 5.

CANADA GOLDENROD

Solidago canadensis var. *canadensis*
Sunflower Family (Asteraceae)

Height: To 6'.

Flowers: Bright yellow, tiny, to ⅛" long; on arching stems in a large, loose, terminal cluster.

Leaves: Dark green, narrowly lance-shaped, with 3 prominent veins; to 5" long.

Blooms: July–September.

Elevation: 3,000 to 8,500'.

Habitat: Clearings in ponderosa forests, meadows, fields, and roadsides.

Comments: Perennial herb. Eleven species of *Solidago* in Arizona. Photograph taken at Dead Horse Ranch State Park, September 9.

MISSOURI GOLDENROD

Prairie Goldenrod
Solidago missouriensis var. *tenuissima*
Sunflower Family (Asteraceae)

Height: To 3'.

Flowers: Bright yellow, to ¼" long, ⅛" wide; about 8 in each flower head; in tight cluster, total flowering cluster to 6" long; on slightly arching stem.

Leaves: Dark green, shiny, smooth; alternate, lance-shaped, margins curved upward; basal leaves to 7" long, ¾" wide; shorter, narrower leaves up stem, with small clusters of tiny leaves at leaf axils.

Blooms: June–August.

Elevation: 5,000 to 9,000'.

Habitat: Along streams and clearings in pine forests.

Comments: Perennial herb; has smooth, reddish stem. Native Americans use leaves as salad greens. Eleven species of *Solidago* in Arizona. Photograph taken near Christopher Creek, August 11. Recognizable by smooth leaves and stem, and by close arrangement of flowers on arching stem.

ALPINE GOLDENROD

Solidago multiradiata
Sunflower Family (Asteraceae)

Height: To 16".

Flowers: Yellow, up to 13 rays, each to ⅝" wide, ¾" long; in tight, terminal cluster.

Leaves: Dark green, alternate, smooth; broadly spatula-shaped and tapering to stem; to 4" long Stem leaves are stalkless.

Blooms: July–September.

Elevation: Not available. Photograph taken at 9,500'.

Habitat: Mountain meadows and clearings in moist coniferous forests.

Comments: Has erect stem. Eleven species of *Solidago* in Arizona. Photograph taken in mountains above Greer, August 7.

DWARF GOLDENROD

Mt. Albert Goldenrod
Solidago simplex ssp. *simplex (Solidago decumbens)*
Sunflower Family (Asteraceae)

Height: Below 10".

Flowers: Yellow rays, yellow disks; to ⅜" wide; in dense cylindrical, terminal cluster.

Leaves: Dark green, alternate, smooth; often toothed at rounded tip; spatula-shaped but variable; to 3" long at base; upper leaves smaller.

Blooms: July–August.

Elevation: 8,000 to 9,500'.

Habitat: Clearings in coniferous forests and mountain meadows.

Comments: Has reddish stems, which often creep along the ground for a few inches before growing erect. Often grows in small patches. Nine Eleven species of *Solidago* in Arizona. Photograph taken in mountains above Greer, August 7.

SPARSE-FLOWERED GOLDENROD

Solidago velutina (Solidago sparsiflora)
Sunflower Family (Asteraceae)

Height: To 2'.

Flowers: Bright yellow, to ¼" wide, ⅛" long; about a dozen in each flower head; in clusters to 1¼" wide on one side of arching stem; total flowering clusters to 7" long.

Leaves: Dull grayish green, alternate, rough; lance-shaped to linear, with 3 prominent veins; to 3" long.

Blooms: June–October.

Elevation: 2,000 to 8,500'.

Habitat: Roadsides, chaparral, and clearings in pine forests.

Comments: Perennial herb. Eleven species of *Solidago* in Arizona. Photograph taken near McNary, August 10. Recognizable by its rough-textured leaves and loose, sparse flowers on arching stem.

COMMON SOWTHISTLE

Annual Sowthistle
Sonchus oleraceus
Sunflower Family (Asteraceae)

Height: To 5'.

Flowers: Yellow, dandelion-like flower head to 1¼" wide; in sparse clusters; followed by seeds ribbed lengthwise, attached to soft white hairs forming a miniature parachute.

Leaves: Dark green, thin, deeply lobed into 1 to 3 lobes on each side; tip lobe broadly triangular; clasping stem; to 7" long.

Blooms: February–November.

Elevation: 150 to 7,000'.

Habitat: Waste areas, roadsides, and disturbed places.

Comments: Fleshy annual. Naturalized from Europe. Birds feed on seeds. Two species of *Sonchus* in Arizona. Photograph taken near Granite Reef Dam, March 1. A similar species, also exotic, is **Spiny Sowthistle** (*Sonchus asper*), which has prickly toothed leaves.

COMMON DANDELION

Taraxacum officinale
Sunflower Family (Asteraceae)

Height: Flower stalks to 15".

Flowers: Golden yellow flower head, straplike ray flowers toothed at tips; flower head to 1½" wide; solitary on hollow stalk; followed by downy, globular mass with seeds attached to parachutelike hairs.

Leaves: Dark green, lance-shaped, deeply cut into triangular-shaped sections; in basal rosette; to 10" long.

Blooms: April–September.

Elevation: 100 to 9,000'.

Habitat: Roadsides, lawns, meadows, and fields.

Comments: Perennial weed, with deep taproot and milky stem juice. From Europe; now naturalized. Used for food and medicine. Three species of *Taraxacum* in Arizona. Photograph taken near Greer, July 4.

GRAY FELT THORN

Spineless Horsebrush
Tetradymia canescens
Sunflower Family (Asteraceae)

Height: Rounded shrub to 2' tall, 5' wide.

Flowers: Yellow, petalless, 4 disk flowers per head; 4 woolly bracts surrounding head; to ⅝" long; in terminal clusters on branches; followed by tan bristles to ¾" long.

Leaves: Grayish green, stiff, woolly, narrow; to ½" long; growing along entire stem.

Blooms: June–October.

Elevation: 5,000 to 8,000'.

Habitat: Rocky, sandy, dry soils in woodlands; clearings in ponderosa pine forests; and roadsides.

Comments: Has a woody base and many branches. Native Americans use plants medicinally. Safely browsed by cattle, but often fatal to sheep when consumed in large quantities. Four species of *Tetradymia* in Arizona. Photograph taken near Aripine, August 4.

HOPI-TEA GREENTHREAD

Colorado Greenthread
Thelesperma megapotamicum
Sunflower Family (Asteraceae)

Height: To 3'.

Flowers: Yellow disk flowers surrounded by green bracts; rayless; to ½" wide, ½" long; terminal on long, leafless stems.

Leaves: Dark green, pinnate, linear; 3 to 7 thread-like segments; mainly basal, opposite, leaf to 4" long.

Blooms: May–October.

Elevation: 4,000 to 7,500'.

Habitat: Open woodlands and forest, plains, roadsides, and mesas.

Comments: Annual herb. Hopi Indians used flowers and young leaves for making tea. A dye extracted from plant is used for textiles and basketry. Three species of *Thelesperma* in Arizona. Photograph taken at Black Canyon Lake, September 29.

NEEDLELEAF DOGWEED

Thymophylla acerosa (Dyssodia acerosa)
Sunflower Family (Asteraceae)

Height: To 1'.

Flowers: Yellow rays (usually 8), slightly darker yellow disks; flower bracts dotted with yellowish glands; to ¾" wide, terminal on branches.

Leaves: Dark green, narrowly linear, needle-like, opposite or alternate, dotted with yellowish glands; to ½" long, all along stems.

Blooms: March–October.

Elevation: 3,500 to 6,000'.

Habitat: Washes; dry, rocky slopes; and mesas.

Comments: Perennial herb. Many-branched, rounded bush; woody at base. Has glands on pinkish brown stems. Three species of *Thymophylla* in Arizona. Photograph taken at Dead Horse Ranch State Park, October 3.

DOGWEED

Thymophylla pentachaeta (Dyssodia pentachaeta)
Sunflower Family (Asteraceae)

Height: To 8".

Flowers: Bright yellow rays and disks; bracts dotted with orangish brown glands; to ½" wide; on leafless stalks above leaves.

Leaves: Dark green, mostly opposite, stiff, pinnately cleft into very narrow lobes with spiny tips; to ½" long.

Blooms: March–September.

Elevation: 2,500 to 4,500'.

Habitat: Desert and dry slopes.

Comments: Plant forms low mound. Foliage has disagreeable odor when handled. Three species of *Thymophylla* in Arizona. Photograph taken at Catalina State Park, April 15.

YELLOW SALSIFY

Meadow Salsify
Tragopogon dubius
Sunflower Family (Asteraceae)

Height: To 3'.

Flowers: Lemon-yellow, individual ray flowers to ½" long; 10 to 13 bracts longer than ray flowers; open in morning, closed by noon; flower head to 2½" wide; followed by a large-spherical seed head resembling a giant dandelion.

Leaves: Grayish green, long, grasslike; clasping stem; to 10" long.

Blooms: June–September.

Elevation: 3,500 to 7,000'.

Habitat: Dry roadsides, fields, and vacant lots.

Comments: Perennial herb. Introduced from Europe; now naturalized. Stems release milky sap when broken. Native Americans used plant for food and medicine. Five species of *Tragopogon* in Arizona. Photograph taken near Prescott, May 26.

TRIXIS

Trixis californica var. *californica*
Sunflower Family (Asteraceae)

Height: Sprawling shrub to 3½'.

Flowers: Bright yellow, rayless, surrounded by leaflike bracts; flower heads composed of 9 to 15 flowers; to ¾" wide, at branch ends; followed by seeds with straw-colored bristles.

Leaves: Dark green, lance-shaped, smooth-edged to fine-toothed; numerous; to ½" wide, 2" long.

Blooms: February–October.

Elevation: Below 5,000'.

Habitat: Rocky slopes, along washes, among other bushes.

Comments: Browsed by cattle. One species of *Trixis* in Arizona. Photograph taken at Usery Mountain Recreation Area, March 1.

SILVER PUFFS

Starpoint
Uropappus lindleyi (Microseris linearifolia)
Sunflower Family (Asteraceae)

Height: Flowering stem to 1½' .

Flowers: Yellow; all rays, dandelion-like; flower head bracts extend beyond rays as sharp points; to 1" wide; followed by delicate, silvery, pufflike seed head to 1½" wide.

Leaves: Grayish green to dark green; linear to partly linear, or pinnately lobed; to 5" long.

Blooms: March–June.

Elevation: Below 6000'.

Habitat: Pine forests down to foothills, plains, and mesas.

Comments: Hollow stemmed. One species of *Uropappus* in Arizona. Photograph taken at Saguaro National Park West, April 17.

GOLDEN CROWNBEARD
Cowpen Daisy
Verbesina encelioides
Sunflower Family (Asteraceae)

Height: To 3′.

Flowers: Yellow rays, each notched into 3 lobes; yellow disk flowers; to 2″ wide.

Leaves: Grayish green, triangular, with toothed margins; to 4″ long.

Blooms: Early germinations, March–July; later germinations, July–December.

Elevation: Below 7,000′.

Habitat: Roadsides, waste areas, and washes.

Comments: Annual. Flattened seed head covered with grayish brown hairs gave rise to name, "crownbeard." Rodents and birds eat seeds. Native Americans and early settlers used plant to treat skin diseases and boils. Hopi Indians use water of steeped plant for treating spider bites. Three species of *Verbesina* in Arizona. Photograph taken near Salome, March 28.

ARIZONA MULES EARS
Wyethia arizonica
Sunflower Family (Asteraceae)

Height: To 2′.

Flowers: Bright yellow, 10 to 15 rays, yellow disk flowers; to 2½″ wide; solitary at tip of stem.

Leaves: Dark green, with white midstripe; hairy, with wavy margins; oblong or elliptical; basal leaves to 1½′ long, 3″ wide; stem leaves much shorter.

Blooms: June–August.

Elevation: 7,000 to 9,000′.

Habitat: Slopes and canyons in ponderosa pine forests.

Comments: Perennial herb. One species of *Wyethia* in Arizona. Photograph taken in Mormon Lake area, June 2.

YELLOW SPINY DAISY

Slender Goldenweed
Xanthisma gracilis (Machaeranthera gracilis)
Sunflower Family (Asteraceae)

Height: To 1'.

Flowers: Yellow rays and disks; to 1¼" wide; terminal on upper branches.

Leaves: Grayish green, hairy, narrow; angled upward, toothed, with spiny bristle at tip of each tooth; lowest leaves have a few lobes; to ¾" long.

Blooms: February–December.

Elevation: Below 7,000'.

Habitat: Dry plains, mesas, and rocky slopes.

Comments: Annual herb. Six species of *Xanthisma* in Arizona. Photograph taken at Apache Lake, March 19. Recognizable by spiny bristle at tip of each leaf tooth.

PRAIRIE ZINNIA

Rocky Mountain Zinnia
Zinnia grandiflora
Sunflower Family (Asteraceae)

Height: To 1'.

Flowers: Bright yellow, 3 to 6 ray flowers; reddish disk flowers in center; to 1½" wide; occurring singly on tips of branches.

Leaves: Grayish green, linear, very narrow; somewhat curled; to 1" long; all along stem.

Blooms: May–October.

Elevation: 4,000 to 6,500'.

Habitat: Mesas, dry plains, roadsides, and pinyon-juniper woodlands.

Comments: Spreading, many-branched herb forming rounded clump. Three species of *Zinnia* in Arizona. Photograph taken near Wupatki National Monument, September 8.

FREMONT BARBERRY
Berberis fremontii
Barberry Family (Berberidaceae)

Height: Shrub to 10′.

Flowers: 6 yellow petals, flower to 1″ wide, in clusters of 3 to 9; followed by loose cluster of dark blue, ½″-wide berries.

Leaves: Dark grayish-green, leathery, pinnately compound, with 3 to 7 leaflets to 3″ long; leaf to 5″ long.

Blooms: April–July.

Elevation: 4,000 to 7,000′.

Habitat: Pinyon-juniper and pine woodlands.

Comments: Evergreen. Fragrant. Hopi Indians use wood for crafts and roots for yellow dye. Berries used for jams and jellies. Contains the drug berberine. Six species of *Berberis* in Arizona. Photograph taken at Vernon, June 13.

RED BARBERRY
Algerita
Berberis haematocarpa
Barberry Family (Berberidaceae)

Height: To 6′.

Flowers: Fragrant, to ½″ wide, with 6 yellow petals and stamens; in loose few-flowered cluster; followed by red, juicy berry to ⅜″ in diameter.

Leaves: Bluish green covered with a whitish bloom, pinnate, 3 to 5 leaflets; leaflets unstalked, leathery, stiff, to ¾″ wide, each tapering to a sharp, terminal spine, terminal leaflet longest, pointed lobes on leaflets ending in sharp spines; leaf to 4″ long.

Blooms: February–May.

Elevation: 3,000 to 5,000′.

Habitat: Desert grasslands and oak woodlands. Native to upper desert slopes and to chaparral in central Arizona.

Comments: Red jelly made from fruits; root and bark used in making a yellow dye. Six species of *Berberis* in Arizona. Photograph taken at Desert Botanical Garden, Phoenix, March 3.

KOFA MOUNTAIN BARBERRY
Berberis harrisoniana
Barberry Family (Berberidaceae)

Height: Small shrub to 3'.

Flowers: Yellow, with 6 large petals and sepals; to ½" wide; in large, loose cluster; followed by bluish black, slightly oval fruits, to ¼" in diameter.

Leaves: Grayish green, palmate, with 3 similar leaflets; thick, leathery, and stiff, ending in stout, sharp spines; leaf to 2" long.

Blooms: February–March.

Elevation: 2,500 to 3,500'.

Habitat: Rocky slopes in Ajo and Kofa Mountains.

Comments: Forms a sprawling bush. Six species of *Berberis* in Arizona. Photograph taken at Palm Canyon in Kofa Mountains, February 22.

CREEPING BARBERRY
Berberis repens
Barberry Family (Berberidaceae)

Height: Low, creeping shrub to 1'.

Flowers: Fragrant, to ¾" wide; 6 yellow petals; in dense cluster, followed by cluster of ¼" bluish purple berries.

Leaves: Dark green, shiny, leathery, hollylike; to 10" long; pinnately compound; 3 to 7 wavy leaflets with spiny margins, leaflets to 3" long.

Blooms: April–June.

Elevation: 5,000 to 10,000'.

Habitat: Open coniferous forests and wooded slopes.

Comments: Evergreen shrub. Excellent ground cover and erosion fighter. Stems root when they come in contact with soil. In fall, leaves turn shades of red, yellow, or purple. Berries are used for jelly and eaten by wildlife; twigs and leaves are used medicinally. Yellow dye is made from roots. Six species of *Berberis* in Arizona. Photograph taken at Sharp Creek northeast of Christopher Creek, April 22. A similar shrub, **Holly Leaf Grape** (*Berberis wilcoxii*), is taller and has fewer than 10 coarse teeth on each leaflet. Photograph taken at Cave Creek, Portal, April 23.

YELLOW TRUMPET BUSH
Yellow Bells
Tecoma stans var. *angustata*
Bignonia Family (Bignoniaceae)

Height: Small shrub to 15′.

Flowers: Bright yellow, trumpet-shaped, with ruffled lobes; to 2″ long; in clusters; followed by a narrow capsule to 8″ long, ¼″ wide.

Leaves: Dark green, glossy, lance-shaped to oval; pointed at tips, toothed; to 6″ long; pinnately compound, with 5 to 13 leaflets.

Blooms: May–October.

Elevation: 2500 to 5,500′.

Habitat: Dry, gravelly hillsides in southeastern Arizona.

Comments: Evergreen in frost-free areas. Browsed by bighorn sheep. Cultivated as an ornamental. One species of *Tecoma* in Arizona. Photograph taken at Mesa, May 9.

MOUNTAIN GROMWELL
Lithospermum cobrense
Forget-me-not Family (Boraginaceae)

Height: To 1′.

Flowers: Pale yellow, funnel-shaped, 5-lobed; to ¾″ wide; clustered together on erect coil.

Leaves: Gray, lancelike, covered with short hairs; to 2″ long.

Blooms: July–August.

Elevation: 5,000 to 9,000′.

Habitat: Ponderosa pine forests.

Comments: Biennial or short-lived perennial. Six species of *Lithospermum* in Arizona. Photograph taken at Vernon, June 13.

FRINGED GROMWELL
Lithospermum incisum
Forget-me-not Family (Boraginaceae)

Height: To 16".

Flowers: Yellow, fringed, and trumpet-shaped; 5 united petals; to 1¼" long, ¾" wide.

Leaves: Grayish green, narrowly lanceolate, to 2½" long.

Blooms: March–May.

Elevation: 4,000 to 7,500'.

Habitat: Foothills, open plains, and slopes.

Comments: Used by the Hopi Indians to produce medicine. Six species of *Lithospermum* in Arizona. Photograph taken at Chiricahua National Monument, April 24.

MANYFLOWER STONESEED
Lithospermum multiflorum
Forget-me-not Family (Boraginaceae)

Height: To 2'.

Flowers: Yellow to yellowish orange, funnellike, with 5 short, rounded petal lobes; to ½" wide; in nodding, coiled, terminal clusters.

Leaves: Grayish green, sandpapery, and covered with stiff hairs; linear to slightly lance-shaped; to 2" long.

Blooms: June–September.

Elevation: 6,000 to 9,500'.

Habitat: Slopes, flats, and clearings in pinyon-juniper woodlands and ponderosa pine forests.

Comments: Perennial, with hairy stems. Native Americans obtained purple dye from roots. Six species of *Lithospermum* in Arizona. Photograph taken at Mormon Lake, June 2.

SAHARA MUSTARD
Brassica tournefortii
Mustard Family (Brassicaceae)

Height: To 3'.

Flowers: Yellow, with 4 petals; to ¼" wide; followed by cylindrical seed pod to 2½" long on 1"-long stem.

Leaves: Dark green to yellowish green, pinnately lobed, warty-looking; hairy, toothed, with clasping stem; to 10" long at base, smaller on upper stem.

Blooms: Late winter to early spring.

Elevation: Below 4,000'. Photograph taken at 1,700'.

Habitat: Roadsides and fields.

Comments: Hairy-stemmed. A native of Europe; rapidly taking over in some areas of Arizona. Four species of *Brassica* in Arizona. Photograph taken in Mesa, February 22.

MOUNTAIN TANSY MUSTARD
Descurainia incisa ssp. *incisa (Descurainia richardsonii)*
Mustard Family (Brassicaceae)

Height: To 40".

Flowers: Bright yellow, with 4 petals; to ¹⁄₁₆" wide; in terminal raceme: followed by ½"-long slender seed capsule containing 1 row of seeds.

Leaves: Grayish green, slightly hairy, pinnate; to 2½" long.

Blooms: July–August.

Elevation: 6,500 to 9,500'.

Habitat: Roadsides and fields.

Comments: Annual. Native Americans used seeds for making pinole, a ground meal. Six species of *Descurainia* in Arizona. Photograph taken near Nutrioso, August 3.

DRABA

Draba asprella
Mustard Family (Brassicaceae)

Height: To 5".

Flowers: Yellow, with 4 petals; to ¼" wide; in long, terminal cluster or raceme on leafless stem.

Leaves: Green, very hairy, oval to elliptical; in basal rosette; to ½" long.

Blooms: Starting in February.

Elevation: 5,000 to 8,000'.

Habitat: Pine forests.

Comments: Plants vary in hairiness of foliage. Fifteen species of *Draba* in Arizona. Photograph taken at Black Canyon Lake area, June 4.

GOLDEN DRABA

Draba aurea
Mustard Family (Brassicaceae)

Height: To 9".

Flowers: Golden yellow, with 4 petals; hairy, tiny; to ¼" long, 3⁄16" wide, in loose, terminal cluster; followed by small, hairy, flat, nearly erect seed pod to ½" long (including stem) ending in persistent style to 1⁄16" long.

Leaves: Grayish green, covered with short, starlike hairs; elliptical-shaped on stem, spatula-shaped in basal rosette; faintly toothed; to 1½" long.

Blooms: July–August.

Elevation: 5,000 to 12,000'.

Habitat: Clearings in coniferous forests.

Comments: Perennial herb. Erect to sprawling stems are covered with short hairs. Fifteen species of *Draba* in Arizona. Photograph taken at Luna Lake, August 5.

WESTERN WALLFLOWER
Erysimum capitatum var. *purshii (Erysimum asperum)*
Mustard Family (Brassicaceae)

Height: To 32".

Flowers: Bright yellow (occasionally purplish, with 4 petals; to ¾" wide; in cluster on rounded, terminal raceme; followed by very slender, erect 4-sided pod to 4" long.

Leaves: Grayish green, lance-shaped, toothed margins; in basal rosette; to 5" long. Stem leaves are narrow with small teeth.

Blooms: March–September.

Elevation: 2,500 to 9,500'.

Habitat: Roadsides, open flats, slopes, and dry, stony banks.

Comments: Biennial or perennial. Three species of *Erysimum* in Arizona. Photograph taken near Globe, March 29.

ARIZONA BLADDERPOD
Physaria arizonica (Lesquerella arizonica)
Mustard Family (Brassicaceae)

Height: To 6".

Flowers: Bright yellow, with 4 petals; to ½" wide; in short flower cluster; followed by a thick, oval fruit to ¼" in diameter.

Leaves: Silvery gray, woolly, linear; to 2" long.

Blooms: April–May.

Elevation: 3,500 to 7,000'.

Habitat: Rocky slopes and mesas.

Comments: Has unbranched, erect stem. Eleven species of *Physaria* in Arizona. Photograph taken on San Carlos Indian Reservation, April 20.

GORDON'S BLADDERPOD

Physaria gordonii (Lesquerella gordonii)
Mustard Family (Brassicaceae)

Height: To 16".

Flowers: Bright yellow, with 4 petals; to ⅓" wide; in terminal, loose raceme; followed by ⅛" diameter spherical pod tipped with slender point.

Leaves: Green, but often appearing silvery because of hairs; narrow, lance- or spatula-shaped; basal leaves often lobed; to 2" long.

Blooms: February–May.

Elevation: 100 to 5,000'.

Habitat: Desert flats, dry plains, and among desert shrubs on mesas.

Comments: An annual; erect or spreading. Forage for cattle. Eleven species of *Physaria* in Arizona. Photograph taken at Alamo Lake, February 26.

NEWBERRY'S TWINPOD

Physaria newberryi
Mustard Family (Brassicaceae)

Height: To 4" in circular tuft.

Flowers: Yellow, with 4 petals; to ½" wide; in terminal raceme; followed by pale pinkish, twin, bladderlike seed capsules deeply notched between the 2 cells; twin pods to 1" wide, in cluster above leaves.

Leaves: Silvery gray, roundish to squarish, and downy; with upwardly curved margins; toothed; basal; to 2" long.

Blooms: May.

Elevation: 5,000 to 7,000'.

Habitat: Dry, rocky slopes; often in volcanic cinders.

Comments: At first glance the pink bladders resemble flowers. Eleven species of *Physaria* in Arizona. Photograph in pod taken at Sunset Crater National Monument, June 5.

TUMBLE MUSTARD

Sisymbrium altissimum
Mustard Family (Brassicaceae)

Height: To 4'.

Flowers: Pale yellow, with 4 petals; to ½" wide; in loose, terminal cluster on branch; followed by slender, 4"-long seed capsule that spreads from stem.

Leaves: Dark green, hairy, deeply lobed; to 11" long; upper leaves are pinnately lobed to linear, to 4" long.

Blooms: April–September.

Elevation: 5,000 to 7,000'.

Habitat: Roadsides, fields, and waste places.

Comments: Many-branched and spreading. Introduced from Europe. Four species of *Sisymbrium* in Arizona. Photograph taken at Dead Horse Ranch State Park, May 30.

LONDON ROCKET

Sisymbrium irio
Mustard Family (Brassicaceae)

Height: To 3'.

Flowers: Yellow, with 4 petals; to ⅛" long; on slender stalk, in small, terminal cluster on stem; followed by long, slender seed pod thicker than flower stalk, to 2" long.

Leaves: Dark green, fleshy, large, and pointed; with terminal lobe, 1 to 4 pairs of smaller lobes below; to 8" long.

Blooms: December–April.

Elevation: 100 to 4,500'.

Habitat: Irrigated fields, roadsides, and waste places.

Comments: Annual. Introduced from Europe. The flower stem gradually elongates as the seed pods mature. Four species of *Sisymbrium* in Arizona. Photograph taken near Granite Reef Dam, February 6.

PRINCE'S PLUME
Stanleya pinnata
Mustard Family (Brassicaceae)

Height: To 6′.

Flowers: Yellow, with long stamens and pistil, 4 narrow petals, and 4 sepals; to 1¼″ long; in dense, terminal spike to 2′ long; blooming first at base and progressing up spike; followed by slender, drooping seed pods to 2″ long.

Leaves: Grayish green, narrow; to 7″ long at base, shorter on upper stem; some pinnately divided at base and on stem.

Blooms: May–September.

Elevation: 2,500 to 7,000′.

Habitat: Mesas, dry plains, and sagebrush areas.

Comments: Perennial herb. Native Americans used seeds for mush and plant as a potherb. One species of *Stanleya* in Arizona. Photograph taken at Sunset Crater National Monument, September 8.

CHAPARRAL HONEYSUCKLE
Lonicera interrupta
Honeysuckle Family (Caprifoliaceae)

Height: Trailing.

Flowers: Creamy white, tubular, with swollen throat; upper lobe rolled backward, lower lobe in watchspring-shaped curl; 4 very long stamens; to ¾″ long; in elongated, terminal cluster.

Leaves: Dark green above, lighter green beneath; opposite, oval; to 1¾″ long.

Blooms: May–June.

Elevation: 4,000 to 6,000′.

Habitat: Along streams in juniper-oak woodlands.

Comments: Has reddish stems. Twelve species of *Lonicera* in Arizona. Photograph taken north of Payson along East Verde River, June 3.

FOUR-WING SALTBUSH

Chamiso
Atriplex canescens var. *canescens*
Goosefoot Family (Chenopodiaceae)

Height: To 8', but more commonly to 4'.

Flowers: Pale yellow, tiny, inconspicuous, in clusters along the stems; male and female on different plants; followed on the female plant by bunches of small, burlike seeds encased in 4 papery, light green, winglike bracts; to ½" wide; bracts dry to pale brown.

Leaves: Gray-green, narrow; to 2" long.

Blooms: July–August.

Elevation: 2,000 to 8,000'.

Habitat: Sandy, sometimes saline soil, from creosote bush to pinyon to ponderosa belts.

Comments: This salt-tolerant shrub is the most widely spread species of *Atriplex* in the U.S. Its deep roots help control erosion. Its foliage tastes salty. Some female bushes become a mass of fruits. A browse shrub for livestock, deer, and antelope, its seeds provide food for birds and small rodents. Over twenty-five species of *Atriplex* in Arizona. Photograph taken at Wupatki National Monument, September 8. **Littleleaf Saltbush** (*Atriplex polycarpa*) resembles the four-wing saltbush but its leaves are small and it lacks the prominent, 4-winged fruits.

YELLOW BEE PLANT

Peritoma jonesii (Cleome lutea var. *jonesii)*
Cleome Family (Cleomaceae)

Height: To 2½'.

Flowers: Yellow, 6 long stamens, 4 petals; to ⅜" wide, ¼" long, in cluster at top of stem; followed by slender seed pod to 1½" long, hanging downward on long stalk.

Leaves: Green, palmately compound; usually 5 lance-shaped leaflets, each to 3" long.

Blooms: May–September.

Elevation: 2,000 to 6,000'.

Habitat: Along streams and in other moist areas.

Comments: The Hopi Indians used immature plants as potherbs. Two species of *Peritoma* in Arizona. Photograph taken in Kayenta area, June 27.

JACKASS CLOVER

Wislizenia refracta ssp. *refracta*
Cleome Family (Cleomaceae)

Height: To 4'.

Flowers: Tiny and yellow, with 4 petals; ⅛" long; in dense, terminal raceme, followed by ⅛" pod on a sharply bent stalk.

Leaves: Light green, with 3 segments, elliptical leaflets; to 1¼" long.

Blooms: April–November.

Elevation: 1,000 to 6,500'.

Habitat: Roadsides, dry streambeds, and other sandy areas.

Comments: Annual. One species of *Wislizenia* in Arizona. Photograph taken at Organ Pipe Cactus National Monument, November 14.

ROCK ECHEVERIA

Live-Forever
Dudleya saxosa ssp. *collomiae*
Orpine Family (Crassulaceae)

Height: Flower stalk to 1½'.

Flowers: 5 yellow petals, reddish orange sepal; to ½" long, spaced at end of curved flower stem.

Leaves: Grayish green, succulent; flat on upper surface, rounded beneath; linear, tapering to reddish-tipped point; to 6" long, ¾" wide, ⅛" thick; in basal rosette.

Blooms: April–June.

Elevation: 3,000 to 5,000'.

Habitat: Dry, rocky, desert slopes.

Comments: Reddish stems. Two species of Dudleya in Arizona. Photograph taken in vicinity of Roosevelt Lake, April 29.

FINGER-LEAVED GOURD
Cucurbita digitata
Gourd Family (Cucurbitaceae)

Height: Trailing vine.

Flowers: Yellow, bell-shaped tube, to 1½" long; followed by a dark green, roundish gourd, with vertical, whitish stripes and blotches; to 3½" in diameter. Matures to pale yellow.

Leaves: Grayish green with central silvery white markings on 5 narrow, fingerlike segments; very hairy beneath; side lobes vary in shape and size; to 10" long including stem.

Blooms: June–October.

Elevation: Below 5,000'.

Habitat: Sandy washes, mesas, and dry plains.

Comments: Has very hairy stems. Three species of *Cucurbita* in Arizona. Photograph taken at Catalina State Park, November 9.

BUFFALO GOURD
Calabazilla
Cucurbita foetidissima
Gourd Family (Cucurbitaceae)

Length: Prostrate to 20' long.

Flowers: Yellow, funnel-shaped, to 4" long; followed by smooth, round, light and dark green striped 4" gourd that turns yellow when mature.

Leaves: Gray-green above, whitish beneath; triangular, finely toothed, long-stalked; to 1' long.

Blooms: May–August.

Elevation: 1,000 to 7,000'.

Habitat: Roadsides and dry or sandy areas.

Comments: Perennial, with huge taproots and foul-smelling leaves. Flowers open very early in day. Pollinated by bees. Gourds are edible before they dry. Native Americans used oil extracted from seeds for cooking, and dried gourds for ceremonial rattles. Three species of *Cucurbita* in Arizona. Photograph taken in vicinity of Vernon, August 4. The **Coyote Melon** (*Cucurbita palmata*) has similar flowers and round gourds, but its leaves are fingerlike instead of triangular.

RUSSIAN OLIVE

Elaeagnus angustifolia
Oleaster Family (Elaeagnaceae)

Height: To 25′.

Trunk: To 4″.

Bark: Grayish brown, fissured, shredding in long strips.

Flowers: Pale yellow inside, silver on outer surface; petalless, bell-shaped, very fragrant; to ⅜″ long; growing from leaf bases; followed by yellowish brown, silver-scaled, elliptical, ½″-long, berrylike fruit.

Leaves: Grayish green above, gray beneath; velvety, toothless, lance-shaped or oblong; to 3½″ long.

Blooms: Early summers.

Elevation: 3,000 to 7,000′.

Habitat: Moist soils along streams and ponds.

Comments: Deciduous, with reddish brown spines on branches and twigs to 2″ long. Native of Europe and western Asia; now naturalized in areas of Arizona, crowding out native species. Fruits eaten by birds; occasionally used for making jelly. One species of *Elaeagnus* in Arizona. Photograph taken at Lyman Lake, August 4.

RUSSET BUFFALO BERRY

Shepherdia canadensis
Oleaster Family (Elaeagnaceae)

Height: To 8′.

Flowers: Yellowish, inconspicuous, petalless; to 1⁄16″ wide; followed by reddish orange, fleshy, juicy, pockmarked, oval, berrylike fruit to ¼″ long.

Leaves: Dull green and dusty-scaled above; silvery with rusty patches beneath; elliptical to oval; to 3″ long.

Blooms: April–June.

Elevation: 7,000 to 9,000′.

Habitat: Moist coniferous forests.

Comments: Thornless, sprawling shrub with grayish brown bark. Fruit tastes bitter; eaten by birds. Male and female flowers on separate plants. Three species of *Shepherdia* in Arizona. Photograph taken at Greer, July 5.

HORNED SPURGE
Euphorbia brachycera (Euphorbia lurida)
Spurge Family (Euphorbiaceae)

Height: To 3'; usually shorter.

Flowers: Yellowish bracts, petalless, male and female flowers occur together in a tiny cup formed by joined bracts; to 1⁄16" wide.

Leaves: Light green, fleshy, numerous; more wide than long, to 3⁄8" wide, 1⁄4" long.

Blooms: April–August.

Elevation: 3,500 to 7,500'.

Habitat: Roadsides, fields, and clearings.

Comments: Stems have milky juice. More than three dozen species of *Euphorbia* in Arizona. Photograph taken northeast of Superior, April 3.

PHYSICNUT
Sangre-De-Drago
Jatropha cuneata
Spurge Family (Euphorbiaceae)

Height: To 6'.

Flowers: Pale yellowish; with deep yellow stamens; tubular, 5-lobed, narrow bell with rim curved backward; to 1⁄4" long, 1⁄8" wide; cluster at tips or on sides of branches.

Leaves: Dark green, thick, fleshy, smooth; alternate, oval with pointed base; creased down middle; to 3⁄4" long; in small clusters on branches. Usually leafless until after the summer rains.

Blooms: July–August.

Elevation: 1,000 to 2,000'.

Habitat: Dry mesas, plains, and slopes in southwestern Arizona.

Comments: Fleshy stems and leaves store water for drought periods. Pinkish stems. Tannin in roots was once used for tanning hides. Reddish sap in roots was used to produce dye and medicine. Three percent rubber can be extracted from dry stems. Four species of *Jatropha* in Arizona. Photograph taken at Organ Pipe Cactus National Monument, August 25.

WHITE-THORN ACACIA

Acacia constricta (Vachellia constricta)
Pea Family (Fabaceae)

Height: Shrub to 10'.

Flowers: Yellowish orange, very fragrant, crowded into ½" balls; followed by curved, reddish brown pod to 5" long, to ¼" wide.

Leaves: Light green, bipinnately compound; to 2" long, 1" wide.

Blooms: May–August.

Elevation: 2,500 to 5,000'.

Habitat: Mesas, dry slopes, washes, shallow caliche soil, and desert grassland.

Comments: Semi-deciduous; leafless during winter months. Branches armed with pairs of white spines to 2" long. Golden stamens (up to 40 per ball) give ball its color. Jackrabbits feed on young growth. Six species of *Acacia* in Arizona. Photograph taken near Saguaro Lake, August 26.

SWEET ACACIA

Huizache
Acacia farnesiana (Acacia smallii, Vachellia farnesiana)
Pea Family (Fabaceae)

Height: Spiny shrub, or small tree to 20'.

Trunk: To 1' in diameter.

Bark: Reddish brown, ridged, scaly, and thin.

Flowers: Golden yellow, ball-shaped, very fragrant; to ⅜" in diameter; followed by a purplish pod to 3" long.

Leaves: Bright green, bipinnately compound; to 4" long; leaflets to ¼" long.

Blooms: November–May.

Elevation: 2,500 to 4,000'.

Habitat: Occasionally found naturally in washes in southern Arizona.

Comments: Deciduous to semi-deciduous. Has white, straight, inch-long spines at base of leaves. Foliage and pods eaten by livestock. When mature has round, widely spreading crown. Although native in Sonora, Mexico, and probably in parts of southern Arizona, sweet acacia is widely used in landscapes and can be found escaping into nearby drainages. Six species of *Acacia* in Arizona. Photograph taken at Tortilla Flat, November 11.

CATCLAW ACACIA

Una De Gato
Acacia greggii (Senegalia greggii)
Pea Family (Fabaceae)

Height: Shrub, or small tree to 23′.

Trunk: To 8″ in diameter.

Bark: Gray to brown; scaly.

Flowers: Cream to pale yellow, dense, fragrant, cylindrical spike, to 2½″ long; followed by flattened, twisted pod to 6″ long, ½″ wide.

Leaves: Grayish green, bipinnately compound; to 3″ long.

Blooms: April–October, with heaviest blooms in April and May.

Elevation: Below 5,000′.

Habitat: Slopes, canyons, desert grasslands, and along washes and streams.

Comments: Deciduous. Branches have short, sharp, ¼″ long, recurved spines that resemble cats' claws and can rip clothing and flesh if brushed against. Wax-coated seeds delay germination for several years. Flowers attract bees and other insects. String bean–like fruits are ground into meal by Native Americans. Wood used for tool handles and fuel. Six species of *Acacia* in Arizona. Photograph taken near Saguaro Lake, August 26.

YELLOW BIRD-OF-PARADISE

Caesalpinia gilliesii
Pea Family (Fabaceae)

Height: Shrub to 10′.

Flowers: Yellow petals, 10 red stamens; to 3″ long; in terminal raceme; followed by a flattened bean, to 5″ long, splitting and curling when mature.

Leaves: Grayish green, twice pinnately compound; to 14 primary leaflets, up to 10 pairs of oval secondary leaflets; leaf to 6″ long, 3″ wide.

Blooms: April–September.

Habitat: Roadsides and waste areas in warmer counties of Arizona.

Comments: Open, deciduous shrub to 6′ wide. Native of South America and Mexico, now naturalized in the Southwest and grown as a landscape shrub. Flowers and foliage have an unpleasant odor. Seed pods are thought to be poisonous. Three species of *Caesalpinia* in Arizona. Photograph taken in vicinity of Scottsdale, May 7.

COURSETIA

Baby Bonnets
Coursetia glandulosa (Coursetia microphylla)
Pea Family (Fabaceae)

Height: To 20'; usually less.

Flowers: Cream-colored upper petals, yellow lower petals, pealike; to ½" long, ¼" wide; in short, glandular raceme; followed by a brown, thick-walled seed pod, to 2" long, constricted between seeds.

Leaves: Grayish green, hairy, pinnately compound; to 1" long, singly or in clusters. Elliptical to oblong leaflets to ¼" long, and in pairs; no terminal leaflet.

Blooms: March–April.

Elevation: Below 4,000'.

Habitat: Canyons and dry, rocky slopes.

Comments: Plant has grayish bark; no spines, and hairy stems. Two species of *Coursetia* in Arizona. Photograph taken at Saguaro National Monument West, April 17.

BIRDSFOOT LOTUS

Ground Honeysuckle
Lotus corniculatus
Pea Family (Fabaceae)

Height: Sprawling stems, tips upright to 18".

Flowers: Bright yellow, pealike, with upright upper hood; 5-petaled; to ⅝" long, ⅜" wide; on axil stem, in flat-topped, loose cluster of up to 12 flowers; cluster to 1¼" wide; followed by 1"-long, narrow pod with a hair at tip; pods radiating outward in half-circle.

Leaves: Dark green, pinnate, with 3 oblong leaflets, each to ¾" long.

Blooms: June–September.

Elevation: Not available. Photograph taken at 5,600'.

Habitat: Roadsides and meadows.

Comments: Perennial. Introduced from Eurasia. Configuration of pods suggests a bird's foot. Fifteen species of *Lotus* in Arizona. Photograph taken in vicinity of Christopher Creek, August 11.

GREEN'S LOTUS

Lotus greenei
Pea Family (Fabaceae)

Height: Creeper to 6" high.

Flowers: Yellow, with orangish backs; pealike; banner (upper) petal very large; to ⅜" wide, on long flower stem.

Leaves: Grayish green, densely haired, pinnate, 4 to 5 spatula-shaped leaflets in fanlike cluster, to ¼" long.

Blooms: March–May.

Elevation: 3,000 to 5,000' in southern Arizona.

Habitat: Roadsides, rocky hillsides, and mesas.

Comments: Fifteen species of *Lotus* in Arizona. Photograph taken at Sonoita, April 26.

DESERT ROCK PEA

Shrubby Deer Vetch
Lotus rigidus
Pea Family (Fabaceae)

Height: To 3'.

Flowers: Yellow-tinged with orange; pealike, banner (upper) petal flares backward; on long flower stem; flower to 1" long; followed by long, narrow seed pod to 1½" long.

Leaves: Grayish green, pinnate, widely spaced on stem; 3 to 5 narrow leaflets; leaf to 1⅛" long.

Blooms: February–May.

Elevation: Below 5,000'.

Habitat: Dry, rocky slopes, in canyons and in deserts.

Comments: The most drought-resistant lotus in Arizona. Several erect, wiry, gray stems. Fifteen species of *Lotus* in Arizona. Photograph taken near Superior, April 3. Recognizable by wide spacing on stems between leaves, and by its very gray stems and foliage.

HAIRY LOTUS

Desert Lotus

Lotus strigosus var. *tomentellus*

Pea Family (Fabaceae)

Height: Prostrate to 10".

Flowers: Yellow and pealike; to ⅛" wide; 1 to 2 on long flower stem.

Leaves: Grayish, hairy, and pinnate, with 5 to 6 blunt leaflets; to 1" long.

Blooms: March–May.

Elevation: Below 3,000'.

Habitat: Sandy deserts.

Comments: Fifteen species of *Lotus* in Arizona. Photograph taken in Superstition Mountains, March 15.

WRIGHT'S DEERVETCH

Lotus wrightii

Pea Family (Fabaceae)

Height: To 16".

Flowers: Yellow, tinged with orange or red with age; pealike; to ½" long; in leaf axils; followed by a slender pod to 1" long.

Leaves: Dark green; pinnately compound with 3 to 5 leaflets, each to ½" long.

Blooms: May–September.

Elevation: 4,500 to 9,000'.

Habitat: Roadsides and dry, open pine forests and juniper woods.

Comments: Fifteen species of *Lotus* in Arizona. Photograph taken in Willow Springs Lake area, August 9.

BLACK MEDICK

Nonesuch

Medicago lupulina

Pea Family (Fabaceae)

Height: Trailing to 2′ long.

Flowers: Bright yellow and pealike; to ⅛″ long; clustered together on short, dense spike; followed by tiny, 1-seeded, kidney-shaped pod, to ⅛″ long.

Leaves: Dark green, divided into 3 leaflets with rounded tips; to 1″ long including stalk.

Blooms: April–September.

Elevation: 2,500 to 9,000′.

Habitat: Roadsides, fields, lawns, and waste areas.

Comments: Annual; a forage plant. Native of Eurasia; now naturalized in U.S. Five species of *Medicago* in Arizona. Photograph taken at Black Canyon Lake, September 29.

BUR CLOVER

Medicago polymorpha (Medicago hispida)

Pea Family (Fabaceae)

Height: Prostrate to semi-erect, to 2′.

Flowers: Yellow and pealike; to ⅛″ wide; 3 to 5 in a cluster on short stalk; followed by coiled, burlike seed pod with curved prickles on the sharp edge of a spiral; to 5⁄16″ in diameter.

Leaves: Bright green, divided into 3 wedge-shaped, toothed leaflets with slightly indented tips; to 1″ wide.

Blooms: March–May; a longer period of time under moist conditions.

Elevation: 100 to 5,500′.

Habitat: Fields, lots, and disturbed areas.

Comments: Annual. Introduced from Europe; now naturalized in the U.S. Five species of *Medicago* in Arizona. Photograph taken on Apache Trail, March 19.

YELLOW SWEET CLOVER
Honey Clover
Melilotus officinalis
Pea Family (Fabaceae)

Height: To 5'.

Flowers: Yellow and pealike; to ¼" long; in long, spikelike raceme to 6" long.

Leaves: Light green, pinnately divided into 3 lance-shaped, toothed leaflets, each to 1" long.

Blooms: June–August.

Elevation: To 8,000'.

Habitat: Roadsides and fields.

Comments: From Eurasia; now naturalized in the U.S. Biennial. Roots bind soil and enrich it with nitrogen. An excellent honey producer; browsed by deer. Three species of *Melilotus* in Arizona. Photograph taken near Greer, June 18.

MEXICAN PALOVERDE
Horse Bean
Parkinsonia aculeata
Pea Family (Fabaceae)

Height: To 40'.

Trunk: To 1' in diameter.

Bark: Yellowish green, smooth, scaly at base, and brown when larger.

Flowers: Golden yellow; 5 petals, the largest orange or yellow with orange spots; to 1" wide, in cluster to 7" long; followed by dark brown pod narrowing between seeds, to 4" long.

Leaves: Bright green, twice pinnately compound; to 20" long; up to 30 pairs of leaflets, each to 3/16" long. Leaves fall off during drought or cold conditions.

Blooms: April–May.

Elevation: To 4,500'.

Habitat: Desert valleys.

Comments: Gradually becoming naturalized in Arizona. Foliage and seeds eaten by wildlife. A favorite of bees. Three species of *Parkinsonia* in Arizona. Photograph taken in Mesa, May 3. Easily recognizable by its branches of long streamers; spines on branches grouped in threes.

BLUE PALOVERDE

Parkinsonia florida (Cercidium floridum)
Pea Family (Fabaceae)

Height: To 30′.

Trunk: To 1½′ in diameter.

Bark: Blue-green, smooth, and thin, becoming brown and scaly on large, older trunks.

Flowers: Bright yellow, 5-petaled (all petals yellow); to 1″ wide; 4 or 5 in 2″-long cluster; followed by flat, yellowish brown pod, short-pointed at ends; to 3″ long.

Leaves: Dull bluish green when present; to 1½″ long; bipinnately compound with oblong leaflets to 3⁄16″ long.

Blooms: April–May.

Elevation: 500 to 4,000′.

Habitat: Along washes, valleys, flood plains, desert slopes, and desert grasslands.

Comments: Blue paloverde is widely spreading, with a very open crown. Photosynthesis takes place in the bark, in addition to the drought-deciduous leaves. Twigs often bear a ¼″ spine at each node. Three species of *Parkinsonia* in Arizona. Blue paloverde is distinguished from **Foothill Paloverde** (*Parkinsonia microphylla*) (at right) by its larger leaves, 5 yellow petals, bluish bark, and ¼″-long spines at the nodes. Photo taken at Usery Mountain Recreation Area, May 3.

FOOTHILL PALOVERDE

Littleleaf Paloverde
Parkinsonia microphylla (Cercidium microphyllum)
Pea Family (Fabaceae)

Height: To 25′.

Trunk: To 1′ in diameter.

Bark: Yellow-green and smooth.

Flowers: Pale yellow, with 5 petals (the largest petal is creamy white); ½″ wide; in 1″ cluster, followed by cylindrical pod with long, narrow points; pod to 3″ long, ¼″ in diameter.

Leaves: Yellowish green, bipinnately compound; 5 to 7 pairs of elliptical leaflets, each to 1⁄16″ long; leaf to ¾″ long.

Blooms: March–May.

Elevation: 500 to 4,000′.

Habitat: Dry, rocky hillsides and mesas, plains, and deserts.

Comments: During the dry season, when leaves are dropped, photosynthesis is carried on by the chlorophyll in the bark. The tree has a widely spreading, open crown. Twigs end in stout, stiff spines, to 2″ long. Three species of *Parkinsonia* in Arizona. Distinguished from **Blue Paloverde** (*Parkinsonia florida*) (at left) by its smaller leaves, 1 whitish petal, yellowish bark, and spine-tipped branchlets with no spines at the nodes. Photo taken at Usery Mountain Recreation Area, May 3.

WESTERN HONEY MESQUITE

Prosopis glandulosa var. *torreyana*
Pea Family (Fabaceae)

Height: To 20′.

Trunk: To 1′ in diameter.

Bark: Brown and smooth, becoming rough with age.

Flowers: Creamy yellow and fragrant; to ¼″ long; in narrow cluster to 3″ long; followed by spiraled or straight beanlike pod to 8″ long.

Leaves: Yellowish green, bipinnately compound; to 8″ long; leaflets narrow, oblong, to 1¼″ long, ⅛″ wide.

Blooms: May.

Elevation: Below 4,500′.

Habitat: Plains, hillsides, and along washes.

Comments: A favorite of bees. Native Americans used pods for meal. A single or pair of white to yellowish spines, to 3½″ long, at large nodes on branches. Three species of *Prosopis* in Arizona. Photograph taken at Desert Botanical Garden, Phoenix, May 20. **Texas Honey Mesquite** (*Prosopis glandulosa* var. *glandulosa*) has reddish spines and larger leaflets. Native to Texas and New Mexico, it is being spread along highways in Arizona by droppings from cattle transport trucks.

SCREWBEAN MESQUITE

Tornillo
Prosopis pubescens
Pea Family (Fabaceae)

Height: Shrub, or small tree to 20′.

Trunk: To 8″ in diameter.

Bark: Light brown to reddish and smooth when young; when mature, separates into long, fibrous strips.

Flowers: Yellow and small; to 3⁄16″ long; in dense, narrow, cylindrical cluster to 2″ long; followed by a light greenish, tightly coiled, spiral pod to 2″ long; on stalk in crowded cluster.

Leaves: Yellowish green, slightly hairy, bipinnately compound; to 3″ long; 5 to 8 pairs of oblong leaflets, to ⅜″ long, ⅛″ wide.

Blooms: May–August.

Elevation: Below 4,000′.

Habitat: Flood plains and bottomlands along rivers.

Comments: Deciduous. Branches are twisted and spiny. Pods used by Native Americans for meal and also eaten by desert animals. Wood used for fence posts and fuel. Three species of *Prosopis* in Arizona, with several varieties. Photograph taken in vicinity of Scottsdale, August 25.

VELVET MESQUITE

Prosopis velutina
Pea Family (Fabaceae)

Height: A large shrub, or small tree to 30' or more.

Trunk: To 2' in diameter.

Bark: Dark brown, rough, separating into long, narrow strips.

Flowers: Cream to yellow, small and fragrant, in slender, cylindrical spikes to 4" long, followed by narrow seed pods (brownish when mature) to 8" long.

Leaves: Yellow-green, bipinnately compound; leaflet to ½" long, leaf to 6" long.

Blooms: April; often again in August.

Elevation: Below 5,000'.

Habitat: Desert washes, alongside streams, and in areas where water table is reasonably high.

Comments: Deciduous shrub or tree, with straight 2" spines on branches. A legume that restores nitrogen to the soil. Roots penetrate ground to 60' in search of water. Often grows in dense thickets.

A favorite of bees and other insects; its flowers are an excellent honey source. Ripened seed pods are eaten by livestock and wild animals; 80 percent of a coyote's diet in late summer and fall is mesquite beans.

Along some desert rivers, such as the Verde, mesquite *bosques* (Spanish for "small forests") are still found; they provide excellent nesting sites for desert birds and habitat for mammals. Three species of *Prosopis* in Arizona. Photograph taken at Organ Pipe Cactus National Monument, April 9.

ROSARY BEAN

Rhynchosia senna (Rhynchosia texana)
Pea Family (Fabaceae)

Height: To 8", but mostly trailing.

Flowers: Yellow to peachy orange, pealike flower with sickle-shaped keel; to ⅛" wide, ¼" long; occurring singly or in small cluster in leaf axils; followed by green, flat pea pod to ¾" long.

Leaves: Dark green, pinnately divided into 3 elliptical segments; leaflet to 1¼" long, leaf to 2¼" long.

Blooms: May–September.

Elevation: 3,500 to 5,500'.

Habitat: Dry plains, mesas, and slopes.

Comments: Perennial herb. Ground cover, often controlling erosion. Three species of *Rhynchosia* in Arizona. Photograph taken at Lynx Lake, September 11.

DESERT SENNA

Senna covesii (Cassia covesii)
Pea Family (Fabaceae)

Height: To 2'.

Flowers: Rusty yellow, with 5 petals; to 1" wide; in terminal cluster; followed by slightly curving, woody seed pod, to 1¼" long.

Leaves: Grayish green, with fine, white hairs; pinnate; 3 pairs of elliptical leaflets; to 2" long.

Blooms: April–October.

Elevation: 1,000 to 3,000'.

Habitat: Dry, rocky slopes, mesas, and roadsides.

Comments: A bushy perennial. Nine species of *Senna* in Arizona. Photograph taken at Usery Mountain Recreation Area, March 1.

PINE THERMOPSIS

Golden Pea
Thermopsis divericarpa (Thermopsis pinetorum)
Pea Family (Fabaceae)

Height: To 3'.

Flowers: Bright yellow and pealike; to 1" long; in terminal, erect cluster; followed by long, slender pod, to 3" long.

Leaves: Bright green, compound, with 3 broad, lance-shaped leaflets; to 4" long.

Blooms: April–July.

Elevation: 6,000 to 9,500'.

Habitat: Meadows and clearings in pine forests.

Comments: Perennial herb; grows in clumps. Two species of *Thermopsis* in Arizona. Photograph taken near Hannagan Meadow, June 24.

GOLDEN CORYDALIS

Scrambled Eggs
Corydalis aurea
Fumitory Family (Fumariaceae)

Height: To 1½', but, because it is weak-stemmed, often supported or prostrate.

Flowers: Golden yellow, irregularly shaped; to ¾" long; in spikelike cluster; followed by a flat-sided, curved, narrow seed capsule to 1" long.

Leaves: Silvery bluish green, finely dissected, slightly succulent; to 6" long.

Blooms: February–June (sometimes late summer).

Elevation: 1,500 to 9,500'.

Habitat: Disturbed areas, washes, and pastures.

Comments: Short-lived perennial herb. Plant contains poisonous alkaloid. One species of *Corydalis* in Arizona. Photograph taken on Mount Graham, May 3.

SPURRED GENTIAN

Halenia recurva
Gentian Family (Gentianaceae)

Height: To 20".

Flowers: Yellowish and pointed, with cylindrical corolla, 4 erect lobes, 4 short spurs at base; to ½" long, ½" wide; at leaf intervals along stem.

Leaves: Yellowish green, linear, in opposite pairs at intervals on erect stem; to 1½" long.

Blooms: August–September.

Elevation: 7,500 to 10,000'.

Habitat: Mountain meadows and moist coniferous forests.

Comments: Annual herb. One species of *Halenia* in Arizona. Photograph taken near Willow Springs Lake, September 13.

WOLF'S CURRANT

Black Currant
Ribes wolfii
Currant Family (Grossulariaceae)

Height: Straggly shrub to 15'.

Flowers: Yellowish white, small, and bell-shaped, in small cluster. Followed by round currant with soft hairs tipped with brown, to ¼" in diameter; in clusters, blackish with a whitish bloom when mature.

Leaves: Green, maplelike, 3-lobed; toothed, deeply veined; to 3" long, 2½" wide.

Blooms: May–August.

Elevation: 8,500 to 11,500'.

Habitat: Moist coniferous forests and areas of springs.

Comments: Often sends shoots up into nearby trees. Browsed by elk. Birds and small mammals feed on fruits. Though tart, fruits are used for baking and for jelly. Ten species of *Ribes* in Arizona. Photograph taken at Greer, August 10.

WHISPERING BELLS

Emmenanthe penduliflora var. *penduliflora*
Waterleaf Family (Hydrophyllaceae)

Height: To 20".

Flowers: Pale yellow, bell-shaped, and nodding; with 5 untied petals; to ½" long; in loosely branched cluster.

Leaves: Grayish green, long, narrow; pinnately lobed; to 4" long.

Blooms: March–June.

Elevation: Below 4,000'.

Habitat: Desert washes, slopes, and along streams.

Comments: Annual herb. Stems are covered with sticky hairs. So-named because when it is dry, paper-thin flowers make rustling sounds in the wind. One species of *Emmenanthe* in Arizona. Photograph taken in the Superstition Mountains, March 15.

SOUTHWESTERN ST. JOHN'S WORT

Hypericum formosum
St. John's Wort Family (Hypericacae)

Height: To 28".

Flowers: 5 bright yellow petals above, reddish orange markings beneath; tiny, black dots on petal edges; numerous yellow stamens; to 1" wide.

Leaves: Dull green, often tinged with pink; oblong to oval; paired around stem; black-dotted on margins (translucent dots all over leaf are oil and pigment glands); to 1" long.

Blooms: July–September.

Elevation: 5,000 to 9,500'.

Habitat: Along mountain streams and moist meadows in coniferous forests.

Comments: Perennial herb. Reddish green erect stem, often with branches. Plant is named for St. John the Baptist. It was once believed if plants were hung in a home, inhabitants were protected from witches and thunder. Three species of *Hypericum* in Arizona. Photograph taken at Greer, July 5.

COMMON BLADDERWORT

Greater Bladderwort
Utricularia macrorhiza (Utricularia vulgaris)
Bladderwort Family (Lentibulariaceae)

Height: To 8" above surface of water.

Flowers: Yellow, 2-lipped, snapdragon-like; large palate faintly striped with red; spurred; to ½" wide, ¾" long; in a sparsely flowered, terminal cluster on a stout, leafless stem.

Leaves: Dark green, finely dissected, hairlike; floating or submerged; provided with scattered, ⅛"-wide bladders; to 2" long.

Blooms: July–August.

Elevation: 8,000 to 9,500'.

Habitat: Mainly shallow water of ponds and lakes.

Comments: Only insectivorous plant in Arizona. The bladders resemble minute bubbles. Each is equipped with a tiny door and a sensitive trigger, which when touched opens the door. The vacuum inside the bubble sucks an organism in, then digestive enzymes disintegrate the prey. Water fleas and mosquito larvae are common victims. Eight species of *Utricularia* in Arizona. Photograph taken at Carnero Lake near Greer, July 11.

DESERT MARIPOSA

Calochortus kennedyi var. *munzii*
Lily Family (Liliaceae)

Height: To 2′.

Flowers: Yellow petals with purple to black markings; membranes at base; short-stemmed when growing in open, long-stemmed among shrubs; to 3″ wide.

Leaves: Grayish green, narrow, grasslike, few; to 8″ long.

Blooms: March–May, but usually April.

Elevation: Below 5,000′.

Habitat: Open or shrubby areas in dry soil.

Comments: Perennial herb. *Mariposa* means "butterfly" in Spanish. Six species of *Calochortus* in Arizona. Photograph taken at Patagonia Lake State Park, April 26.

NEW MEXICAN YELLOW FLAX

Linum neomexicanum
Flax Family (Linaceae)

Height: To 2′.

Flowers: Yellow and starlike, with 5 petals; to ½″ wide. Many grow on elongated flower stems to 1½″ long.

Leaves: Dark green, narrow; to ½″ long; pointed upward, hugging stem; all along upward-pointing branches.

Blooms: June–September.

Elevation: 4,500 to 9,000′.

Habitat: Pine forests.

Comments: Nine species of *Linum* in Arizona. Photograph taken in vicinity of Woods Canyon Lake, August 3.

VENUS BLAZING STAR
Jones' Blazing Star
Mentzelia jonesii (Mentzelia nitens)
Stick Leaf Family (Loasaceae)

Height: To 2', often prostrate.

Flowers: Bright yellow, 5 rounded petals, to ¼" wide; in small, hairy clusters at tip of branches.

Leaves: Grayish green, rough, hairy; pinnately and deeply cleft into very narrow lobes; lower leaves to 6" long.

Blooms: February–May.

Elevation: Below 3,000'.

Habitat: Sandy deserts, often along rivers.

Comments: Several varieties of this species. Over twenty species of *Mentzelia* in Arizona. Photograph taken along Bill Williams River below Alamo Lake Dam, February 26.

DESERT BLAZING STAR
Adonis Blazing Star
Mentzelia multiflora (Mentzelia pumila)
Stick Leaf Family (Loasaceae)

Height: To 3'.

Flowers: Bright yellow, star-shaped; outer stamens have broad, flattened filaments and resemble additional petals; to 2" wide; occurring at ends of branches; followed by bullet-shaped seed capsule.

Leaves: Grayish green, sandpapery, long, narrow, many-lobed; to 4" long.

Blooms: February–October.

Elevation: 100 to 8,000'.

Habitat: Roadsides, dry stream beds, pinyon-juniper woods, and ponderosa forest clearings.

Comments: Plant has whitish stems. Flowers open in late afternoon. This species has a bullet-shaped ovary below flower head. Leaves and stems cling to fabric like Velcro, due to hooked hairs. Native Americans ground seeds for meal. Over twenty species of *Mentzelia* in Arizona. Photograph taken at Painted Rocks Dam, March 31.

DESERT VINE

Janusia gracilis

Malpighia Family (Malpighiaceae)

Height: Twining, tangled vine climbing over cacti and trees up to 6', vine to 9' long horizontally.

Flowers: Yellow, with 5 spoon-shaped petals; to ½" wide; occurring singly or in small cluster; followed by a 2- or 3-winged, reddish fruit similar to maple samara.

Leaves: Grayish green, narrow, very hairy above and beneath; opposite, linear; to 1¼" long.

Blooms: April–October.

Elevation: 1,000 to 5,000'.

Habitat: Washes; dry, rocky slopes; and desert flats.

Comments: Very slender stems. One species of *Janusia* in Arizona. Photograph taken at Saguaro National Monument West, April 16.

BALLOON MALLOW

Herissantia crispa

Mallow Family (Malvaceae)

Height: Straggly and weak-stemmed, to 2' long.

Flowers: 5 very pale peachy yellow petals, orangish center, to ¾" wide; followed by an angular, hairy, inflated, balloon-shaped fruit, to ½" wide, ⅜" high.

Leaves: Grayish green, heart-shaped, toothed, hairy above and beneath; to 2" long.

Blooms: Almost year-round.

Elevation: Below 3,500'.

Habitat: Dry slopes.

Comments: Often vinelike. One species of *Herissantia* in Arizona. Photograph taken at Saguaro National Park West, March 31.

DESERT ROSE MALLOW

Coulter's Hibiscus
Hibiscus coulteri
Mallow Family (Malvaceae)

Height: To 4', but usually less.

Flowers: Butter yellow to whitish with red basal spot, 5 rounded petals; cup-shaped; stamens joined at bases form tube surrounding style; to 2" wide.

Leaves: Dark green and reddish-margined; toothed, hairy, glandular; 3-lobed upper leaves to 1" long, undivided lower leaves to 1½" long.

Blooms: Periodically throughout the year.

Elevation: 1,500 to 4,000'.

Habitat: Canyons and rocky slopes.

Comments: Straggling shrub with gray, woody stems. Frequented by bees. Three species of *Hibiscus* in Arizona. Photograph taken at Saguaro National Monument West, April 17.

SPREADING FANPETALS

Sida abutifolia (Sida filicaulis)
Mallow Family (Malvaceae)

Height: Sprawling vine.

Flowers: Yellowish to peachy orange, with bright yellow center, 5 flattened petals slightly indented at tips; to ⅞" wide, borne singly from leaf axil.

Leaves: Dark green, hairy, elliptical, with scalloped margins; folded slightly upward; to 1" long.

Blooms: April–October.

Elevation: 2,500 to 6,000'.

Habitat: Plains and mesas in dry, sandy soil.

Comments: Stems are pinkish and hairy. Five species of *Sida* in Arizona. Photograph taken at Portal, April 23.

FRINGED LOOSESTRIFE

Lysimachia hybrida (Lysimachia ciliata var. *validula)*
Myrsine Family (Myrsinaceae)

Height: To 3'.

Flowers: Shiny yellow, flattened; with 5 rounded, crinkly edged, toothed petals pinkish toward base; 5 green sepals showing between petals; 5 stamens; to ¾" wide, on stalks in leaf axils.

Leaves: Light green with pinkish margins; opposite, folded upward, lance-shaped to elliptical; to 2¾" long.

Blooms: July–September.

Elevation: 6,000 to 7,500'.

Habitat: Moist soil of meadows, pondsides, and stream banks.

Comments: Perennial herb. Many-branched, with squarish stems. One species of *Lysimachia* in Arizona. Photograph taken in vicinity of McNary, July 7.

YELLOW MENODORA

Twinfruit
Menodora scabra
Olive Family (Oleaceae)

Height: To 1½'.

Flowers: Bright yellow and tubular, with 5 or 6 spreading lobes; to ¾" wide; in loose clusters at branch tips; reddish buds, followed by 2 translucent, round fruits side by side, each to ¼" in diameter.

Leaves: Grayish green, thick, rough, and lance-shaped; to 1½" long; occurring along the length of entire stem.

Blooms: April–September.

Elevation: 1,500 to 7,500'.

Habitat: Dry mesas and rocky slopes.

Comments: Perennial herb. Many-branched; browsed by wildlife. Two species of *Menodora* in Arizona. Photograph taken at Dead Horse Ranch State Park, May 30.

YELLOW EVENING PRIMROSE

Calylophus hartwegii ssp. *hartwegii*
Evening Primrose Family (Onagraceae)

Height: To 1'.

Flowers: Bright yellow, with 4 very crinkly petals; to 2½" wide; large buds are reddish with green stripes.

Leaves: Gray-green, hairy, narrow, with wavy margins; to 2" long.

Blooms: April–June.

Elevation: 3,000 to 7,000'.

Habitat: Roadsides, hillsides, and plains.

Comments: Perennial herb. Four species of *Calylophus* in Arizona. Photograph taken in vicinity of Safford, April 20.

YELLOW CUPS

Sundrop
Camissonia brevipes
Evening Primrose Family (Onagraceae)

Height: To 22".

Flowers: Bright yellow, with 4 petals; cup-shaped; to 1½" wide; in broad raceme; followed by slender pod, to 3" long.

Leaves: Green with reddish tinge, mostly in basal rosette; coarse, oval or pinnately lobed; to 5" long.

Blooms: February–May.

Elevation: Below 4,500'.

Habitat: Desert slopes and washes.

Comments: Annual. Blooms at sunrise instead of sunset, like most evening primroses. Around two dozen species of *Camissonia* in Arizona. Photograph taken at Alamo Lake, February 13.

MUSTARD EVENING PRIMROSE

Camissonia californica
Evening Primrose Family (Onagraceae)

Height: To 3′.

Flowers: Yellow, 4-petaled; often with pinkish spots at bases of petals; to ¾″ wide; followed by a long, very narrow, mustardlike seed pod to 2½″ long.

Leaves: Dark green, narrow; linear on stems; lance-shaped with irregular margins toward base of plant; to 4″ long.

Blooms: February–June.

Elevation: Below 4,500′.

Habitat: Washes, dry slopes, and plains.

Comments: Many-branched. Resembles members of the Mustard Family. Around two dozen species of *Camissonia* in Arizona. Photograph taken in Superstition Mountains, March 22.

MINIATURE SUNCUPS

Camissonia micrantha (Oenothera micrantha)
Evening Primrose Family (Onagraceae)

Height: To 2′.

Flowers: Bright yellow, with orangish buds, 4 petals; to ½″ wide.

Leaves: Gray-green, leathery with hairs; arrow-shaped; base clasps stem; to 2″ long at base, smaller up along stem.

Blooms: March–May.

Elevation: Below 4,500′.

Habitat: Washes and desert flats.

Comments: Around two dozen species of *Camissonia* in Arizona. Photograph taken in vicinity of Saguaro Lake, May 20.

HOOKER'S EVENING PRIMROSE

Yellow Flowered Evening Primrose
Oenothera elata ssp. *hirsutissima (Oenothera hookeri)*
Evening Primrose Family (Onagraceae)

Height: To 4'.

Flowers: Yellow, becoming pink to orange the following day; 4 broad petals; to 3" wide; in a simple or branching raceme; followed by a slender pod to 2" long.

Leaves: Green, long, lance-shaped or elliptical; to 9" long, graduating to smaller from base to top of stem.

Blooms: July–October.

Elevation: 3,500 to 9,500'.

Habitat: Roadsides, pinyon-juniper woodlands, and ponderosa pine clearings.

Comments: Biennial herb. Seeds eaten by Native Americans. Flowers open in late afternoon and close by noon the following day. Twenty-one species of *Oenothera* in Arizona. Photograph taken near Willow Springs Lake, September 13.

YELLOW PRIMROSE

Oenothera flava ssp. *taraxacoides (Oenothera taraxacoides)*
Evening Primrose Family (Onagraceae)

Height: To 6".

Flowers: Bright yellow (pink when faded), with 4 quilted petals, yellow stamens, long, yellowish green stigma with 4 threadlike branches; to 3½" wide.

Leaves: Dark green, very finely haired, deeply lobed; to 6" long.

Blooms: May–August.

Elevation: 5,000 to 9,500'.

Habitat: Sandy or moist soil along roadsides and in pine forests.

Comments: Twenty-one species of *Oenothera* in Arizona. Photograph taken at Crescent Lake, July 2.

BOTTLE EVENING PRIMROSE

Sundrop
Oenothera primiveris
Evening Primrose Family (Onagraceae)

Height: Prostrate, to 4".

Flowers: Yellow, 4 petals, notched; to 2" wide, opening in the evening, closing following morning.

Leaves: Greenish gray, pinnate, broad, rounded lobes, basal; to 4" long.

Blooms: Mid-February–May.

Elevation: Below 4,500'.

Habitat: Dry and open deserts.

Comments: Twenty-one species of *Oenothera* in Arizona. Photograph taken at Organ Pipe Cactus National Monument, February 28.

STRIPED CORAL ROOT

Corallorhiza striata var. *vreelandii*
Orchid Family (Orchidaceae)

Height: To 20".

Flowers: Pale brownish, yellowish, or whitish, striped with purplish to brownish, tiny orchids to 1" wide; oval lip to ½" long is bent downward; in racemes on erect, reddish purple stems.

Leaves: Nearly leafless, with only a few scalelike vestiges of leaves on lower stems.

Blooms: June–July.

Elevation: 7,000 to 9,000'.

Habitat: Ponderosa pine and spruce-fir forests.

Comments: Saprophytic orchid lacking chlorophyll. Receives nourishment from a fungus that decomposes dead plant material. Has corallike underground stem. Three species of *Corallorhiza* in Arizona. Photograph taken in vicinity of Forest Lakes, June 13.

YELLOW LADY'S SLIPPER

Yellow Moccasin Flower
Cypripedium parviflorum var. *pubescens (Cypripedium calceolus* var. *pubescens)*
Orchid Family (Orchidaceae)

Height: To 2′.

Flowers: Golden yellow, inflated slipperlike lip petal to 2″ long; 2 spirally twisted side petals ranging in color from yellow to yellowish brown with reddish markings. Two greenish yellow, lance-shaped sepals with reddish markings, one above and one below lip petal; usually 1 bloom but sometimes twin blooms per stem.

Leaves: Dark green, clasping stem; hairy, oval to elliptical, deeply veined; usually 3 to 5 per stem; to 8″ long.

Blooms: June–July.

Elevation: 6,000 to 9,000′.

Habitat: Rich soil in well-shaded locations of moist coniferous forests.

Comments: Perennial herb. Extremely rare in Arizona. One species of *Cypripedium* in Arizona. (Because of its rarity many have never seen this orchid growing in Arizona. The author of the first edition was, however, very familiar with the species and included here a photograph taken in 1982 in her wildflowers garden in the Northeast where her plant multiplied year after year.)

MEXICAN CANCER-ROOT

Conopholis alpina var. *mexicana*
Broomrape Family (Orobanchaceae)

Height: To 12″.

Flowers: Yellowish, curved with protruding stigmas; in several rows on elongated spike; to ½″ long.

Leaves: Yellow (lack chlorophyll); scalelike.

Blooms: April–June.

Elevation: 5,000 to 6,000′.

Habitat: The humus in pine, oak, cypress, and madrone areas.

Comments: Plant resembles a large pine cone. A saprophyte on decaying vegetation. One species of *Conopholis* in Arizona. Photograph taken in vicinity of Cave Creek, Portal, April 22.

CLUSTERED BROOMRAPE
Cancer-Root
Orobanche fasciculata
Broomrape Family (Orobanchaceae)

Height: To 4".

Flowers: Yellowish with slight pinkish color on lobes; tubular, 5-lobed; pointed, hairy calyx lobes; glandular hairs on flower and calyx; 2 pair of very hairy, yellow stamens; flower to ⅜" wide, 1" long.

Leaves: Cream-colored scales.

Blooms: May–August.

Elevation: 4,000 to 8,000'.

Habitat: Chaparral and coniferous forests, frequently in volcanic cinders.

Comments: Cream-colored, hairy, separate stems in cluster. A root parasite. Six species of *Orobanche* in Arizona. Photograph taken at Sunset Crater National Monument, June 4.

CREEPING WOOD SORREL
Little Yellow Sorrel
Oxalis corniculata
Oxalis Family (Oxalidaceae)

Height: Creeping stems to 8" long.

Flowers: Golden yellow, with 5 petals; to ½" wide; in cluster of 1 to 5 on slender stalk from leaf axil; followed by yellowish green, cylindrical, 5-angled, pointed, 1"-long seed pod.

Leaves: Dark green (sometimes tinged with purple); palmately compound with 3 notched leaflets; to 1" wide; folding together at night.

Blooms: February–November.

Elevation: 100 to 8,000'.

Habitat: Lawns, gardens, and fields.

Comments: Common garden weed; originated in Europe, now naturalized in the U.S. Roots at joints. Mature seed pods burst open, throwing seeds in all directions. Eight species of *Oxalis* in Arizona. Photograph taken at Chiricahua National Monument, April 24.

LITTLE GOLD POPPY

Pygmy Poppy
Eschscholzia minutiflora
Poppy Family (Papaveraceae)

Height: To 20".

Flowers: Yellow-orange, 4 petals; to ¾" wide; followed by a long, slender seed capsule.

Leaves: Bluish green, fernlike, divided into narrow segments; on long stalks at stem joints; to 2" near the ground, decreasing in size upward.

Blooms: February–May.

Elevation: Below 4,500'.

Habitat: Lower deserts in sandy soil.

Comments: A rather bushy plant. Three species of *Eschscholzia* in Arizona. Photograph taken in Kofa Mountain area, March 29.

CREAM CUPS

Platystemon californicus
Poppy Family (Papaveraceae)

Height: To 1'.

Flowers: Pale yellow or cream; normally 6 petals, occasionally more (as in photograph); terminal on stem; to 1" wide.

Leaves: Grayish green, softly haired, linear to narrowly lance-shaped; to 3" long; mainly on lower stems.

Blooms: March–May.

Elevation: 1,500 to 4,500'.

Habitat: Rocky slopes, hillsides, and along streams.

Comments: Annual. One species of *Platystemon* in Arizona. Photograph taken in vicinity of Roosevelt Dam, March 23.

YELLOW LINANTHUS

Desert Gold
Leptosiphon aureus (Linanthus aureus)
Phlox Family (Polemoniaceae)

Height: To 4".

Flowers: Golden yellow, with bright orange center; 5-lobed, funnel-shaped, upright; to ½" wide.

Leaves: Green, divided into 3 to 7 linear lobes; to ¼" long; in rings at well-spaced intervals on stem.

Blooms: March–June.

Elevation: 2,000 to 6,000'.

Habitat: Dry plains, mesas, and oak woodlands.

Comments: Annual. Threadlike, with reddish stalks. Three species of *Leptosiphon* in Arizona. Photograph taken on Mount Graham, April 21.

RIGID SPINY-HERB

Devil's Spiny-Herb
Chorizanthe rigida
Buckwheat Family (Polygonaceae)

Height: To 4".

Flowers: Yellowish green, surrounded by 3 long spines.

Leaves: Dark green, basal, broadly oval, long-stemmed; to 1½" long. Stem leaves are bractlike.

Blooms: March–May.

Elevation: Below 2,500'.

Habitat: Hot, coarse gravels of lower deserts.

Comments: Annual; sprouts up by the thousands after wet winter, then dies. Dried, blackened plants remain in desert soil for a year or more like spiny tufts. Four species of *Chorizanthe* in Arizona. Photograph taken south of Gila Bend, March 30.

WINGED BUCKWHEAT

Eriogonum alatum var. *alatum*
Buckwheat Family (Polygonaceae)

Height: To 5'.

Flowers: Yellowish green to yellowish brown; tubular, petalless, small (to 3/16" wide); in numerous loose clusters on upper branches, followed by small, hard, winged, triangular seeds.

Leaves: Grayish green, spatula-shaped, very hairy; mainly basal; to 7" long.

Blooms: July–September.

Elevation: 5,500 to 9,500'.

Habitat: Roadsides and forest clearings.

Comments: Erect and many-branched, with hairy stems. Navajo and Hopi Indians used plant medicinally to ease pain. Over fifty species of *Eriogonum* in Arizona. Photograph taken in vicinity of Woods Canyon Lake, August 2.

WILD BUCKWHEAT

Eriogonum corymbosum var. *glutinosum (Eriogonum aureum)*
Buckwheat Family (Polygonaceae)

Height: To 2'.

Flowers: Grayish-white to yellowish, tiny, and numerous; in rounded clusters to 3/4" wide; on many-stemmed branches.

Leaves: Light green, oval; to 1 3/4" long, 1 1/8" wide.

Blooms: July–October.

Elevation: 4,500 to 8,000'.

Habitat: Clearings in ponderosa pine forests.

Comments: Over fifty species of *Eriogonum* in Arizona. Photograph taken at Sunset Crater National Monument, September 6.

DESERT TRUMPET

Bladderstem

Eriogonum inflatum

Buckwheat Family (Polygonaceae)

Height: To 3'.

Flowers: Yellow, tiny; on threadlike stalks at stem divisions.

Leaves: Dark green, oval, long-stemmed; to 2" long; in basal rosette.

Blooms: February–October.

Elevation: Below 3,500'.

Habitat: Rocky or sandy desert slopes.

Comments: Perennial herb. Stems above nodes are inflated. Some species of wasps drill holes in the hollow stems, fill the hollows with captured insect larvae, then lay their eggs within the stems, thus ensuring food for their young. Over fifty species of *Eriogonum* in Arizona. Photograph taken at Cattail Cove State Park, February 24. The inflated stems are common to numerous species of *Eriogonum.*

YELLOW-FLOWERED ERIOGONUM

Eriogonum sp.

Buckwheat Family (Polygonaceae)

Height: To 1½'.

Flowers: Bright yellow, with 3 petals, 3 sepals, to 3⁄16" wide; in flattish, terminal cluster to 3" wide.

Leaves: Grayish-green, hairy, lance-shaped; mainly basal; to 6" long.

Blooms: May–September.

Elevation: 5,000 to 9,000'.

Habitat: Clearings in ponderosa pine forests and pinyon-juniper woodlands.

Comments: A variable species with over 20 varieties. Has a leafless flower stem, and grows in clumps. Almost all species of *Eriogonum* are difficult to identify, even for the expert botanist. For the amateur, simply recognizing wild buckwheat as such is an accomplishment. Over fifty species of *Eriogonum* in Arizona. Photograph taken at Walnut Canyon National Monument, September 7.

COMMON PURSLANE

Portulaca oleracea
Purslane Family (Portulacaceae)

Height: Normally prostrate to 2' long; occasionally erect to 6" high.

Flowers: Yellow, with 5 petals; to ¼" wide; singly or in small clusters in leaf axils or at stem tips; followed by a small, round capsule.

Leaves: Bronze-green with reddish margins, succulent, thick, smooth, and shiny; wedge-shaped, rounded at tip; to 1½" long.

Blooms: June–September.

Elevation: 1,000 to 8,500'.

Habitat: Clearings in ponderosa forests, overgrazed areas, meadows, and cultivated areas.

Comments: Annual. Reddish on stems. Joints produce roots when in contact with soil. Introduced from Europe; now naturalized. Its iron content is very high, so it is eaten as a salad green and as a potherb. Six species of *Portulaca* in Arizona. Photograph taken in Pine, September 2.

YELLOW COLUMBINE

Golden Columbine
Aquilegia chrysantha
Buttercup Family (Ranunculaceae)

Height: To 4'.

Flowers: Canary yellow, with 5 yellow petals with 2"-long spurs projecting backward; flower horizontal or upward pointing; to 3" wide.

Leaves: Bluish green, mostly basal, divided into threes; leaflets to 1½" long about as wide.

Blooms: April–September.

Elevation: 3,000 to 11,000'.

Habitat: Alongside streams and in rich, moist soil in shady forests.

Comments: Perennial herb. Seven species of *Aquilegia* in Arizona. Photograph taken at Willow Springs Lake, September 9.

HEARTLEAF BUTTERCUP

Ranunculus cardiophyllus
Buttercup Family (Ranunculaceae)

Height: To 16".

Flowers: Shiny yellow, with 5 waxy petals; to 1½" wide.

Leaves: Dark green, heart-shaped, with scalloped margins; to 2" long.

Blooms: June–July.

Elevation: 7,000 to 9,500'.

Habitat: Moist meadows in pine and spruce-fir belts.

Comments: Buttercups contain a cardiac poison and are poisonous if eaten. Over twenty species of *Ranunculus* in Arizona. Photograph taken near Greer, June 21.

AQUATIC BUTTERCUP

Ranunculus hydrocharoides
Buttercup Family (Ranunculaceae)

Height: To 3".

Flowers: Golden yellow, with 5 petals; waxy-looking; to ¼" wide; grow singly at end of stalk.

Leaves: Green, shiny, oval to lance-shaped; to 1½" long.

Blooms: June–September.

Elevation: 7,000 to 9,500'.

Habitat: Springs, marshes, streams, and wet meadows in mixed coniferous forests.

Comments: Aquatic perennial. Over twenty species of *Ranunculus* in Arizona. Photograph taken near Hannagan Meadow, June 24.

MACOUN'S BUTTERCUP

Ranunculus macounii
Buttercup Family (Ranunculaceae)

Height: To 3'.

Flowers: Golden yellow, with 5 rounded petals; to ⅝" wide.

Leaves: Dark green, finely haired above, lighter green beneath; triangular-shaped but divided and cleft; to 8" long, including stem; leaves smaller on upper stems.

Blooms: July–August.

Elevation: 6,000 to 8,000'.

Habitat: In mud along streams and in marshes in coniferous forests.

Comments: Perennial. A sprawling plant with either hairy or smooth stems. Over twenty species of *Ranunculus* in Arizona. Photograph taken at Greer, July 5.

ROADSIDE AGRIMONY

Agrimonia striata
Rose Family (Rosaceae)

Height: To 6'.

Flowers: Yellow, with 5 petals, yellow stamens; to 5⁄16" wide; on long, slender spike.

Leaves: Dark green and hairy above, pale green and hairy beneath; pinnate, divided into large and small leaflets to 2" long, with pointed tips, toothed; up to 11 leaflets per leaf.

Blooms: July–September.

Elevation: 6,500 to 8,500'.

Habitat: Rich soil in pine forests and along streams.

Comments: Perennial herb with very hairy stem. Two species of *Agrimonia* in Arizona. Photograph taken at Greer, July 5.

COMMON SILVERWEED

Argentina anserina (Potentilla anserine)
Rose Family (Rosaceae)

Height: Strawberrylike runners, with flower stalk to 12".

Flowers: Yellow, with 5 petals; to ¾" wide.

Leaves: Silvery green, featherlike; upper surface is silky-haired, lower surface woolly haired; pinnately compound, with 9 to 31 lancelike, sharply toothed leaflets; to 10" long.

Blooms: May–August.

Elevation: 5,600 to 9,500'.

Habitat: Open, moist ground in ponderosa pine and spruce-fir forests.

Comments: Perennial herb, introduced from Eurasia. At joints on runners, roots form and enter soil; leaves then develop and new plants form. One species of *Argentina* in Arizona. Photograph taken at Luna Lake, July 23.

BIG-LEAF AVENS

Largeleaf Avens
Geum macrophyllum
Rose Family (Rosaceae)

Height: To 4'.

Flowers: Yellow, wavy with 5 petals as long or longer than sepals; numerous stamens and pistils; flower to ½" wide, in loose cluster on upper branches.

Leaves: Dark green, pinnately compound; bristly haired, toothed, and large; roundish segment at tip with small segments toward main stem; basal leaf to 18" long, progressively shorter toward flowers.

Blooms: July–September.

Elevation: 7,000 to 9,000'.

Habitat: Mountain stream banks and clearings in moist coniferous forests.

Comments: Perennial herb, with bristly, hairy stems. Four species of *Geum* in Arizona. Photograph taken in Mount Baldy Wilderness, August 13.

TALL CINQUEFOIL

Potentilla arguta
Rose Family (Rosaceae)

Height: To 40".

Flowers: Pale yellow, with 5 rounded petals; 5 long sepals visible between petals; yellow-centered stamens; to ¾" wide.

Leaves: Dark green, very hairy, pinnate; 5 to 9 broadly oval, toothed leaflets; unequally divided leaf to 8" long.

Blooms: June–July.

Elevation: 5,000 to 8,000'.

Habitat: Meadows and hillsides.

Comments: Stems are reddish, hairy, and sticky. Twenty species of *Potentilla* in Arizona. Photograph taken at Black Canyon Lake, June 4.

VARILEAF CINQUEFOIL

Meadow Cinquefoil
Potentilla diversifolia
Rose Family (Rosaceae)

Height: To 20".

Flowers: Golden yellow and buttercup-like, with 5 notched petals; 5 sepals showing between petals; orangish stamens; to ¾" wide; in loose, terminal cluster.

Leaves: Dark green and silky-haired above, whitish beneath; palmate, to 7 leaflets; sharply toothed, with hairs extending beyond teeth; to 2¾" wide, 2" long.

Blooms: June–September

Elevation: 8,000 to 12,000'.

Habitat: Roadsides and moist, rocky areas.

Comments: *Cinquefoil* means "five leaves" in French. Twenty species of *Potentilla* in Arizona. Photograph taken south of Alpine, June 30.

SHRUBBY CINQUEFOIL
Potentilla fruticosa
Rose Family (Rosaceae)

Height: To 3′.

Flowers: Bright golden yellow, with 5 broad petals, up to 30 stamens; to 1¼″ wide.

Leaves: Grayish green above, paler beneath; very hairy, pinnately divided into 3 to 7 leaflets (normally 5) tipped with a red dot and with margins rolled under; leaflet to ¾″ long, leaf to 1⅛″ long; cover entire shrub.

Blooms: June–August.

Elevation: 7,000 to 9,500′.

Habitat: Moist meadows, pine forest clearings, streamsides, and plains.

Comments: Only shrubby *Potentilla* growing in Arizona. This erosion fighter has reddish brown, shreddy bark. Browsed by livestock and deer. Twenty species of *Potentilla* in Arizona. Photograph taken in Greer area, July 3.

CLUBLEAF CINQUEFOIL
Potentilla subviscosa
Rose Family (Rosaceae)

Height: To 4″ tall in leafy, rather flat rosette at higher elevations.

Flowers: Bright yellow, with 5 slightly notched petals; sepals shorter than petals; tack-shaped, glandular hairs on sepals and buds; to ⅝″ wide.

Leaves: Dark green, very hairy; toothed; 3 large segments with 2 smaller segments; to ¾″ wide, 1¾″ long, including stem.

Blooms: April–June.

Elevation: 6,500 to 12,000′.

Habitat: Mountain meadows and coniferous forests; often blooming at edges of snowbanks.

Comments: Hairy-stemmed. Twenty species of *Potentilla* in Arizona. Photograph taken on Mount Graham at 8,300′, May 3.

STANSBURY CLIFF-ROSE

Buckbrush

Purshia stansburiana (Cowania mexicana var. *stansburiana)*

Rose Family (Rosaceae)

Height: Shrub to 8′, or small tree to 20′.

Trunk: To 8″ in diameter for a tree.

Bark: Reddish brown, shredding.

Flowers: Creamy white to pale yellow, 5-petaled with gold centers, very fragrant; to 1″ wide; each flower followed by 5 to 10 ¼″-long fruits, each with a 2″-long feathery plume attached.

Leaves: Dark green above, white, woolly hairs beneath; leathery, glandular-dotted, wedge-shaped; divided into 3 to 5 narrow lobes; edges rolled under; to 1″ long.

Blooms: April–September.

Elevation: 3,500 to 8,000′.

Habitat: Dry, rocky hillsides and plateaus in upper desert, grasslands, and oak-pinyon-juniper areas.

Comments: Evergreen. Excellent winter browse for deer, sheep, and cattle despite its bitter taste. Native Americans used stringy bark for mats and clothing and wood for arrows. Six species of *Purshia* in Arizona. Photograph taken at North Rim of Grand Canyon National Park, June 25. Unlike the **Apache Plume** (*Fallugia paradoxa*), cliff-rose is treelike, with a single trunk, and has waxy, hairless leaves, cream-colored flowers, and sparser plumes. Another similar shrub, **Antelopebrush** (*Purshia tridentata*), has wider leaves and smaller, yellowish flowers.

HOPBUSH

Dodonaea viscosa

Soapberry Family (Sapindaceae)

Height: Shrub to 12′.

Flowers: Yellowish, small, without petals; in small, terminal raceme; followed by 2 to 4 creamy to pinkish, broad-winged fruits.

Leaves: Green and oblong; to 4″ long, ¾″ wide.

Blooms: February–October.

Elevation: 2,000 to 5,000′.

Habitat: Dry, rocky slopes and in canyons.

Comments: Fruits used as substitute for hops. Contains saponin, a poisonous substance, used as laundry soap. One species of *Dodonaea* in Arizona. Photograph taken in Superstition Mountains, March 15.

MOGOLLON INDIAN PAINTBRUSH

Painted-Cup
Castilleja sulphurea (Castilleja mogollonica)
Figwort Family (Scrophulariaceae)

Height: To 14".

Flowers: Pale yellow; 3-toothed bracts, each 1" long, with pinkish, pointed tips, center tooth wider than others; very hairy, clustered on 4"-long erect spike. Flowers inconspicuous and generally hidden within bracts.

Leaves: Light green, finely haired, narrow; to 1¼" long.

Blooms: July–August.

Elevation: Around 9,500'.

Habitat: Wet alpine meadows.

Comments: Stem is very hairy. Seventeen species of *Castilleja* in Arizona. Photograph taken in mountain meadow above Greer, August 8.

WRIGHT'S BIRDBEAK

Club-Flower
Cordylanthus wrightii
Figwort Family (Scrophulariaceae)

Height: To 2'.

Flowers: Yellowish green, beaklike, narrow; upper and lower lips nearly equal; surrounded by long bracts; to 1¼" long, in clusters at branch ends.

Leaves: Light green, tinged with pink; hairlike, divided into very narrow, curly segments; to 2" long; occurring all along stems.

Blooms: June–October.

Elevation: 5,000 to 7,500'.

Habitat: Roadsides, sandy mesas, and flats, often growing among junipers.

Comments: Annual. Partially root-parasitic. Many-branched, spindly, and bushlike. Four species of *Cordylanthus* in Arizona. Photograph taken north of St. Johns, August 4.

BUSH PENSTEMON

Keckiella antirrhinoides ssp. *microphylla (Penstemon microphyllus)*
Figwort Family (Scrophulariaceae)

Height: Sprawling shrub to 8'.

Flowers: Bright yellow, snapdragon-like, with 2 upper lobes, 3 lower lobes; 4 yellow stamens curved upward in throat, upper surface of fifth stamen (sterile, lacking anther) heavily bearded; flower to 1" long, ⅜" wide; all along stems and side branches.

Leaves: Light green, hairy, elliptical to oblong; to ¾" long; in small clusters along branches.

Blooms: March–May.

Elevation: 1,500 to 5,000'.

Habitat: Rocky slopes.

Comments: Only occasionally browsed by livestock. One species of *Keckiella* in Arizona. Photograph taken south of Superior, April 9.

BUTTER AND EGGS

Dalmatian Toadflax
Linaria dalmatica
Figwort Family (Scrophulariaceae)

Height: To 4'.

Flowers: Pale yellow, snapdragon-like; long, slender, basal spur; 2 lobes of upper lip pointing upward, 2 lobes of lower lip pointing downward; orange palate (in throat); to 2" long; in elongated cluster along stem.

Leaves: Bluish green, broad, leathery; oval to lance-shaped; clasping stem at regular intervals; to 3" long.

Blooms: May–September.

Elevation: 5,500 to 7,500'.

Habitat: Roadsides, fields, and waste areas.

Comments: Perennial herb. Orange palate serves as a honey guide for bees. Naturalized from Eurasia. Four species of *Linaria* in Arizona. Photograph taken in Lynx Lake area, May 26.

SEEP MONKEY FLOWER
Yellow Monkey Flower
Mimulus guttatus
Figwort Family (Scrophulariaceae)

Height: To 3′.

Flowers: Bright yellow, hairy throat spotted with reddish pink; 2 lobes of upper lip point upward, 3 lobes of lower lip point downward; to 1½″ long, 1¼″ wide; in upper leaf axils.

Leaves: Dark green, oval, opposite; margins toothed; to 4″ long, upper leaves lack stalks.

Blooms: March–September.

Elevation: 500 to 9,500′.

Habitat: Along brooks, springs, and other wet places.

Comments: Perennial herb with hollow stems. Variable in size from tall and spindly to large and bushy. Native Americans used leaves for salad greens. Fourteen species of *Mimulus* in Arizona. Photograph taken at Organ Pipe Cactus National Monument, April 1.

GHOST FLOWER
Mohavea confertifolia
Figwort Family (Scrophulariaceae)

Height: To 16″.

Flowers: Pale cream-colored to yellowish, with pinkish purple dots on inside surface of roughly fringed lobes; cuplike; lower petal has reddish purple spot with 2 bright yellow stamens curving upward over spot; lower lip deeply indented on underside by hairy calyx; flower to 1½″ long; in clusters among leaves.

Leaves: Light green, succulent, hairy on upper surface; elliptical to lance-shaped; to 4″ long.

Blooms: February–April.

Elevation: Below 2,500′.

Habitat: Sandy desert washes and rocky talus slopes.

Comments: Annual. One main stem with many side stems. Translucent appearance of flower gives it its common name. Two species of *Mohavea* in Arizona. Photograph taken at Cattail Cove State Park, March 8.

YELLOW OWL'S CLOVER

Buttered Owl's Clover
Orthocarpus luteus
Figwort Family (Scrophulariaceae)

Height: To 16".

Flowers: Golden yellow; 2-lipped, with upper lip forming short beak, saclike lower lip; to ½" long; in axils of hairy, 3-lobed bracts on single spike; flowering section of spike to 3" long.

Leaves: Dark green to reddish green, alternate, spiraling up stem; linear or at times narrowly 3-lobed; to 1" long.

Blooms: July–September.

Elevation: 7,000 to 9,500'.

Habitat: Coniferous forests and moist meadows and hillsides.

Comments: Annual. Erect stem. Six species of *Orthocarpus* in Arizona. Photograph taken in vicinity of Greer, August 8.

GRAY'S LOUSEWORT

Fern-Leaf
Pedicularis procera (Pedicularis grayi)
Figwort Family (Scrophulariaceae)

Height: To 5'.

Flowers: Yellowish with red-brown lines; 2-lipped, tubular, short-beaked; to ¾" long, in a dense, bracted, terminal spike.

Leaves: Dark green, fernlike, twice pinnate; to 1½' long.

Blooms: July–August.

Elevation: 8,000 to 10,000'.

Habitat: Rich soil in coniferous forests.

Comments: Eight species of *Pedicularis* in Arizona. *Pediculus* means "louse" in Latin; in Roman times seeds were used to kill lice. Photograph taken in Greer area, July 21.

YELLOW RATTLE

Rhinanthus minor ssp. *minor (Rhinanthus rigidus)*
Figwort Family (Scrophulariaceae)

Height: To 8".

Flowers: Yellow, with arched upper lip, 3-lobed lower lip; to ½" log; in dense, elongated cluster; followed by enlarged, bladderlike, oval calyx, to ⅜" thick, ⅝" long, ½" wide.

Leaves: Light green streaked with reddish purple; opposite, lance-shaped, toothed; to 2" long.

Blooms: August–September.

Elevation: 9,000 to 9,500'.

Habitat: Streamsides and moist mountain meadows.

Comments: Annual. Erect, 4-angled stems. Plant used as an insecticide. One species of *Rhinanthus* in Arizona. Photograph taken in Mount Baldy Wilderness, August 13.

MOTH MULLEIN

Verbascum blattaria
Figwort Family (Scrophulariaceae)

Height: To 5'.

Flowers: Yellow (sometimes white), with 5 rounded, slightly unequal lobes; flat; red-orange anthers with purplish hairs on stamen filaments; to 1¼" wide; clustered along slender, erect spike.

Leaves: Dark green, straplike, lobed or toothed; to 5" long in basal rosette, decreasing in length toward flowers.

Blooms: August–September.

Elevation: 6,000 to 7,000'.

Habitat: Fields, roadsides, and wastelands.

Comments: Biennial weed. Introduced from Europe; now naturalized in U.S. Hairy stamen filaments resemble a moth's antennae; the flattened flowers on the stem resemble resting moths; hence the name "moth mullein." Five species of *Verbascum* in Arizona. Photograph taken near Prescott, September 11.

COMMON MULLEIN

Woolly Mullein
Verbascum thapsus
Figwort Family (Scrophulariaceae)

Height: To 6'.

Flowers: Yellow, with 5 slightly unequal lobes; 5 orange-tipped stamens; to 1" wide; in tightly wedged, spikelike cluster; to 20" long.

Leaves: Grayish green, oblong, feltlike, grayish-haired; basal leaves in rosette; to 16" long, shorter on stem.

Blooms: June–September.

Elevation: 5,000 to 7,000'.

Habitat: Roadsides, open areas, and disturbed places.

Comments: Biennial; naturalized from Europe. Seeds eaten by birds. Leaves used as wicks. For warmth, colonists and Native Americans lined footwear with leaves. Various parts of plant used medicinally. Five species of *Verbascum* in Arizona. Photograph taken in vicinity of Woods Canyon Lake, August 7.

HAIRY FIVE-EYES

Chamaesaracha sordida
Nightshade Family (Solanaceae)

Height: To 1'; usually much less because plant hugs the ground.

Flowers: Pale yellowish or whitish above, purplish streaks beneath; flat, with 5 spreading lobes; 5 greenish bands radiating outward from center; eyelike markings at base of each lobe; to ½" wide, in upper leaf axils.

Leaves: Dull green, glandular, very sticky, and hairy; lance-shaped to oblong, often pinnately cleft or toothed, with wavy margins; to 1½" long.

Blooms: March–October.

Elevation: 3,500 to 5,500'.

Habitat: Dry mesas and plains.

Comments: Because of stickiness, foliage is often covered with pollen dust or soil particles. Three species of *Chamaesaracha* in Arizona. Photograph taken in vicinity of Fort Bowie, May 8.

TREE TOBACCO

Nicotiana glauca
Nightshade Family (Solanaceae)

Height: To 20′.

Flowers: Pale yellow, tubular; to 2″ long; in loose clusters at ends of branches.

Leaves: Bluish green, long-stalked, oval, and smooth; to 7″ long.

Blooms: Throughout the year.

Elevation: Below 3,000′.

Habitat: Roadsides, washes, hillsides, and rocky canyons.

Comments: Grows as an open shrub or small tree. Contains the poisonous alkaloid nicotine. Plant also contains anabasine, another potent poison. An insecticide for aphids is brewed by soaking plant parts in water, and then using solution on infected plant. Native to South America, now naturalized in Arizona. The name *Nicotiana* is in honor of Jean Nicot, who introduced tobacco plants to the French royalty. Four species of *Nicotiana* in Arizona. Photograph taken in vicinity of Crown King, May 4.

THICK-LEAVED GROUND CHERRY

Physalis crassifolia
Nightshade Family (Solanaceae)

Height: To 2′.

Flowers: Tawny yellow, bell-shaped, with spreading lobes; petals joined; to ½″ wide; followed by a silvery, inflated, ¾″-long, lantern-shaped calyx containing the berry of seeds.

Leaves: Dark green, triangular to heart-shaped; to 1″ wide.

Blooms: February–October.

Elevation: Below 3,000′.

Habitat: Dry, rocky slopes and washes.

Comments: Perennial herb. Plant forms small bush. Fourteen species of *Physalis* in Arizona. Photograph taken at Cattail Cove State Park, March 8.

BUFFALO-BUR
Texas Thistle
Solanum rostratum
Nightshade Family (Solanaceae)

Height: To 2½'.

Flowers: Bright yellow, starlike, with 5 crinkly lobes, 5 yellow anthers form cone in center; to 1" wide; followed by spiny bur to 1" wide.

Leaves: Dark green, deeply cut into 5 to 7 pinnate lobes; stalks and vein backs covered with spines; to 6" long.

Blooms: May–August.

Elevation: 1,000 to 7,000'.

Habitat: Roadsides, fields, and disturbed areas.

Comments: Annual herb. Probably the original host of the Colorado potato beetle. Entire plant is covered with straight, sharp, straw-colored spines, to ½" long, which cause pain if touched. Leaves and seed pods are poisonous. Fifteen species of *Solanum* in Arizona. Photograph taken at Dead Horse Ranch State Park, May 29.

CREOSOTE BUSH
Hediondilla
Larrea tridentata var. *tridentata (Larrea divaricata)*
Caltrop Family (Zygophyllaceae)

Height: Many-branched shrub to 10'.

Flowers: Yellow, solitary, 5-petaled; to 1" in diameter; followed by a globe-shaped, fuzzy, white, dry capsule, to ¼" in diameter.

Leaves: Dark green to yellowish green, waxy, and strong-scented; resinous, 2 leaflets are joined at base; to ⅜" long.

Blooms: Periodically; peaks March–April and November–December.

Elevation: Below 4,500'.

Habitat: Dry plains and mesas.

Comments: Evergreen. Certain creosote plants are thought to be oldest living plants, at over 11,000 years old. Gives off musty odor after rain. When pollinated, petals twist 90 degrees. Varnishlike coating on leaves conserves water by slowing evaporation. During extreme drought plant grows tougher, smaller leaves. Resin from branches was used by Native Americans as a glue; other parts of plant were used medicinally. One species of *Larrea* in Arizona. Photograph taken at Cattail Cove State Park, February 24.

PUNCTURE VINE
Goat's Head
Tribulus terrestris
Caltrop Family (Zygophyllaceae)

Height: Prostrate, to 2" high, 5' long.

Flowers: Yellow, with 5 broad petals, to ½" wide; followed by hard, star-shaped fruit which separates into 5 brownish gray nutlets, each bearing a pair of ¼"-long spines.

Leaves: Dark green, opposite, pinnately compound; to 2" long; with leaflets to ½" long.

Blooms: March–October.

Elevation: Below 7,000'.

Habitat: Fields, wastelands, roadsides, and desert.

Comments: Annual weed. Two-spined segment of fruit resembles goat's head with horns. Native of Mediterranean region; now naturalized in Southwest. Spines injure livestock, puncture bicycle tires, are very painful to bare feet, and become embedded in fur and fabric. One species of *Tribulus* in Arizona. Photograph taken at Saguaro Lake, October 18.

ANGIOSPERMS: ORANGE FLOWERS

DESERT HONEYSUCKLE

Anisacanthus thurberi
Acanthus Family (Acanthaceae)

Height: Shrub to 6′.

Flowers: Vermillion or orange, with long slender tube; to 1¼″ long.

Leaves: Dark green, short-stemmed, elliptical; to 2½″ long, ¾″ wide.

Blooms: Chiefly in spring, but also other times of year.

Elevation: 2,500 to 5,500′.

Habitat: Along sandy washes and in canyons.

Comments: Has woody stems and shreddy bark. Browsed by sheep and cattle; pollinated by hummingbirds. One species of *Anisacanthus* in Arizona. Photograph taken at Patagonia, April 27.

BUTTERFLY WEED

Asclepias tuberosa
Milkweed Family (Asclepiadaceae)

Height: To 3′.

Flowers: Brilliant orange to yellow (page 109); to ½″ wide, ½″ long; 5 small sepals, 5 petals (bent back) and 5 hoods, in flat-topped, erect, terminal cluster to 3″ wide; followed by a narrow, tapered pod to 5″ long.

Leaves: Light green, narrowly arrow-shaped; to 4½″ long.

Blooms: May–September.

Elevation: 4,000 to 8,000′.

Habitat: Dry, open grasslands and open areas in pine forests.

Comments: Perennial, bushy herb with stems hairy. Seeds have white, silky hairs. Unlike most milkweeds, sap of this species is not milky. Twenty-nine species of *Asclepias* in Arizona. Photograph taken at Oak Creek Canyon, June 18.

ORANGE AGOSERIS

Orange Mountain-Dandelion
Agoseris aurantiaca
Sunflower Family (Asteraceae)

Height: To 2′.

Flowers: Burnt orange, all rays, with shorter rays toward center; stamens stand upright; flower head to 1″ wide; solitary on erect, leafless flower stem; followed by seeds topped with silvery bristles.

Leaves: Grayish green, narrow, widest above middle; with or without teeth or lobes; to 10″ long.

Blooms: June–August.

Elevation: 5,000 to 9,500′.

Habitat: Meadows and clearings in coniferous forests.

Comments: Perennial herb; has pinkish stem. Produces milky sap. Three species of *Agoseris* in Arizona. Photograph taken in Greer area, August 8.

FIDDLENECK

Amsinckia menziesii var. *intermedia (Amsinckia intermedia)*
Forget-me-not Family (Boraginaceae)

Height: To 20″.

Flowers: Yellow-orange, funnel-shaped; to ⅛″ wide; in coiled spike.

Leaves: Green, narrow, with bristly hairs; 6″ at base, graduating upward to ¾″ at tips of stem.

Blooms: March–May.

Elevation: Below 4,000′.

Habitat: Fields, roadsides, and dry, open areas.

Comments: Annual. At start of blooming, flower head is coiled in a manner resembling the neck of a violin; as individual flowers open, the coil uncurls. Two species of *Amsinckia* in Arizona. Photograph taken at Usery Mountain Recreation Area, March 7.

WHEELER'S WALLFLOWER

Erysimum capitatum var. *capitatum (Erysimum wheeleri*
Mustard Family (Brassicaceae)

Height: To 32".

Flowers: Deep orange to orange-yellow, with 4 petals; to ¾" wide; in cluster on rounded, terminal raceme; followed by very slender, erect 4-sided pod to 4" long.

Leaves: Grayish green, lance-shaped, toothed margins; in basal rosette; to 5" long. Stem leaves are narrow with small teeth.

Blooms: March–September.

Elevation: Above 7,000'.

Habitat: Coniferous forests.

Comments: Biennial or perennial. Three species of *Erysimum* in Arizona. Photograph taken south of Alpine, July 23.

BEARBERRY HONEYSUCKLE

Twinberry Honeysuckle
Lonicera involucrata
Honeysuckle Family (Caprifoliaceae)

Height: To 7'.

Flowers: Yellow tinged with red; tubular; to ½" long; in pairs, with 2 large bracts at base; followed by a pair of shiny, purplish black, pea-sized berries.

Leaves: Dark green, glandular-dotted, oval, and hairy; to 5" long.

Blooms: June–July.

Elevation: 7,500 to 10,500'.

Habitat: Along streams and in moist coniferous forests.

Comments: Often forms thickets. Frequented by hummingbirds. Birds and mammals eat its sour berries. Twelve species of *Lonicera* in Arizona. Photograph taken in Greer, July 20.

ORANGE GOOSEBERRY
Ribes pinetorum
Currant Family (Grossulariaceae)

Height: Sprawling to 6′.

Flowers: Reddish orange, with 5 petals; hairy; to ¾″ long; on short branchlet; followed by very spiny, ½″ round berry, maturing to dark reddish purple.

Leaves: Dark green, alternate, roundish; 5-lobed, toothed, glandular-hairy; to 2½″ long, 2″ wide; in clusters on branches.

Blooms: April–September.

Elevation: 7,000 to 10,000′.

Habitat: Coniferous forests.

Comments: The most abundant species of gooseberry in mountains of southern Arizona. Tan spines, to ⅜″ long, with 1 to 3 at notes, curve downward from branches. Ten species of *Ribes* in Arizona. Photograph taken on Mount Lemmon, May 13.

TINKER'S PENNY
Hypericum anagalloides
St. John's Wort Family (Hypericacae)

Height: Prostrate, to 8″ long, 2″ high.

Flowers: Yellow to orange, with 5 petals, 5 sepals, many stamens; to ¼″ wide at branch tips.

Leaves: Light green, opposite; round, slightly oval to elliptical; to ⅝″ long; all along stems.

Blooms: June–August.

Elevation: 7,500 to 8,000′.

Habitat: Lakeshores and other wet areas.

Comments: Perennial herb. Forms leafy mats. Prostrate stems root along their length. Three species of *Hypericum* in Arizona. Photograph taken at Woods Canyon Lake, July 7.

YELLOW-EYED GRASS

Sisyrinchium arizonicum
Iris Family (Iridaceae)

Height: To 2′ tall.

Flowers: Orange to yellowish orange; 6 pointed segments, all alike; 3 yellow stamens tipped with black; 3-branched style; flower to 1½" wide, followed by an oblong capsule to ½" long.

Leaves: Dark green, linear, prominently veined; grasslike, flattened; to 10" long, ¼" wide, basal and up along stem.

Blooms: July–August.

Elevation: 6,000 to 9,500′.

Habitat: Coniferous forests.

Comments: Perennial herb. Stems erect and branching. Six species of Sisyrinchium in Arizona. Photograph taken at Woods Canyon Lake, August 3.

DESERT MARIPOSA

Calochortus kennedyi var. *kennedyi*
Lily Family (Liliaceae)

Height: To 2′.

Flowers: Bright orange petals with purple to black markings; membranes at base; short-stemmed when growing in open, long-stemmed among shrubs; to 3" wide.

Leaves: Grayish green, narrow, grasslike, few; to 8" long.

Blooms: March–May, but usually April.

Elevation: Below 5,000′.

Habitat: Open or shrubby areas in dry soil.

Comments: Perennial herb. *Mariposa* means "butterfly" in Spanish. Six species of *Calochortus* in Arizona. Photograph taken near Superior, April 12.

TORREY'S CRAG LILY
Echeandia flavescens (Anthericum torreyi)
Lily Family (Liliaceae)

Height: To 16".

Flowers: Yellowish orange, starlike, with 3 wide petals, 3 narrower sepals; petals and sepals have greenish to brownish vertical veins; to 1" wide; along slender flower stalk, followed by an oblong capsule.

Leaves: Dark green, basal, pointing upward; grasslike, margins curved inward; to ⅛" wide, to 8" long.

Blooms: August.

Elevation: 6,000 to 9,000'.

Habitat: Coniferous forests and canyons.

Comments: Leafless stem. One species of *Echeandia* in Arizona. Photograph taken at Woods Canyon Lake, August 3.

LEMMON'S STAR
Triteleia lemmoniae
Lily Family (Liliaceae)

Height: To 9".

Flowers: Yellow-orange, star-shaped, 6 segments, in terminal cluster on leafless stem; to ¾" wide.

Leaves: Dark green, 1 or 2, grasslike, to 6" long.

Blooms: May–August.

Elevation: 5,000 to 7,700'.

Habitat: Partial shade among ponderosa pines.

Comments: This species is unique to Arizona. One species of *Triteleia* in Arizona. Photograph taken near Willow Springs Lake, June 16.

PLAINS FLAX

Yellow Flax
Linum puberulum
Flax Family (Linaceae)

Height: To 15".

Flowers: Pale orangish with reddish and white inner rings, 5 petals; to 2" wide.

Leaves: Grayish green, wiry, and very narrow; to 1" long; at intervals along the stem.

Blooms: April–October.

Elevation: 2,000 to 6,500'.

Habitat: Desert and mesas.

Comments: Nine species of *Linum* in Arizona. Photograph taken at Tortilla Flat, April 1.

INDIAN MALLOW

Pelotazo
Abutilon incanum
Mallow Family (Malvaceae)

Height: Bush to 8'.

Flowers: Orange-yellow, with 5 petals, reddish brown center, bright yellow stamens; to ⅜" wide; followed by a short-pointed, hairy, round fruit with up to 7 segments.

Leaves: Grayish green, woolly, velvetlike, heart-shaped, toothed; to 3" long.

Blooms: March–May.

Elevation: 1,000 to 3,000'.

Habitat: Dry, rocky slopes and deserts.

Comments: Ten species of *Abutilon* in Arizona. Photograph taken in Superstition Mountains, March 26.

PALMER'S INDIAN MALLOW
Abutilon palmeri
Mallow Family (Malvaceae)

Height: To 6'.

Flowers: Orangish yellow, with 5 broad petals; to 1½" wide; followed by roundish, segmented fruit with very short point.

Leaves: Grayish green above, paler beneath; very finely haired, somewhat oval, with tapered point; scalloped or toothed; to 3" long.

Blooms: March–October.

Elevation: 1,000 to 4,000'.

Habitat: Dry slopes.

Comments: Ten species of *Abutilon* in Arizona. Photograph taken below Horse Mesa Dam, Apache Lake, March 23.

DESERT GLOBEMALLOW
Sore-Eye Poppy
Sphaeralcea ambigua
Mallow Family (Malvaceae)

Height: To 40".

Flowers: Bright orange, with variants of white, pink, purplish, or reddish-maroon hues, with 5 petals; to 1½" wide; in clusters along upper stems.

Leaves: Grayish, maplelike, 3-lobed; with scalloped edges; covered with starlike hairs; to 2½" long; equally wide.

Blooms: Year-round. (Each plant establishes its own time.)

Elevation: Below 3,500'.

Habitat: Roadsides, banks of sandy washes, and flats.

Comments: Perennial herb that grows in large clumps. The most drought-tolerant of the globemallows. Browsed by bighorn sheep, domesticated sheep and goats, and desert tortoises; attracts bees. Leaf hairs are an eye irritant to some people. Sixteen species of *Sphaeralcea* in Arizona. Photograph taken at Apache Junction, March 22.

LITTLELEAF GLOBEMALLOW

Sphaeralcea parvifolia
Mallow Family (Malvaceae)

Height: To 3'.

Flowers: Orange-red, with 5 petals; to 1" wide.

Leaves: Green, broad, 3-lobed or none; whitish hairs, to 1½" long.

Blooms: May–September.

Elevation: 4,000 to 7,000'.

Habitat: Roadsides and dry slopes.

Comments: Perennial herb. Whitish hairs on stems. Sixteen species of *Sphaeralcea* in Arizona. Photograph taken at Wupatki National Monument, September 8.

MEXICAN GOLD POPPY

Amapola Del Campo
Eschscholzia californica ssp. *mexicana (Eschscholzia mexicana)*
Poppy Family (Papaveraceae)

Height: To 16".

Flowers: Orange, (rarely cream with orange spots at petal bases) with 4 petals forming a cup; to 1½" wide on a single stalk, with numerous flowers per plant; followed by a slender seed capsule to 4" long.

Leaves: Fernlike, pale bluish green; to 2½" long.

Blooms: Mid-February–May, starting earliest in warmer desert areas.

Elevation: Below 4,500'.

Habitat: Slopes, plains, foothills, and mesas.

Comments: Annual; remains open only in full sunlight. Whole areas of desert become a sea of gold from these poppies after abundant winter rainfalls. Dr. Eschscholtz, for whom genus is named, was a Russian surgeon, naturalist, and traveler. Three species of *Eschscholzia* in Arizona. Photograph taken in Superstition Mountains, February 22. The **Desert Gold** or **Mohave Poppy** (*Eschscholzia glyptosperma*) has orange-yellow flowers on long, naked stems, with all the leaves in a basal cluster.

ARIZONA CALTROP

Summer Poppy
Kallstroemia grandiflora
Caltrop Family (Zygophyllaceae)

Height: Sprawling to 3'.

Flowers: Brilliant orange, crimson in center, with 5 broad, rounded petals forming bowl; deep orange stamens, hairy sepals; to 1" wide; facing upward on very hairy stems.

Leaves: Grayish green, pinnately compound; opposite, to 2½" long; oval leaflets, very hairy.

Blooms: July–October.

Elevation: Below 5,000'.

Habitat: Roadsides, open plains, mesas, and desert slopes.

Comments: Annual. Four species of *Kallstroemia* in Arizona. Photograph taken in Phoenix area, August 16. This species distinguished from **Mexican Gold Poppy** (*Eschscholzia californica* ssp. *mexicana*) (page 222) mainly by 5 petals instead of 4, by its very hairy, compound leaves, and by its later blooming period.

ANGIOSPERMS: RED FLOWERS

CHUPAROSA

Beloperone
Justicia californica
Acanthus Family (Acanthaceae)

Height: To 6′.

Flowers: Dull red, tubular; 2-lobed upper and 3-lobed lower lip; anther with large, white point at tip; to 2″ long, in terminal clusters; followed by 2-celled capsule.

Leaves: (When present) grayish green, covered with soft hairs; oval to egg-shaped; to 1″ long; falling during drought or cold.

Blooms: On and off throughout year.

Elevation: 1,000 to 2,500′.

Habitat: Rocky slopes and along washes.

Comments: Stems are soft and hairy. A favorite with hummingbirds (*Chuparosa* is Spanish for "hummingbird"). Three species of *Justicia* in Arizona.

Photograph taken at Usery Mountain Recreation Area, March 7. Distinguished from **Red Jacobinia** (*Justicia candicans*) (at right) by terminal clusters of flowers; more narrow tube without prominent white throat markings, and by presence of white point at anther tip.

RED JACOBINIA

Justicia candicans (Jacobinia ovata)
Acanthus Family (Acanthaceae)

Height: Sprawling shrub to 5′.

Flowers: Deep red with white markings in throat; tubular, 2-lobed upper and 3-lobed lower tube; petals united for about two-thirds of tube; lower lobes wide, notched, and curving downward; anther lacks large white point at tip; flower to 1½″ long, in upper leaf axils; followed by 2-celled capsule.

Leaves: Grayish green with deep purplish brown markings; covered with soft hairs; oval to sharply pointed; to 1½″ long, ¾″ wide.

Blooms: On and off throughout year.

Elevation: 1,500 to 3,500′.

Habitat: Rocky foothills, washes, and canyons.

Comments: Three species of *Justicia* in Arizona. Photograph taken below Horse Mesa Dam, Apache Lake, March 23.

ARIZONA THISTLE

Cirsium arizonicum
Sunflower Family (Asteraceae)

Height: To 4'.

Flowers: Bright red or carmine, with slender flower heads; to 2" long.

Leaves: Grayish green, very hairy, alternate; spiny-toothed and tipped; to 2" long.

Blooms: May–October.

Elevation: 3,000 to 7,000'.

Habitat: Roadsides, chaparral, and clearings in ponderosa pine forest.

Comments: Native Americans use thistles medicinally. Around two dozen species of *Cirsium* in Arizona. Photograph taken at Lynx Lake, October 5.

BLANKETFLOWER

Firewheel
Gaillardia pulchella
Sunflower Family (Asteraceae)

Height: To 2'.

Flowers: Reddish purple, wedge-shaped ray flowers tipped with bright yellow; broad ends of rays divided into 3 sharp lobes; reddish purple disk flowers in dome-shaped center; to 3" wide.

Leaves: Green; upper leaves oblong, lower ones lobed; to 3" long.

Blooms: April–September.

Elevation: 3,500 to 5,500'.

Habitat: Roadsides, fields, and clearings in pinyon-juniper woodlands and ponderosa forests.

Comments: Annual. Five species of *Gaillardia* in Arizona. Photograph taken at Payson, September 12.

CARDINAL FLOWER

Scarlet Lobelia
Lobelia cardinalis
Bellflower Family (Campanulaceae)

Height: To 5'.

Flowers: Bright red, tubular, with 2 small upper lobes, 3 larger lower lobes; stamens united in column; to 1½" long; in elongated cluster on erect stalk.

Leaves: Dark green, oblong, toothed; to 5" long.

Blooms: June–October.

Elevation: 3,000 to 7,500'.

Habitat: Wet areas and along streams.

Comments: Perennial herb. Attracts hummingbirds. Six species of *Lobelia* in Arizona. Photograph taken in Tucson area, August 23.

ARIZONA HONEYSUCKLE

Madreselva
Lonicera arizonica
Honeysuckle Family (Caprifoliaceae)

Height: A stiff, trailing vine to 3' or more.

Flowers: Red outside, with orange throat inside; trumpet-shaped; in small, terminal, whorled cluster arising from 2 joined leaves; followed by cluster of red berries.

Leaves: Bluish green above, paler beneath; very finely haired; oval; joined together in pairs; to 2" wide, 2¾" long.

Blooms: June–July.

Elevation: 6,000 to 9,000'.

Habitat: Open coniferous forests.

Comments: Has scaly, brown trunk. A favorite of hummingbirds. Berries eaten by birds and small mammals. Twelve species of *Lonicera* in Arizona. Photograph taken near Willow Springs Lake, June 11.

MEXICAN PINK

Mexican Campion
Silene laciniata ssp. *greggii*
Pink Family (Caryophyllaceae)

Height: To 3′.

Flowers: Cardinal red, with 5 petals, each cut at tips into 4 segments; flower head to 1½″ wide.

Leaves: Dark green, narrowly lance-shaped, sticky, and fine-haired; to 6″ long.

Blooms: July–October.

Elevation: 5,500 to 9,000′.

Habitat: Coniferous forests.

Comments: Perennial herb. "Pinked" petals give family its name. Nine species of *Silene* in Arizona. Photograph taken at Greer, September 11.

SCARLET CREEPER

Scarlet Morning Glory
Ipomoea cristulata (Ipomoea coccinea)
Morning Glory Family (Convolvulaceae)

Height: Long, twining vine.

Flowers: Bright reddish orange, tubular, and narrow; with white-tipped stamens; to 1″ long, ½″ wide.

Leaves: Dark green, heart-shaped or 3- to 5-lobed; to 2¼″ wide, 2½″ long, not including stem.

Blooms: May–October.

Elevation: 2,500 to 6,000′.

Habitat: Canyons, woodlands, hillsides, and along streams.

Comments: Frequented by hummingbirds. Eighteen species of *Ipomoea* in Arizona. Photograph taken at Clear Creek near Camp Verde, September 30.

SOUTHWESTERN CORAL BEAN

Chilicote

Erythrina flabelliformis

Pea Family (Fabaceae)

Height: Shrub, or small tree to 15', usually shorter.

Flowers: Bright red or reddish orange, with whitish, waxy calyx, and long, narrow corolla; pealike; to 3" long, ¼" wide; several in terminal cluster, followed by large, thick-walled pod, to 10" long, ¾" wide; seeds to ½" long.

Leaves: (When present) grayish green, pinnate, with 3 triangular leaflets, each to 3" long, 4" wide.

Blooms: May–July (sometimes in late summer).

Elevation: 3,000 to 5,000'.

Habitat: Dry, rocky hillsides.

Comments: Has short, hooked spines on stems and leafstalks. Flowers appear before leaves. Bright red to brown seeds are very poisonous; used in Mexican jewelry. One species of *Erythrina* in Arizona. Photograph taken in Tucson area, May 12.

OCOTILLO

Coachwhip

Fouquieria splendens ssp. *splendens*

Ocotillo Family (Fouquieriaceae)

Height: To 20' tall.

Flowers: Red, tubular, about 1" long; in clusters to 10" long, at tips of canes.

Leaves: Green, oval, to 2" long.

Blooms: March–June.

Elevation: Below 5,000'.

Habitat: Desert, especially on rocky, well-drained slopes.

Comments: Named for Pierre Fouquier, a French professor of medicine. Relative of the boojum tree of Baja. Despite spines on its stem, the ocotillo is not a cactus. Most of the year its canes are leafless, but after a heavy rain bright green leaves appear on the long stems. When arid conditions return, the leaves change to brown and fall. This drought-responsive process may be repeated several times during the warmer months. Sections of ocotillo planted in rows soon become living fences. Mature plants have up to 75 slender branches. One species of *Fouquieria* in Arizona. Flower close-up taken at Usery Mountain Recreation Area, March 18.

TEXAS BETONY

Scarlet Sage
Stachys coccinea
Mint Family (Lamiaceae)

Height: To 3′.

Flowers: Scarlet, tubular; 2-lipped, with upper lip erect, lower lip 3-lobed and spreading; to 1¼″ long; in whorls around stem.

Leaves: Grayish green, oval to triangular, toothed; hairy, with netlike surface; opposite; to 3″ long.

Blooms: March–October.

Elevation: 1,500 to 8,000′.

Habitat: Slopes and canyons in rich soil.

Comments: Perennial herb. Square stems. Five species of *Stachys* in Arizona. Photograph taken below Kitt Peak, April 18.

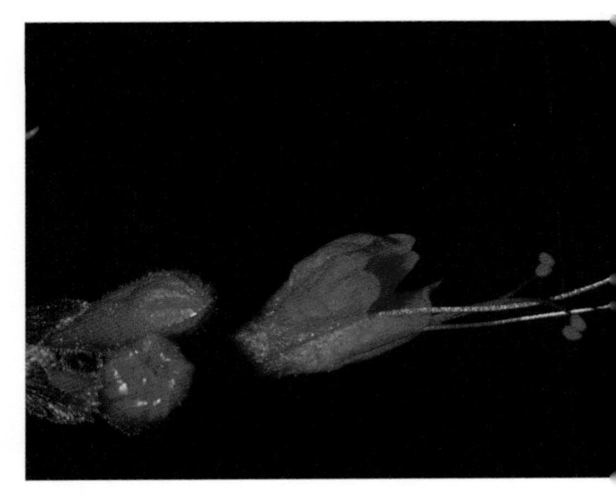

SCARLET FOUR O'CLOCK

Mirabilis coccinea (Oxybaphus coccineus)
Four O'Clock Family (Nyctaginaceae)

Height: To 3½′.

Flowers: Deep carmine-red, tubular, with very long stamens; bright purple on tip of longest stamen; to 1¾″ including stamens; in long-stemmed clusters of 3.

Leaves: Grayish green, very narrow, almost threadlike; to 4″ long; no basal leaves, few leaves on stalk.

Blooms: May–August.

Elevation: 4,000 to 6,500′.

Habitat: Grassy slopes and clearings in ponderosa pine forests.

Comments: Produces a brilliant flower that opens only at night. Thirteen species of *Mirabilis* in Arizona. Photograph taken at Lynx Lake, May 27.

HUMMINGBIRD TRUMPET

Epilobium canum ssp. *latifolium (Zauschneria californica* ssp. *latifolia)*
Evening Primrose Family (Onagraceae)

Height: To 2'.

Flowers: Reddish orange, tubular, crinkly, with 4 lobes notched in center; long stamens; to 1½" long, hanging in clusters.

Leaves: Dark green, hairy, narrowly elliptical, toothed; to 1" long, growing all along stems.

Blooms: June–December.

Elevation: 2,500 to 7,000'.

Habitat: Canyons, along washes, and in other damp areas.

Comments: Perennial herb; shredding bark. Thirteen species of *Epilobium* in Arizona. Photograph taken near Christopher Creek, September 27.

SKYROCKET

Scarlet Gilia
Ipomopsis aggregata (Gilia aggregata)
Phlox Family (Polemoniaceae)

Height: Flowering stem to 3'.

Flowers: Bright red to deep pink, mottled with cream color; funnel-shaped; 5 flaring, pointed lobes; to 1½" long; in leaf axils at tops of nearly leafless stems.

Leaves: Dark green; pinnately divided into very narrow segments; mostly basal; to 2" long.

Blooms: May–September.

Elevation: 5,000 to 9,000'.

Habitat: Roadsides and openings in coniferous forests.

Comments: Biennial; first year, produces rosette of leaves; in the second year, blooms. Has sticky stems; bruised leaves smell skunky. Attracts hummingbirds. Browsed by deer, pronghorn, and livestock. Native Americans used plant medicinally and for ceremonies. Fourteen species of *Ipomopsis* in Arizona. Photograph taken south of Alpine, July 23.

ARIZONA GILIA

Ipomopsis arizonica (Ipomopsis aggregata var. *arizonica)*
Phlox Family (Polemoniaceae)

Height: Flowering stem to 16" with red stems.

Flowers: Bright red trumpet-shaped flowers to 1" wide and ¾" long; in leaf axils at tops of nearly leafless stems.

Leaves: Dark green and hairy, basal, pinnately divided leaves to 1" long.

Blooms: May–September.

Elevation: 5,000 to 9,000'.

Habitat: Roadsides and openings in coniferous forests. It often grows on volcanic soil.

Comments: Biennial; first year, produces rosette of leaves; in the second year, blooms. Has sticky stems; bruised leaves smell skunky. Attracts hummingbirds. Browsed by deer, pronghorn, and livestock. Native Americans used plant medicinally and for ceremonies. Fourteen species of *Ipomopsis* in Arizona. Photograph taken at Sunset Crater National Monument, May 31.

RED COLUMBINE

Aquilegia desertorum (Aquilegia triternata)
Buttercup Family (Ranunculaceae)

Height: To 1'.

Flowers: 5 reddish petals tipped with yellow; long spurs pointing backward; 5 reddish sepals, stamens projecting and nodding; to 1¾" long.

Leaves: Olive-green, divided into threes; leaflets to 1½".

Blooms: May–October.

Elevation: 4,000 to 10,000'.

Habitat: Moist coniferous forests.

Comments: Perennial herb. Seven species of *Aquilegia* in Arizona. Photograph taken south of Alpine, August 2.

SCARLET CINQUEFOIL

Five-Finger
Potentilla thurberi
Rose Family (Rosaceae)

Height: To 16".

Flowers: Dark red, with 5 petals; in loose cluster; to 1" wide.

Leaves: Dark green, basal leaves with silky hairs on underside and 5 to 7 finely toothed leaflets, each to 2" long; smaller leaves on stem.

Blooms: July–October.

Elevation: 6,000 to 9,000'.

Habitat: Rich soil in coniferous forests.

Comments: Perennial herb. Twenty species of *Potentilla* in Arizona. Photograph taken south of Alpine, July 23.

SMOOTH BOUVARDIA

Bouvardia ternifolia (Bouvardia glaberrima)
Madder Family (Rubiaceae)

Height: Small shrub to 3'.

Flowers: Bright reddish orange, narrow, tubular, flaring into 4 lobes; honeysuckle-like; to 1¼" long, ⁵⁄₁₆" wide; in clusters.

Leaves: Dark green, lance-shaped, hairy; generally in whorls of 3, to 3" long.

Blooms: May–October.

Elevation: 3,000 to 9,000'.

Habitat: Slopes and canyons.

Comments: Visited by hummingbirds. Used medicinally in Mexico. One species of *Bouvardia* in Arizona. Photograph taken at Madera Canyon, May 11.

DESERT PAINTBRUSH

Painted-Cup
Castilleja angustifolia (Castilleja chromosa)
Figwort Family (Scrophulariaceae)

Height: To 16".

Flowers: Reddish or orangish bracts, shorter and wider than leaves; hairy; in terminal cluster. Flowers are inconspicuous and generally hidden within bracts.

Leaves: Reddish green, grayish haired, linear to linear-lobed; to 2" long.

Blooms: March–September.

Elevation: 2,000 to 8,000'.

Habitat: Roadsides, chaparral, and clearings in ponderosa pine forests.

Comments: Removes selenium from soil. Seventeen species of *Castilleja* in Arizona; the species of this genus are difficult to identify. Photograph taken at Mormon Lake, September 3.

WOOLLY PAINTBRUSH

Painted-Cup
Castilleja lanata ssp. *lanata*
Figwort Family (Scrophulariaceae)

Height: To 2'.

Flowers: Red to reddish orange, with somewhat narrow bracts, some lobed, very hairy, in terminal cluster. Flowers inconspicuous and generally hidden within bracts.

Leaves: Reddish green, hairy, margins curved upward; to 2" long; linear to linear-lobed; bunches of small leaves in leaf axils.

Blooms: March–August.

Elevation: 2,500 to 7,000'.

Habitat: Arid slopes and desert.

Comments: Seventeen species of *Castilleja* in Arizona; the species of this genus are difficult to identify. Photograph taken in desert area below Superstition Mountains, March 15.

TWINING SNAPDRAGON

Snapdragon Vine
Maurandya antirrhiniflora (Maurandella antirrhiniflora)
Figwort Family (Scrophulariaceae)

Height: Vine; twining over bushes to 8'.

Flowers: Red or purple, snapdragon-like; white in throat with pinkish lines; side lobes flare outward, upper lobes paired and flare upward; to 1" long, ⅝" wide.

Leaves: Dark green, arrow-shaped; leathery, strong network; rounded at tip on lower leaves, pointed at tip on younger leaves; to 1¼" wide, 2" long.

Blooms: April–October.

Elevation: 1,500 to 6,000'.

Habitat: Among shrubs in washes, on rocky slopes, and in pinyon-juniper woodlands.

Comments: Perennial herb. Stems of flowers and leaves twist in all directions. Three species of *Maurandya* in Arizona. Photograph taken at Dead Horse Ranch State Park, May 30.

CRIMSON MONKEY FLOWER

Scarlet Monkey Flower
Mimulus cardinalis
Figwort Family (Scrophulariaceae)

Height: To 3'.

Flowers: Crimson red to reddish orange, tubular; 2-lipped, with 3 lobes of lower lip notched, upper lip arched upward; yellow stamens, hairy sepals; to 2" long, 1" wide; terminal, on 2"-long stem.

Leaves: Dark green, hairy, opposite; oblong to oval, coarsely toothed, sticky; clasping stem; to 4¼" long.

Blooms: March–October.

Elevation: 1,800 to 8,000'.

Habitat: Seeps, springs, along flowing streams, and in wet canyons.

Comments: Stems hairy. Fourteen species of *Mimulus* in Arizona. Photograph taken along West Fork of Oak Creek Canyon, October 1.

GOLDEN-BEARD PENSTEMON

Beardlip Penstemon
Penstemon barbatus
Figwort Family (Scrophulariaceae)

Height: To 4'.

Flowers: Scarlet red, narrow, and tubular; with upper lip projecting forward, lower lip bent downward with sides flared backward; to 1½" long; in open, terminal raceme mostly on one side of stem.

Leaves: Gray-green, narrow, smooth; to 5" long.

Blooms: June–October.

Elevation: 4,000 to 10,000'.

Habitat: Roadsides, oak woods, and coniferous forests.

Comments: Native Americans use plant for medicinal and ceremonial purposes. Pollinated by hummingbirds. More than three dozen species of *Penstemon* in Arizona. Photograph taken south of Alpine, August 2.

FIRECRACKER PENSTEMON

Penstemon eatonii
Figwort Family (Scrophulariaceae)

Height: Flower stalk to 2'.

Flowers: Bright scarlet red, tubular, with width of tube fixed; flaring lobes about equal; to 1" long; in elongated cluster.

Leaves: Dark green, leathery, mainly basal; to 2½" long.

Blooms: February–June.

Elevation: 2,000 to 7,000'.

Habitat: Roadsides, desert slopes, and mesas.

Comments: Stems are purplish. Several subspecies. More than three dozen species of *Penstemon* in Arizona. Photograph taken in Alamo Lake area, February 27. *P. eatonii* is similar to **Scarlet Bugler** (*Penstemon subulatus*) (page 238), but its tubes are shorter.

SCARLET BUGLER

Penstemon subulatus
Figwort Family (Scrophulariaceae)

Height: To 3'.

Flowers: Bright red, tubular, with 5 short lobes rounded at tips; to 1¼" long; on one side all along upper stems.

Leaves: Dark green, smooth, thick; lance-shaped to spatula-shaped, opposite; to 3" long.

Blooms: March–June.

Elevation: 600 to 4,500'.

Habitat: Dry hillsides and cliffs.

Comments: Stem is pinkish. More than three dozen species of *Penstemon* in Arizona. Photograph taken near Apache Lake, March 29.

ANGIOSPERMS: PINK TO PURPLE FLOWERS

PURPLE SCALYSTEM

Elytraria imbricata
Acanthus Family (Acanthaceae)

Height: To 10".

Flowers: Blue to lavender; 2 upright, earlike lobes; 2 cleft side lobes; 1 cleft lower lobe; to 5/16" long, 5/16" wide; on long narrow, scaly flower stem.

Leaves: Dark green, scalelike.

Blooms: April–September.

Elevation: 3,500 to 5,000'.

Habitat: Among rocks on slopes or mesas.

Comments: Numerous upright or spreading flower stems forming a clump. One species of *Elytraria* in Arizona. Photograph taken at Patagonia Lake State Park, May 10.

MACDOUGAL'S BISCUITROOT

Lomatium foeniculaceum ssp. *macdougalii (Lomatium macdougalii)*
Carrot Family (Apiaceae)

Height: To 1'.

Flowers: Purplish or yellowish tinged, tiny, in umbel; followed by flat, oval fruit to 1/4" long, 3/16" wide on 3/8"-long stem; in cluster on long stem above leaves.

Leaves: Grayish green, fernlike, much-dissected, hairy on both surfaces; very aromatic, triangular, basal; to 3" long.

Blooms: March–June.

Elevation: 4,500 to 8,000'.

Habitat: Oak-juniper woodlands and clearings in ponderosa forests.

Comments: Perennial. Has hairy stems. Nine species of *Lomatium* in Arizona. Photograph taken in vicinity of Ashurst Lake, June 2.

SPREADING DOGBANE

Honey Bloom

Apocynum androsaemifolium

Dogbane Family (Apocynaceae)

Height: To 20".

Flowers: Pale pink with deeper pink stripes within; bell-shaped; 5 flared lobes on rim, often curved backward; to ⅜" long, 5⁄16" wide; in loose, terminal cluster or in leaf axil; followed by slender pod to 7" long.

Leaves: Dark green above, paler or whitish beneath; oval, in drooping pairs; to 3½" long.

Blooms: June–August.

Elevation: 7,000 to 9,500'.

Habitat: Clearings in ponderosa and mixed conifer forests.

Comments: Perennial herb. Many-branched, with milky sap. Poisonous to livestock; once believed this plant was poisonous to dogs, hence its common name. Native Americans used stem fiber for making cloth, cordage, and fishing nets. Three species of *Apocynum* in Arizona. Photograph taken at Greer, July 21.

INDIAN HEMP

Dogbane

Apocynum cannabinum

Dogbane Family (Apocynaceae)

Height: To 5'.

Flowers: White to very pale pink, with 5 pointed lobes; 3⁄16" long, 3⁄16" wide, in open, terminal, upright cluster; followed by a slender, cylindrical pod to 6" long.

Leaves: Light green above, paler beneath; lance-shaped, opposite; on short leaf stalk; with whitish midvein; to 4" long.

Blooms: June–September.

Elevation: To 7,500'.

Habitat: Clearings and borders of woodlands.

Comments: Perennial herb. Reddish stems. A variable species, as it interbreeds with other species of *Apocynum*. A low-grade heart stimulant was produced from the root. Three species of *Apocynum* in Arizona. Photograph taken at Oak Creek Canyon, Sedona, June 9.

MYRTLE
Bigleaf Periwinkle
Vinca major
Dogbane Family (Apocynaceae)

Height: Trailing evergreen, often rooting at tips.

Flowers: Bluish lavender, tubular, with 5 flat lobes twisted to left; solitary, stalked; to 2" wide, in axils of leaves.

Leaves: Dark green, shiny, evergreen; oval, heart-shaped at base; to 3" long.

Blooms: March–July.

Elevation: Not available. Photograph take at approximately 5,000'.

Habitat: Light shade under trees.

Comments: Perennial herb. Native of Europe. An escapee from cultivation in certain areas of Arizona. Becomes a weedy pest. Shoots used as styptic to stop bleeding. Photograph taken in Oak Creek Canyon, June 18.

SHOWY MILKWEED
Asclepias speciosa
Milkweed Family (Asclepiadaceae)

Height: To 4'.

Flowers: Dull pink, hairy, and starlike, with 5 darker pink sepals; 5 pink petals bent backward; 5 pink; pointed, wide-spreading hoods with incurved horns; flower to 1" wide; in clusters to 3" wide on stems in upper leaf axils and terminal; followed by white-woolly pod, to 4" long, with soft spines.

Leaves: Light green, short-stalked, thick, and leathery; shiny above, hairy beneath; prominent, whitish midvein; opposite; oblong to lance-shaped; to 8" long.

Blooms: June–August.

Elevation: 6,000 to 9,000'.

Habitat: Clearings in coniferous forests and roadsides.

Comments: Perennial herb. Produces milky sap. Has unbranched, woolly stems. Twenty-nine species of *Asclepias* in Arizona. Photograph in flower taken in vicinity of McNary, July 7.

RAMBLING MILKWEED

Funastrum hirtellum (Sarcostemma hirtellum)
Milkweed Family (Asclepiadaceae)

Height: Climbs over shrubs and up into trees.

Flowers: Pale pink to purplish, creamy-colored in center; hairy, starlike, fragrant; 5 petals, 5 sepals; to ½" wide; in cluster, followed by velvety, long-tapering, plump seed pod to 3½" long.

Leaves: Green, very narrow, linear to lance-shaped; to 2" long.

Blooms: March–October.

Elevation: Below 5,500'.

Habitat: Along desert washes.

Comments: Perennial. Seed pods smell like onion. Produces a milky juice. Its numerous stems often twist into a bluish green, living rope. Tohono O'Odham Indians consumed fruits. Three species of *Funastrum* in Arizona. Photograph taken at Usery Mountain Recreation Area, April 6.

DESERT HOLLY

Acourtia nana
Sunflower Family (Asteraceae)

Height: To 10", but usually 4 to 5".

Flowers: Whitish to pinkish, tiny, 15 to 24 per head; with purplish, diamond-shaped bracts; flower head to 1" long; borne singly.

Leaves: Grayish green, hollylike, stalkless, spiny-toothed; somewhat round, stiff, brittle, and alternate; to 2" long, 2" wide.

Blooms: March–June.

Elevation: Below 6,000'.

Habitat: Slopes and dry plains.

Comments: Perennial herb. Flowers smell like violets. Three species of *Acourtia* in Arizona. Photograph taken in vicinity of Portal, May 5.

BROWNFOOT
Acourtia wrightii
Sunflower Family (Asteraceae)

Height: To 3'.

Flowers: Pinkish lavender, to ¾" wide, in loosely branched, terminal cluster; followed by numerous tawny bristles.

Leaves: Dark green, leathery, oval or oblong; wrinkled; toothed with spiny teeth; base of leaf clasping stem; to 5" long.

Blooms: January–June.

Elevation: Below 6,000'.

Habitat: Foothills and canyons.

Comments: Perennial herb. Attracts butterflies and bees. Three species of *Acourtia* in Arizona. Photograph taken at Catalina State Park, April 30.

RUSSIAN KNAPWEED
Hardheads
Acroptilon repens (Centaurea repens)
Sunflower Family (Asteraceae)

Height: To 3'.

Flowers: Lavender, tubular; in spineless, thistlelike flower head, to ¾" wide, 1" long; surrounded by overlapping, silvery bracts; solitary at tips of leafy branches, followed by seeds with white bristles on top.

Leaves: Grayish green, rough, thick, deeply lobed at base, to 4" long; narrowly oblong, sharp-pointed, toothed; to 2½" long on stems.

Blooms: May–October.

Elevation: 1,000 to 7,000'.

Habitat: Roadsides, waste places, fields, lots, and farmland.

Comments: Perennial weed. Introduced from Europe and Asia; now naturalized throughout the West. Difficult to control due to horizontal roots that produce new plants. One species of *Acroptilon* in Arizona. Photograph taken at Dead Horse Ranch State Park, May 28.

NODDING THISTLE

Carduus nutans
Sunflower Family (Asteraceae)

Height: To 3'.

Flowers: Pink to reddish lavender, rayless, and nodding; flower head surrounded by series of spiny-toothed bracts; to 2½" wide; single; followed by seeds with long, white bristles at top.

Leaves: Green, deeply lobed, spiny, stalkless, to 8" long.

Blooms: June–October.

Elevation: 5,000 to 9,000'.

Habitat: Roadsides, fields, and rangeland.

Comments: Biennial. Native to Europe; now naturalized in certain areas of U.S. Three species of *Carduus* in Arizona. Photograph taken in vicinity of Mexican Hay Lake, July 21.

BRISTLEHEAD

Carphochaete bigelovii
Sunflower Family (Asteraceae)

Height: To 2'.

Flowers: Light pink to lavender, darker pink toward center; tubular, with 5 pointed lobes; to ½" wide; in terminal cluster to 1¼" wide.

Leaves: Grayish green, gland-dotted, hairy; opposite, linear to elliptical; clasping stem; to ¾" long.

Blooms: March–July.

Elevation: 4,000 to 7,000'.

Habitat: Canyons and rocky slopes.

Comments: Woody at base with reddish brown stems. Browsed by deer and livestock. One species of *Carphochaete* in Arizona. Photograph taken northeast of Superior, April 3.

COMMON CHICORY

Succory
Cichorium intybus
Sunflower Family (Asteraceae)

Height: To 4'.

Flowers: Light blue to bluish lavender, square-tipped ray flowers, each with 5 small teeth; prickly bracts; flower heads to 2" wide; close, on well-branched stems.

Leaves: Grayish green, wavy; clasping stem, nearly leafless above, lower leaves variable: lance-shaped, pinnately toothed, or lobed, with prickly margins; to 8" long.

Blooms: June–October.

Elevation: Below 7,000'.

Habitat: Fields, roadsides, and waste areas.

Comments: Perennial herb; produces milky sap. Each flower lasts only one day. Introduced from Europe; now naturalized. Immature leaves used as vegetable or for salad. Roots are ground and roasted as coffee additive or substitute. One species of *Cichorium* in Arizona. Photograph taken near Show Low, July 6.

NEW MEXICO THISTLE

Cirsium neomexicanum var. *neomexicanum*
Sunflower Family (Asteraceae)

Height: To 6'.

Flowers: Pink-purple flower head of tiny, tubular, disk flowers; to 2" wide; surrounded by long, spine-tipped bracts, with outer bracts pointing downward.

Leaves: Dark green, shiny, straplike; spiny, coarsely pinnately lobed, to 7" long.

Blooms: March–September.

Elevation: 1,000 to 6,500'.

Habitat: Foothills, mesas, and plains.

Comments: Has spiny stems. Native Americans use thistle plants medicinally. Around two dozen species of *Cirsium* in Arizona. Photograph taken in Superstition Mountains, April 6.

YELLOWSPINE THISTLE

Santa Fe Thistle
Cirsium ochrocentrum
Sunflower Family (Asteraceae)

Height: To 3′.

Flowers: Purple to rose to cream-colored flower head of tiny, tubular, disk flowers; broad flower head to 3″ wide, 2″ tall; usually occurring singly at tips of branches.

Leaves: Grayish green and very woolly above, white and very hairy beneath; thick, deeply lobed, with yellowish spines on edges of leaves; to 7″ long.

Blooms: May–October.

Elevation: 4,500 to 8,000′.

Habitat: Roadsides and clearings in pinyon-juniper woodlands.

Comments: Has woolly haired stem. Around two dozen species of *Cirsium* in Arizona. Photograph taken in vicinity of Heber, August 4.

WAVYLEAF THISTLE

Cirsium undulatum var. *undulatum*
Sunflower Family (Asteraceae)

Height: To 16″.

Flowers: Pink flower head of tiny, tubular, disk flowers; to 2″ long, 2″ wide; long, narrow bracts with yellow spines.

Leaves: Grayish green, wavy, yellow-spined; gray-haired on both surfaces; to 8″ long, becoming smaller on upper stems.

Blooms: June–October.

Elevation: 7,000 to 9,500′.

Habitat: Roadsides and fields.

Comments: Around two dozen species of *Cirsium* in Arizona. Photograph taken in vicinity of Nutrioso, July 23.

BULL THISTLE

Cirsium vulgare
Sunflower Family (Asteraceae)

Height: To 4′.

Flowers: Rose-purple flower head of tiny, tubular, disk flowers; spiny, yellow-tipped bracts; flower head to 2″ wide, 2″ long.

Leaves: Dark green above, white beneath; pinnately lobed, very spiny; clasping stem and extending onto stem, forming spiny wings; to 6″ long.

Blooms: June–September.

Elevation: Not available. Photograph taken at 7,500′.

Habitat: Roadsides, fields, and disturbed areas.

Comments: Introduced biennial. Around two dozen species of *Cirsium* in Arizona. Photograph taken in vicinity of Woods Canyon Lake, September 14.

WHEELER THISTLE

Cirsium wheeleri
Sunflower Family (Asteraceae)

Height: To 4′.

Flowers: Rose-purple or lavender flower head of tiny, tubular, disk flowers; bracts wide and oval-shaped; loose flower head to 1½″ wide, 2″ long; occurring singly, at branch tips.

Leaves: Dark green above, white-woolly beneath; alternate and narrow, with widely spaced, spiny lobes; to 6″ long at base, smaller on upper stem.

Blooms: June–October.

Elevation: 5,000 to 9,000′.

Habitat: Clearings in pine forests.

Comments: Native Americans use thistles medicinally. Around two dozen species of *Cirsium* in Arizona. Photograph taken in vicinity of Christopher Creek, September 27.

COSMOS

Cosmos parviflorus
Sunflower Family (Asteraceae)

Height: To 2½'.

Flowers: Has 8 pale pink ray flowers, bright yellow disk flowers; 8 long, narrow sepals; flower to 1" wide, on long, leafless stem; followed by seeds with stiff bristles.

Leaves: Dark green, linear, very finely divided; to 3" long.

Blooms: July–October.

Elevation: 4,000 to 9,000'.

Habitat: Clearings in ponderosa pine forests and hillsides.

Comments: Annual, with reddish stem. Two species of Cosmos in Arizona. Photograph taken at McNary, August 10.

MACHAERANTHERA

Dietaria asteroides var. *asteroides (Machaeranthera asteroides* var. *asteroides)*
Sunflower Family (Asteraceae)

Height: To 4'.

Flowers: Purple rays, orange disk flowers, to 1" wide.

Leaves: Grayish green, linear, sandpapery, with spines on margins; to 3" long.

Blooms: April–May.

Elevation: 1,500 to 2,500'.

Habitat: Washes and desert flats.

Comments: Three species of *Dieteria* in Arizona. Photograph taken in vicinity of Saguaro Lake, May 20.

MACHAERANTHERA

Dietaria asteroides var. *glandulosa (Machaeranthera asteroides* var. *glandulosa)*
Sunflower Family (Asteraceae)

Height: To 3'.

Flowers: Purple rays, orange disk flowers, to 1¼" wide.

Leaves: Grayish green, linear and sandpapery, toothed, with spines on margins; to 5 ½" long.

Blooms: April–May.

Elevation: 1,500 to 2,500'.

Habitat: Washes and desert flats.

Comments: Three species of *Dieteria* in Arizona. Photograph taken in Superstition Mountains, April 6.

PRINCELY DAISY

Erigeron formosissimus
Sunflower Family (Asteraceae)

Height: To 16".

Flowers: Lavender to blue, with very narrow rays, yellow disks; flower head to 1½" wide.

Leaves: Grayish green, hairy. Spatula-shaped, basal leaves to 3" long. Stem leaves shorter, alternate, lance-shaped, clasping stem, progressively smaller.

Blooms: July–September.

Elevation: 5,500 to 10,500'.

Habitat: Coniferous forests, meadows, and edges of aspen-spruce forests.

Comments: Stems hairy. There are more than two dozen species of *Erigeron* in Arizona. Photograph taken in mountains above Greer, August 13.

ASPEN FLEABANE
Erigeron speciosus (Erigeron macranthus)
Sunflower Family (Asteraceae)

Height: To 3', but usually closer to 2'.

Flowers: Pale bluish lavender rays, very long and narrow; orangish yellow disks (large for a fleabane); to 1½" wide; numerous flower heads.

Leaves: Dark green, hairy, wavy; lance-shaped, alternate, to 6" long.

Blooms: July–October.

Elevation: 6,000 to 9,500'.

Habitat: Oak thickets and pine forests.

Comments: Hairy stems. Fleabanes are usually difficult to identify; they come in all sizes and range in color from white to pink to bluish lavender. Some bloom in spring, others in summer and fall. In general, they are easily distinguished from asters by their numerous, very narrow ray flowers. There are more than two dozen species of *Erigeron* in Arizona. Photograph taken in Greer area, July 5.

WRIGHT BEEFLOWER
Hymenothrix wrightii
Sunflower Family (Asteraceae)

Height: To 3'.

Flowers: Pinkish, with two circles of flower bracts, outer ones narrow, short, and pointed; inner ones broad, long, and blunt; flower to 5⁄16" long; flower head to ½" wide; numerous on widely spread top stems.

Leaves: Dark green, hairy, divided into very narrow segments; to 1½" long, graduating to smaller up stem.

Blooms: June–October.

Elevation: 4,000 to 8,000'.

Habitat: Roadsides, dry slopes in scrub-oak and pine belts.

Comments: Perennial herb. Has a hairy stem. Three species of *Hymenothrix* in Arizona. Photograph taken east of Camp Verde, September 30.

TANSYLEAF ASTER
Tahoka Daisy
Machaeranthera tanacetifolia
Sunflower Family (Asteraceae)

Height: To 16".

Flowers: Bluish purple rays, yellow disks; to 2" wide.

Leaves: Grayish green, sticky-hairy, much-divided, fernlike, pinnate leaves tipped with tiny spines, leaf to 3" long.

Blooms: June–October.

Elevation: 1,000 to 8,000'.

Habitat: Roadsides, fields, and disturbed soil.

Comments: Perennial herb. Hairy, pinkish stems. Two species of *Machaeranthera* in Arizona. Photograph taken near Nutrioso, August 7. Fernlike leaves with tiny spines at leaf tips help identify the species.

SPANISH NEEDLES
Palafoxia arida var. *gigantea (Palafoxia linearis)*
Sunflower Family (Asteraceae)

Height: To 3'.

Flowers: Pinkish, petalless, with 10 to 20 disk flowers in pincushionlike flower head; to ¾" high, ½" wide; followed by seeds with needlelike bristles.

Leaves: Grayish green, hairy, narrow, and linear; to 2½" long.

Blooms: February–November.

Elevation: Below 2,000'.

Habitat: Sandy washes, plains, mesas, and dunes.

Comments: Annual herb. One species of *Palafoxia* in Arizona. Photograph taken north of Yuma, March 29.

MARSH FLEABANE

Arrow Weed
Pluchea sericea (Pluchea purpurascens var. *purpurascens)*
Sunflower Family (Asteraceae)

Height: To 10'.

Flowers: Reddish purple to lavender, petalless, to ¼" wide; in clusters at branch ends; followed by whitish, tufted seed heads.

Leaves: Grayish, silvery-haired, linear to lance-shaped; to 1¼" long, ¼" wide, occurring all along stem.

Blooms: May.

Elevation: Below 3,000'.

Habitat: Along rivers and streams.

Comments: Willowlike; forms dense thickets. Twigs covered with silvery, soft, flat hairs. Rank-smelling. Browsed by horses, cattle, and deer. Source of honey. Native Americans used stems for building huts and making baskets. One species of *Pluchea* in Arizona. Photograph taken near Scottsdale, May 7.

ODORA

Hierba Del Venado
Porophyllum gracile
Sunflower Family (Asteraceae)

Height: To 28".

Flowers: Purplish to purplish white, streaked with dark purple lines; petalless; flower head to ⅝" wide, ¾" log, single at ends of branches; followed by numerous, gray bristles.

Leaves: Dark green, narrowly linear to threadlike, to 2" long.

Blooms: March–October.

Elevation: Below 4,000'.

Habitat: Dry, rocky slopes, mesas, and washes.

Comments: Perennial; many-branched; rank odor when crushed. Browsed by cattle and deer. Two species of *Porophyllum* in Arizona. Photograph taken at Saguaro National Park East, April 15.

MILK THISTLE
Silybum marianum
Sunflower Family (Asteraceae)

Height: Flower stem to 6′.

Flowers: Purplish red flower head, to 2″ wide; curved spines on bracts surrounding flower head; to 1½″ long on stout, erect stem.

Leaves: Shiny green with white blotches and veins; margins lobed and spiny; to 18″ long, 12″ wide.

Blooms: May–September.

Elevation: Below 3,000′.

Habitat: Roadsides, wasteland, and fields.

Comments: Annual, sometimes biennial, field weed; hard to eliminate. Native of Europe, now naturalized in parts of U.S. One species of *Silybum* in Arizona. Photograph of leaves and undeveloped flowers taken at Hassayampa River Preserve, Wickenburg, May 7.

DESERT STRAW
Brownplume Wire-Lettuce
Stephanomeria pauciflora
Sunflower Family (Asteraceae)

Height: To 2′.

Flowers: Lavender, with deep pink stamens; all rays, toothed, to ¾″ wide; followed by fruits with brownish, feathery bristles.

Leaves: Bluish green with whitish midrib; narrow, smooth, sharply lobed; to 3″ long at base of plant; sparsely leaved on upper stems.

Blooms: Almost throughout the year.

Elevation: 150 to 7,000′.

Habitat: Washes and dry, sandy plains and slopes.

Comments: Perennial herb. Many-branched with bluish green stems. Smooth stems and leaves. Seven species of *Stephanomeria* in Arizona. Photograph taken at Fort Bowie, May 8.

THURBER'S WIRE LETTUCE
Stephanomeria thurberi
Sunflower Family (Asteraceae)

Height: To 3'.

Flowers: Lavender; up to 20 rays, each tipped with 6 shallow notches; purplish stamens; pointed bracts tipped with purple; flower to ⅝" wide, ¾" long, on ½"-long stem, followed by bright white, feathery bristles.

Leaves: Grayish green, mostly basal, alternate; pinnately cleft with lobes turned backward; clasping stem; leaf to 6" long. Upper leaves, few, entire, linear, short.

Blooms: April–August.

Elevation: 4,000 to 8,000'.

Habitat: Clearings in pinyon-juniper and ponderosa pine forests.

Comments: Erect stem, branched above. Seven species of *Stephanomeria* in Arizona. Photograph taken at Oak Creek Canyon, June 8.

LEAFYBRACT ASTER
Leafy Aster
Symphyotrichum foliaceum var. *foliaceum (Aster foliaceus)*
Sunflower Family (Asteraceae)

Height: To 18".

Flowers: Narrow, with lavender to purple rays, yellow disks, and large, leafy bracts around flower head; flower head to 2" wide.

Leaves: Dark green, with hairy margins; lance-shaped at base; to 6" long, ½" wide. Leaves clasping on upper stem, alternate, elongate, to 2" long.

Blooms: August–September.

Elevation: 7,500 to 10,000'.

Habitat: Mountain meadows and coniferous forests.

Comments: Stems are reddish and hairy. More than two dozen species of *Symphyotrichum* in Arizona. Photograph taken at Lee Valley Reservoir in mountains above Greer, August 14.

MOHAVE ASTER

Desert Aster

Xylorhiza tortifolia (Machaeranthera tortifolia)

Sunflower Family (Asteraceae)

Height: To 2½′.

Flowers: Pale lavender to almost white rays; yellow disk flowers; terminal, to 2½″ wide.

Leaves: Grayish green, grayish-haired, narrowly lance-shaped; spiny teeth along margins, leaf tipped with a spine; to 3″ long.

Blooms: March–May.

Elevation: 2,000 to 3,500′.

Habitat: Dry, rocky hills and slopes.

Comments: Perennial herb. As many as 20 flower heads on one plant. One species of *Xylorhiza* in Arizona. Photograph taken at Alamo Lake, February 27.

DESERT WILLOW

Mimbre

Chilopsis linearis

Bignonia Family (Bignoniaceae)

Height: Large shrub, or small tree to 25′.

Trunk: To 1′ in diameter.

Bark: Dark brown, scaly, ridged.

Flowers: White to pink, tinged with lavender and yellow, tubular, orchidlike; to 1½″ long; fragrant, in 3- to 4-inch clusters; followed by very narrow, brown seed capsules, to ¼″ wide, 8″ long.

Leaves: Light green, very narrow, untoothed; to 6″ long, to ⅜″ wide.

Blooms: April–August.

Elevation: 1,500 to 5,000′.

Habitat: Along washes and other waterways in desert, grasslands, foothills, and mesas.

Comments: Deciduous; has spreading crown. Desert willows help control erosion by forming dense thickets along washes. Not a true willow, but related to catalpa, which has very similar flowers and seed capsules (although catalpa leaves are very large and heart-shaped). Wood was used for bows by Native Americans. One species of *Chilopsis* in Arizona. Photograph taken in vicinity of Scottsdale, May 7.

SHRUBBY COLDENIA

Woody Crinklemat
Tiquilia canescens
Forget-me-not Family (Boraginaceae)

Height: To 8".

Flowers: Pinkish to white, single, tubular, and 5-lobed; to ⅜" long, ¼" wide.

Leaves: Grayish green, woolly, oval, and numerous; to ⅜" long.

Blooms: February–May.

Elevation: Below 3,500'.

Habitat: Dry, sunny mesas and slopes.

Comments: Woody shrub often forming mats or mounds. Six species of *Tiquilia* in Arizona. Photograph taken in Kofa Mountains, February 21.

PALMER'S CRINKLE MAT

Tiquilia palmeri
Forget-me-not Family (Boraginaceae)

Height: To 10".

Flowers: Pale lavender to blue, funnel-like, 5-lobed; to ¼" wide, ¼" long.

Leaves: Dark green, oval, and crinkly, with grayish, feltlike hairs; to ¼" long.

Blooms: March–October.

Elevation: Below 2,000'.

Habitat: Sandy flats, rocky ridges, and dry river bottoms.

Comments: Perennial. Forms rounded mounds to 2' wide. Six species of *Tiquilia* in Arizona. Photograph taken near Yuma, March 29.

STIFFARM ROCK CRESS

Boechera perennans (Arabis perennans)
Mustard Family (Brassicaceae)

Height: To 2'.

Flowers: Pink to purplish with darker veins; 4 petals; to ¼" long, ½" wide; in loose terminal cluster; followed by widely spreading, horizontal seed pod to 2" long.

Leaves: Grayish green, toothed, spatula-shaped, and roughly haired; mainly basal, straplike leaves on lower stem; to 3" long.

Blooms: February–October.

Elevation: 2,000 to 8,000.

Habitat: Hot canyons and lower mountain slopes.

Comments: Ten species of *Boechera* in Arizona. Photograph taken in Superstition Mountains, February 4.

PINK WINDMILLS

Hesperidanthus linearifolius (Schoenocrambe linearifolia, Thelypodiopsis linearifolia)
Mustard Family (Brassicaceae)

Height: To 4'.

Flowers: Lavender, with 6 stamens, and 4 separated petals narrower at bases; to 1" long; in terminal cluster, followed by long, narrow, upright seed pod.

Leaves: Light green, smooth, and narrow; linear to lance-shaped; to 1¾" long.

Blooms: May–September.

Elevation: 2,500 to 9,500'.

Habitat: Fields, woodlands, chaparral, and ponderosa pine forests.

Comments: One species of *Hesperidanthus* in Arizona. Photograph taken at Oak Creek Canyon, September 9.

HAREBELL

Bluebell Bellflower
Campanula rotundifolia
Bellflower Family (Campanulaceae)

Height: To 20".

Flowers: Blue to violet, bell-shaped, 5-lobed; drooping on slender stem; to 1" long.

Leaves: Green, roundish, basal; withering before flowers appear; to 3" long. Leaves along stem very narrow.

Blooms: June–September.

Elevation: 8,000 to 12,000'.

Habitat: Meadows and rocky slopes.

Comments: Nine species of *Campanula* in Arizona. Photograph taken at Greer, July 20. Occasionally, harebell will have 8 lobes. The flowers of a similar species, **Parry Bellflower** (*Campanula parryi*), are erect and basal leaves are lance-shaped, tapering at base.

APACHE LOBELIA

Southwestern Blue Lobelia
Lobelia anatina
Bellflower Family (Campanulaceae)

Height: To 18".

Flowers: Deep lavender, tubular; 2 smaller upper lobes, crossed, 3 larger flaring lower lobes; to ¾" wide, ¾" long; with about 8 flowers on slender, erect stem.

Leaves: Dark green, lance-shaped to linear, toothed; to 3" long.

Blooms: July–October.

Elevation: 5,500 to 9,000'.

Habitat: Moist mountain meadows, marshy places, and along streams.

Comments: Perennial herb. Visited by bees and hummingbirds. Six species of *Lobelia* in Arizona. Photograph taken at Luna Lake, August 5.

TWINFLOWER

Northern Twinflower
Linnaea borealis
Honeysuckle Family (Caprifoliaceae)

Height: Trailing to 2', upright flower stem to 4".

Flowers: Pink, bell-like, narrow; nodding, fragrant, tubular, 5-lobed; in twin pair on forked stalk; flower to ¼" wide to ½" long.

Leaves: Dark green to yellowish green, opposite, broadly elliptic; thick; 2 shallow teeth on each leaf margin toward tip; shiny, basal, to ¾" long.

Blooms: June–August.

Elevation: In 9,000' range.

Habitat: Moist spruce-fir forests.

Comments: Evergreen. Eaten by deer and grouse. Genus is named for Carl Linnaeus, the famous botanist. One species of *Linnaea* in Arizona. Photograph taken in mountains above Greer, July 8.

LONG-FLOWERED SNOWBERRY

Desert Snowberry
Symphoricarpos longiflorus
Honeysuckle Family (Caprifoliaceae)

Height: To 40".

Flowers: Pink, trumpetlike, with 5 spreading lobes; to ½" long; followed by white, pea-sized berry.

Leaves: Pale green, with whitish bloom above, prominent veins on underside; opposite, lance-shaped to oval; to ½" long, ¼" wide.

Blooms: April–August.

Elevation: 4,000 to 8,000'.

Habitat: Pine forests, canyons, and foothills.

Comments: Birds and small mammals feed on berries. Seven species of *Symphoricarpos* in Arizona. Photograph taken at North Rim of Grand Canyon National Park, June 25.

MOUNTAIN SNOWBERRY

Symphoricarpos oreophilus

Honeysuckle Family (Caprifoliaceae)

Height: To 4½′.

Flowers: Pinkish, tubular, slender, with 5 spreading lobes; to ½″ long; followed by white, egg-shaped fruit; to ⅜″ long.

Leaves: Pale green, smooth above, pale green below with prominent veins; opposite, oval, thin; to 1″ long.

Blooms: May–August.

Elevation: 5,500 to 9,000′.

Habitat: Pine forests.

Comments: Deciduous. Has grayish bark, smooth twigs. Browsed by deer and livestock. Fruits are eaten by small mammals and birds. Seven species of *Symphoricarpos* in Arizona. Photograph taken south of Alpine, June 30.

ROUNDLEAF SNOWBERRY

Symphoricarpos rotundifolius

Honeysuckle Family (Caprifoliaceae)

Height: To 3½′.

Flowers: Pinkish, tubular, slender, lobed at tip; to ½″ long; in clusters of up to 6 in leaf axils; followed by a round, white berry to ½″ in diameter.

Leaves: Grayish green above, paler beneath; prominent network, thick, finely haired, oval, toothed or untoothed, to 1½″ long, ¾″ wide.

Blooms: May–June.

Elevation: 4,000 to 10,000′.

Habitat: Rocky slopes, and clearings in ponderosa pine forests.

Comments: Twigs are brown and finely haired. Stems pinkish on new growth. Berries remain for most of winter. Foliage browsed by livestock and deer. Seven species of *Symphoricarpos* in Arizona. Photograph taken on Mingus Mountain, Prescott, May 28.

SLEEPY CATCHFLY

Silene antirrhina
Pink Family (Caryophyllaceae)

Height: To 3'.

Flowers: Light pink petals notched at tips; to ¼" wide.

Leaves: Dark green and narrow; to 3" long.

Blooms: March–August.

Elevation: Below 6,000'.

Habitat: Washes and waste areas.

Comments: Sticky areas on stems. Nine species of *Silene* in Arizona. Photograph taken in Superstition Mountains, March 26.

SCOULER'S CATCHFLY

Silene scouleri ssp. *pringlei*
Pink Family (Caryophyllaceae)

Height: To 3'.

Flowers: Pinkish, with 5 deeply forked, very narrowly lobed petals in groups of 2; sticky; to ¾" wide, ¾" long. Sticky calyx distended and striped like a miniature watermelon.

Leaves: Dark green, opposite, hairy, and sticky, with deep midvein; spoon-shaped at base to 5" long; linear and gradually smaller along upper stem.

Blooms: July–September.

Elevation: 5,000 to 9,500'.

Habitat: Moist pine and spruce-fir forests and mountain meadows.

Comments: Entire plant is sticky and hairy. Nine species of *Silene* in Arizona. Photograph taken south of Alpine, August 6.

DESERT HOLLY

Atriplex hymenelytra
Goosefoot Family (Chenopodiaceae)

Height: To 3′.

Flowers: Pinkish tube with 2 dark pink stamens, in terminal clusters of small silvery leaves; followed by large, light green, fan-shaped, compressed, fruiting bractlets.

Leaves: Silvery, succulent, evergreen, and hollylike; rubbery, thick, crinkly; fan-shaped with pointed lobes or teeth; to 1½″ wide, 1¾″ long (including stem).

Blooms: March.

Elevation: Below 1,000′.

Habitat: Dry, sandy soil of washes, hillsides, and plains.

Comments: A rounded shrub whose foliage is often sold as Christmas decorations. Over twenty-five species of *Atriplex* in Arizona. Photograph taken in Kofa Mountains, March 6.

ROCKY MOUNTAIN BEE PLANT

Peritoma serrulata (Cleome serrulata)
Cleome Family (Cleomaceae)

Height: To 4′.

Flowers: Pink to purple, 4-petaled; ½″ long; long stamens tipped with green anthers; raceme 2 to 3″ long (sometimes up to 10″ long) and 2 to 3″ wide; followed by thin, banana-like seed pod, to 2½″ long.

Leaves: Bluish green, palmately compound; leaflets to 3″ long.

Blooms: June–September.

Elevation: 4,500 to 7,000′.

Habitat: Roadsides, plains, foothills, and fields.

Comments: Annual. Seeds eaten by doves. Attracts bees and hummingbirds, and is an excellent source for honey. Boiling the leaves produces a thick, black substance which is dried and later softened to use as a black pottery paint. Two species of *Peritoma* in Arizona. Photograph taken in vicinity of Nutrioso, August 3.

WESTERN SPIDERWORT

Prairie Spiderwort

Tradescantia occidentalis

Spiderwort Family (Commelinaceae)

Height: To 30"

Flowers: Pink to rose or purple; 3 (rarely 4) equal-sized petals; to 1" wide, in clusters at tip of stems.

Leaves: Bluish green, long, narrow, folded lengthwise; drooping, clasping stems at base; to ⅜" wide, to 12" long.

Blooms: April–September.

Elevation: 2,500 to 7,000'.

Habitat: Rocky slopes, mesas, and clearings in pine forests.

Comments: Perennial herb. Stems contain slimy sap. Used by Native Americans as potherbs. Pollinated by bees. Two species of *Tradescantia* in Arizona. Photograph taken at Oak Creek Canyon, June 18. (Notice that instead of the normal 3 petals, one of the flowers in the photograph has 4.)

HOARY BINDWEED

Convolvulus equitans

Morning Glory Family (Convolvulaceae)

Height: Trailing to 3' long.

Flowers: Pale pink, funnel-shaped, 5 veins from center to lobes on edge; to ¾" wide, 1" long.

Leaves: Grayish green, finely haired, narrow, with lobes at base; to 2" long.

Blooms: April–October.

Elevation: 3,000 to 6,000'.

Habitat: Roadsides, dry slopes, and mesas.

Comments: Two species of *Convolvulus* in Arizona. Photograph taken near Sierra Vista, April 26.

BIRD'S FOOT MORNING GLORY

Ipomoea ternifolia var. *leptotoma (Ipomoea leptotoma)*
Morning Glory Family (Convolvulaceae)

Height: Twining or spreading vine.

Flowers: Pink, bluish, or purple (rarely white) trumpet-shaped; to 1½" long.

Leaves: Dark green, palmately cleft into 3 or 5 very narrow lobes; to 3" long.

Blooms: June–November.

Elevation: 2,500 to 4,500'.

Habitat: Washes; dry, grassy plains; and mesas.

Comments: Can be hairy-stemmed. Eighteen species of *Ipomoea* in Arizona. Photograph taken at Catalina State Park, November 10.

QUEEN'S CROWN

Stonecrop
Rhodiola rhodantha (Sedum rhodanthum)
Orpine Family (Crassulaceae)

Height: To 16".

Flowers: Pink to white, with 5 pointed petals forming partly closed, tulip-shaped cup; bright red-tipped stamens; to ⅜" long, ⅛" wide; in rounded, terminal cluster.

Leaves: Light green, rubbery, and succulent; flat, elliptical; alternate, spiraling up stem; to 1" long.

Blooms: July–August.

Elevation: 9,000 to 12,000'.

Habitat: Along mountain streams.

Comments: Plant forms a cluster of stiff, upright stems. One species of *Rhodiola* in Arizona. Photograph taken in Mount Baldy Wilderness, August 12.

COCKERELL'S SEDUM

Stonecrop
Sedum cockerellii
Orpine Family (Crassulaceae)

Height: To 8".

Flowers: White to pale pink, stamens with darker pink or purplish tips; 5-petaled, starlike; to ¼" wide; in small, loose cluster at branch end.

Leaves: Light green (reddish in sunny locations) and succulent; flattened in cross-section; to ½" long; in compact, basal rosette and up flowering stem.

Blooms: June–October.

Elevation: 5,000 to 11,500'.

Habitat: Rocky areas, often among mosses and usually in shady locations.

Comments: Three species of *Sedum* in Arizona. Photograph taken in Greer area, July 29.

FULLER'S TEASEL

Dipsacus fullonum ssp. *sylvestris (Dipsacus sylvestris)*
Teasel Family (Dipsacaceae)

Height: To 6'.

Flowers: Lavender, tubular, 4-lobed; to 1⁄16" wide, ½" long; clustered on thistlelike cone, surrounded by short, spiny bracts and numerous very long, upward-pointing, spiny bracts with straight tips; cluster to 4" long, 2" wide.

Leaves: Light green above, with a row of spines on each side of white midvein; sharp spines beneath on midvein; spined margins; lance-shaped, opposite, basal leaves in rosette, to 14" long, 2½" wide; smaller leaves up stem.

Blooms: July–October.

Elevation: 5,600-8,400'. Photograph taken at 5,800'.

Habitat: Roadsides and old fields.

Comments: Biennial. The first flowers start blooming in a belt around center. Succeeding blooms open daily above and below this band; soon 2 bands of flowers form. Outdoors, dried plants weather rain and snow. Introduced from Europe, plant was once cultivated for its dried, spiny heads which were used in teasing wool to raise nap. Dried stem with spiny flower head used for dried floral arrangements. In various areas escapees are found. Photograph taken along roadside near Christopher Creek, August 11. *Dipsacus* not previously recorded for Arizona.

COMMON PIPSISSEWA

Prince's Pine

Chimaphila umbellata ssp. *umbellata*

Heather Family (Ericaceae)

Height: Flower stem to 8".

Flowers: Pinkish to white, waxy, with 5 widely spreading petals, 10 prominent stamens; bell-shaped, nodding; to ⅝" wide; in loose cluster at top of flower stalk.

Leaves: Dark green, shiny, thick, leathery; evergreen, lance-shaped, with toothed margins from tip to midleaf; to 3" long; whorled around stem.

Blooms: July–August.

Elevation: Above 6,500'.

Habitat: Coniferous forests.

Comments: Perennial herb. An ingredient in root beer. Two species of *Chimaphila* in Arizona. Photograph taken in Greer area, July 5.

ROCKY MOUNTAIN BLUEBERRY

Huckleberry

Vaccinium myrtillus (Vaccinium oreophilum)

Heather Family (Ericaceae)

Height: To 1'.

Flowers: Pink and white, urn-shaped and waxy; to ⅛" long; followed by roundish, juicy berry with flattened base, reddish, ripening to bluish black, to ⅜" in diameter, hanging beneath leaves.

Leaves: Dark green, finely toothed, elliptical, thin; to 1½" long.

Blooms: June–July.

Elevation: 8,000 to 11,000'.

Habitat: Coniferous forests and mountain slopes, often in very mossy forest areas.

Comments: A sprawling shrub, woody at base. Berries used in making jelly, pies, and other recipes. Three species of *Vaccinium* in Arizona. Photograph taken in coniferous forest above Greer, August 12.

FALSE INDIGO

Bastard Indigo
Amorpha fruticosa
Pea Family (Fabaceae)

Height: To 10′.

Flowers: Deep violet-purple, 1-petaled; stamens tipped with bright orange anthers; to ⅜″ long, ⅛″ wide; dense; on long, slender flower spike (spike to 7″ long) in leaf axil, often 2 or 3 spikes per axil.

Leaves: Dark green above, lighter green beneath; compound; to 10 pairs of oblong leaflets, each to 2″ long.

Blooms: May–July.

Elevation: 2,500 to 7,000′.

Habitat: Streamsides, canyons, and other moist locations.

Comments: Deciduous. Two species of *Amorpha* in Arizona. Photograph taken at Cave Creek, Portal, May 5.

FAIRY DUSTER

False Mesquite
Calliandra eriophylla
Pea Family (Fabaceae)

Height: To 4′.

Flowers: Each pink powder puff contains several flowers. Long stamens, white at base and tipped with pink, make up the 2″-wide puff; followed by flat, brown, 2″-long seed pod covered with short, velvety hairs. Some local populations tend toward red or ivory color. Upon ripening, the pod splits and its halves stand widely apart for a long period.

Leaves: Dark green, twice-compound, acacia-like; to 3″ long.

Blooms: At any time of year, but mostly October–May.

Elevation: Below 5,000′.

Habitat: Open hillsides, washes, and arid grasslands.

Comments: Thornless, perennial shrub and soil-binding plant. Many desert animals feed on foliage, and insects and hummingbirds frequent its flowers. During dry spells leaves wilt, but soon revive after rain. Two species of *Calliandra* in Arizona. Photograph taken in vicinity of Apache Lake, March 29.

NARROWLEAF TICK CLOVER

Cologania angustifolia (Cologania longifolia)
Pea Family (Fabaceae)

Height: Trailing or nearly erect to 4′ long.

Flowers: Reddish purple, pealike, notched; to 1″ long, ¾″ wide on ⅛″-long stem; erect upper petal to ⅝″ wide; lower petals curve forward; occurs singly or 2 to 3 in leaf axils, followed by a downy or smooth, flat pod.

Leaves: Dark green, pinnate; to 6″ on long stem; usually 3 but sometimes 5 very narrow leaflets, often folded in half lengthwise, hairy on margins and beneath, to 2½″ long. Leaflets very variable.

Blooms: July–September.

Elevation: 4,000 to 9,000′.

Habitat: Coniferous forests.

Comments: Perennial herb. A highly variable species. Three species of *Cologania* in Arizona. Photograph taken near Willow Springs Lake, September 13.

FEATHERPLUME

Dalea formosa
Pea Family (Fabaceae)

Height: To 16″.

Flowers: Purplish, pealike, largest petal cream-colored, to ½″ long; 2 to 6 in feathery cluster.

Leaves: Dark, green, pinnate; 7 to 15 narrow leaflets dotted with very dark glands; to ⅜″ long.

Blooms: March–June.

Elevation: 2,000 to 6,500′.

Habitat: Rocky hillsides, mountains, and dry plains.

Comments: Many-branched. Browsed by deer; pollinated by bees. Thirty species of *Dalea* in Arizona. Photograph taken north of Safford, April 20.

© MAX LICHER

FOXTAIL DALEA

Dalea leporina

Pea Family (Fabaceae)

Height: To 3'.

Flowers: Bluish or purplish with white bases; to 3⁄16" long; flower spike to 4" long.

Leaves: Dark green, smooth, glandular-dotted; to 3" long, with 19 to 35 leaf segments.

Blooms: June–October.

Elevation: 5,500 to 8,000'.

Habitat: Roadsides and at borders of ponderosa forests.

Comments: Many smooth stems and branches. Thirty species of *Dalea* in Arizona. Photograph taken in vicinity of McNary, August 10.

PEA BUSH

Dalea pulchra

Pea Family (Fabaceae)

Height: To 4'.

Flowers: Rose-purple above, whitish below, pea-like; to ¼" wide, in round head.

Leaves: Grayish green to silvery, velvety-haired, with 3 to 9 segments, each to ¼" long; leaf to ½" long.

Blooms: February–May.

Elevation: 2,500 to 5,000'.

Habitat: Rocky hills and mountainsides.

Comments: Thirty species of *Dalea* in Arizona. Photograph taken at Saguaro National Park East, April 15.

SAN PEDRO TICKREFOIL

Beggar's Ticks
Desmodium batocaulon
Pea Family (Fabaceae)

Height: Climbing or prostrate, to 5'.

Flowers: Rose to purplish, pealike, with large upper petal; lower petals folded lengthwise; column of stamens projecting upward; to ½" wide, ¾" long; followed by seed pods formed by 1-seeded segments covered with barbed hairs; up to 7 segments per pod.

Leaves: Dark green with silvery white, irregularly shaped central markings; pinnately divided into 3 narrow, lancelike segments; prickly haired, leaflets to 2" long, leaf to 3½" long.

Blooms: June–September.

Elevation: 3,500 to 6,500'.

Habitat: Roadsides and pine forests.

Comments: Barbed hairs cause individual seed pods to break off and cling to anything that touches them. Stems are prickly and sticky. Fifteen species of *Desmodium* in Arizona. Photograph taken in vicinity of Christopher Creek, August 2.

GRAHAM'S TICK CLOVER

Desmodium grahami
Pea Family (Fabaceae)

Height: Sprawling, to 2'. Flower stem to 15" tall.

Flowers: Light pink on outer lobes, darker pink on inner lobes; pea-shaped; stamens project forward. Flower to ⅜" long, in upright raceme in leaf axils and terminal, followed by beadlike seed pod with 3 to 5 joints.

Leaves: Grayish green, very hairy on both surfaces, with 3 broadly oval leaflets, slightly curved upward; to 2" long.

Blooms: June–September.

Elevation: 4,500 to 8,000'.

Habitat: Clearings in pine woods and hillsides.

Comments: Stems are reddish and very hairy. Fifteen species of *Desmodium* in Arizona. Photograph taken in vicinity of Prescott, June 7.

SILVERSTEM LUPINE

Silvery Lupine
Lupinus argenteus
Pea Family (Fabaceae)

Height: To 2'.

Flowers: Bluish to lilac, pealike; banner (upper) petal wide and hairy on back; flowers to ½" long, in long, terminal cluster; followed by hairy pod, to 1" long.

Leaves: Dark green above with short hairs, lighter green and hairy beneath; troughlike; to 3" wide, 3" long; palmately divided, with 7 to 9 leaflets.

Blooms: June–October.

Elevation: 7,000 to 10,000'.

Habitat: Clearings in coniferous forests.

Comments: A perennial herb, with silvery-haired stems. Seeds are eaten by birds. Twenty-five species of *Lupinus* in Arizona. Photograph taken south of Alpine, August 6. Its elevation, leaf shape, and silvery-haired stem help to identify this variable species.

ARIZONA LUPINE

Lupinus arizonicus
Pea Family (Fabaceae)

Height: To 2'.

Flowers: Pinkish-lavender, with a yellow center; pealike; to ½" long; in long, terminal cluster.

Leaves: Bright green above, hairy beneath; palmately divided, with up to 9 leaflets; to 1½" long.

Blooms: January–May.

Elevation: Below 3,000'.

Habitat: Roadsides and sandy washes in desert.

Comments: Twenty-five species of *Lupinus* in Arizona. Photograph taken near Hope, March 28.

BAJADA LUPINE
Lupinus concinnus
Pea Family (Fabaceae)

Height: To 6".

Flowers: Pale purple, edged with deeper purple; pealike, in flower cluster among leaves.

Leaves: Grayish green, densely haired; palmate, rounded at tips; to 1" wide.

Blooms: March–May.

Elevation: Below 5,000'.

Habitat: Sandy desert flats and bajadas.

Comments: Low, prostrate stems. Twenty-five species of *Lupinus* in Arizona. Photograph taken at Usery Mountain Recreation Area, March 10.

HILL'S LUPINE
Lupinus hillii
Pea Family (Fabaceae)

Height: To 8".

Flowers: Bluish to purple and pealike; to ¼" long, in an elongated cluster.

Leaves: Grayish green, very hairy; palmate, with 8 to 10 leaflets; to 2½" wide.

Blooms: May–August.

Elevation: 6,000 to 9,000'.

Habitat: Ponderosa pine forests.

Comments: Perennial. Less erect than other lupines. Twenty-five species of *Lupinus* in Arizona. Photograph taken at North Rim of Grand Canyon National Park, June 25.

DESERT LUPINE
Lupinus sparsiflorus
Pea Family (Fabaceae)

Height: To 16".

Flowers: Pale blue to violet; pealike, banner (upper) petal has yellow spot that changes to purplish red when bees contact it; keel (lower) petals are short, wide, and curve upward; flowers to ½" long, in elongated cluster; followed by flattened pod to 1½" long.

Leaves: Dark green, palmately divided; to 3" long; 5 to 9 linear leaflets, each to 1½" long, ⅛" wide.

Blooms: January–May.

Elevation: Below 3,000'.

Habitat: Desert roadsides, slopes, and mesas.

Comments: Perennial herb; a favorite of bees. Like other legumes, it improves the soil. Lupines adjust their leaflets during the day to absorb maximum sunlight. When rains of fall and winter are sufficient, these plants carpet roadsides during the blooming period. Twenty-five species of *Lupinus* in Arizona. Photograph taken in Superstition Mountains, March 22.

PARRY DALEA
Marina parryi (Dalea parryi)
Pea Family (Fabaceae)

Height: Sprawling, to 2½'.

Flowers: Deep blue or purple upper petals, white lower petals, pealike; to ⅜" long, 3⁄16" wide; in terminal, many-flowered, spikelike cluster.

Leaves: Grayish green, hairy, pinnate; leaflets to ⅛" long, leaf to 1" long.

Blooms: March–June.

Elevation: Below 4,000'.

Habitat: Desert flats and washes.

Comments: Perennial. Has very open growth and branches freely. Slender-branched and glandular. Three species of *Marina* in Arizona. Photograph taken at Saguaro National Park West, April 17.

ALFALFA
Medicago sativa
Pea Family (Fabaceae)

Height: To 3′.

Flowers: Violet to purplish, pealike; in short, terminal cluster from leaf axils; followed by downy, spirally coiled seed pod.

Leaves: Dark green, pinnately divided into 3 segments; sharply toothed on margins near tips; to 2″ long.

Blooms: April–October.

Elevation: Not available. Photograph taken at approximately 7,500′.

Habitat: Roadsides and old fields.

Comments: Perennial herb; native of southwestern Asia. Grown as a forage plant; occasionally escapes from cultivation. Has long taproot and enriches soil by adding nitrogen. A favorite of bees and butterflies. Five species of *Medicago* in Arizona. Photograph taken at Nelson Reservoir, August 3.

DESERT IRONWOOD
Palo Fierro
Olneya tesota
Pea Family (Fabaceae)

Height: To 30′.

Trunk: To 1½″ in diameter.

Bark: Gray and smooth on young trees; fissured and dark gray with age.

Flowers: Pink and purplish to white; pealike; 5 unequal petals; to ½″ wide in short cluster; followed by brown, hairy, beanlike pod, to 2½″ long.

Leaves: Bluish green, finely haired, pinnately compound; to 2″ long; 2 to 10 oblong leaflets to ¾″ long.

Blooms: May–June.

Elevation: Below 2,500′.

Habitat: Along desert washes and in sandy canyons.

Comments: Evergreen, with widely spreading crown that can extend up to 30′ across. At base of each leaf is pair of ½″-long, brown-tipped, slightly curved spines. To germinate, the hard coating on seeds must pass through digestive system of deer or cattle. **Desert Mistletoe** (*Phoradendron californicum*) (page 397) is often found growing in branches. A frost-sensitive tree, therefore useful as a gauge for citrus grove selection. Browsed by bighorn sheep. Seeds provide food to birds and other animals and were also an important part of the Native American diet. Ironwood is one of the heaviest woods in the world; one cubic foot weighs 66 pounds. It makes excellent firewood, producing intense heat and lasting coals. One species of *Olneya* in Arizona. Photograph taken at Usery Mountain Recreation Area, May 20.

PURPLE LOCOWEED

Oxytropis lambertii
Pea Family (Fabaceae)

Height: To 16".

Flowers: Reddish purple or white, pealike; 2 keel (lower) petals are pointed; to 1" long; in raceme held above leaves; followed by leathery, beaked pod to 1¼" long.

Leaves: Silvery, silky-haired, basal; to 12" long; pinnately compound, 7 to 17 leaflets, each to 1½" long.

Blooms: June–September.

Elevation: 5,000 to 8,000'.

Habitat: Plains and clearings in ponderosa forests.

Comments: Perennial herb. A dangerous locoweed toxic to livestock. Addictive. Often fatal. Taproot can descend 8'. Two species of *Oxytropis* in Arizona. Photograph taken at Sunset Crater National Monument, September 7.

SLIMLEAF LIMA BEAN

Phaseolus angustissimus
Pea Family (Fabaceae)

Height: Trailing vine to 10" high.

Flowers: Rose-pink, pealike; to ½" wide; followed by lima bean–like pea pod.

Leaves: Dark green, pinnately compound; to 3" long; long, narrow leaflets to 1¼" long. There is a wide range of leaf shapes and sizes within this species.

Blooms: May–October.

Elevation: 3,500 to 7,000'.

Habitat: Mesas, clearings in ponderosa forests, and hillsides.

Comments: Eight species of *Phaseolus* in Arizona. Photograph taken at Sunset Crater National Monument, September 7.

DESERT BEAN

Phaseolus filiformis (Phaseolus wrightii)
Pea Family (Fabaceae)

Height: Trailing vine.

Flowers: Reddish pink and pea-shaped; to ⅜" wide; occurring singly or in small cluster; followed by lima bean–like, hairy pea pod.

Leaves: Green, 3 leaflets with long, rounded lobes; reddish, hairy margins; to 2¼" wide.

Blooms: Throughout the year.

Elevation: 1,000 to 4,000'.

Habitat: Rocky slopes, canyons, and roadsides.

Comments: Eight species of *Phaseolus* in Arizona. Photograph taken near Tortilla Flat, March 19.

GRAY'S LIMA BEAN

Phaseolus grayanus
Pea Family (Fabaceae)

Height: Trailing vine.

Flowers: Deep pink and pealike, rounded upper lobe with shallow cleft; to ½" long, ⅜" wide; in terminal cluster on stem to 3" long; followed by pealike, curved seed pod to 2" long.

Leaves: Grayish green, pinnately compound; 3 leaflets with 3 pointed lobes; silvery central blotches; to 3" long, 3" wide. Center leaflet to 1¾" long; 2 side leaflets to 1½" long; terminal leaflet may fold back between 2 side leaflets.

Blooms: July–September.

Elevation: 5,000 to 8,500'.

Habitat: Clearings in ponderosa pine forests.

Comments: Eight species of *Phaseolus* in Arizona. Photograph taken on Mogollon Rim near Woods Canyon Lake, August 9.

LEMON SCURFPEA

Psoralidium lanceolatum (Psoralea lanceolata)

Pea Family (Fabaceae)

Height: To 16".

Flowers: White with dark purple on lower lip; pealike; to ⅛" wide, ⅛" long; in dense, spikelike, terminal cluster, followed by a roundish, warty pod.

Leaves: Dark green, compound, with 3 linear segments; dotted with prominent glands; to 2" long.

Blooms: May–September.

Elevation: 5,500 to 7,500'.

Habitat: Sandy soil, clearings in pine forests, and mesas.

Comments: Perennial herb. Three species of *Psoralidium* in Arizona. Photograph taken north of St. Johns, June 28.

EMORY INDIGO BUSH

Dyebush

Psorothamnus emoryi (Dalea emoryi)

Pea Family (Fabaceae)

Height: To 3', but straggly.

Flowers: Purple, pealike; to ¼" long; in roundish flower cluster to 1" wide.

Leaves: Grayish white, felty, covered with spreading white hairs and sunken black glands; pinnate; to 2" long; 1 to 13 narrow leaflets; terminal leaflet to 1" long, side leaflets to ¼" long.

Blooms: March–April; occasionally in the fall.

Elevation: Below 500'.

Habitat: Sandy mesas.

Comments: Entire plant is very felty. Seven species of *Psorothamnus* in Arizona. Photograph taken near Yuma, March 29.

SMOKETREE

Psorothamnus spinosus (Dalea spinosa)
Pea Family (Fabaceae)

Height: Shrub, or small tree to 20'.

Trunk: To 1' in diameter.

Bark: Dark, grayish brown, scaly, fissured.

Flowers: Bluish purple, pealike; ½" long; in cluster to 1½" long; followed by egg-shaped, brown, ⅜"-long, gland-dotted seed pod.

Leaves: When present, wedge-shaped, hairy, dotted with glands; to 1" long.

Blooms: May–June.

Elevation: 250 to 1,000'.

Habitat: Sandy washes in frost-free areas.

Comments: Spiny shrub or tree which is smoky gray all over and leafless most of the year. Twigs produce food by performing photosynthesis. Seven species of *Psorothamnus* in Arizona. Photograph taken in Kofa Mountains, February 21.

NEW MEXICO LOCUST

Robinia neomexicana
Pea Family (Fabaceae)

Height: Large, spiny-branched shrub, or small tree to 25'.

Trunk: To 8" in diameter.

Bark: Grayish brown, smooth; becomes furrowed and scaly with age.

Flowers: Purplish pink, pealike, fragrant; to ¾" long; in dense, drooping clusters; followed by reddish brown, bristly haired, flat seed pod to 4½" long.

Leaves: Bluish green, pinnately compound, finely haired; to 12' long; 15 to 21 rounded leaflets, each to 1½" long.

Blooms: May–July.

Elevation: 4,000 to 8,500'.

Habitat: Roadsides, canyons, and coniferous forests.

Comments: At nodes, twigs have reddish brown, stout, curved, ½"-long, paired spines. Forms erosion-controlling thickets. Browsed by cattle and deer. Birds and small mammals eat seeds. Seeds, bark, and roots are poisonous to humans; New Mexican Indians ate flowers raw. Used medicinally by Hopi Indians. Fence posts are made from trunks. Two species of *Robinia* in Arizona. Photograph taken at Lynx Lake, Prescott, May 27.

WHITE MOUNTAIN CLOVER

Trifolium neurophyllum
Pea Family (Fabaceae)

Height: To 2'.

Flowers: Reddish purple and pealike; to ½" long; in large flower head to 1¾" long, 1¼" wide. Fading flowers turn downward on flower head.

Leaves: Dark green above, lighter green and very hairy beneath; heavily veined; to 4½" long, compound, with 3 narrowly elliptical, sharply pointed, finely toothed, stalkless leaflets to 2¼" long.

Blooms: August.

Elevation: 8,000 to 8,500'.

Habitat: Along mountain streams and in moist mountain meadows.

Comments: More than twenty species of *Trifolium* in Arizona. Photograph taken at Luna Lake, August 5. Recognizable by large flower heads and hairy stems.

COW CLOVER

Pine Clover
Trifolium pinetorum
Pea Family (Fabaceae)

Height: To 20".

Flowers: Reddish lavender with whitish tips; pealike; to ⅛" wide, ½" long; flower head to 1¼" wide, has jagged-edged collar.

Leaves: Dark green, toothed, narrowly elliptical; 2½" long; compound, with leaflets usually in threes, to 1¼" long.

Blooms: June–October.

Elevation: 6,500 to 9,000'.

Habitat: Moist soil, stream banks, and wet meadows in coniferous forests.

Comments: More than twenty species of *Trifolium* in Arizona. Photograph taken at Luna Lake, June 29.

RED CLOVER

Trifolium pratense
Pea Family (Fabaceae)

Height: To 2′.

Flowers: Deep pink to pink, pealike; ½″ long; with 3 striped, bractlike stipules below each flower head; ball-shaped flower head, to 1½″ wide.

Leaves: Dark green, hairy; to 4″ wide; compound, with 3 oblong leaflets with lighter central chevron; leaflet to 2½″ long.

Blooms: Summer.

Elevation: Not available. Photograph taken at about 7,500′.

Habitat: Roadsides, fields, and disturbed soil.

Comments: Introduced from Europe, now naturalized in the U.S. Frequented by butterflies and bumblebees. More than twenty species of *Trifolium* in Arizona. Photograph taken at Forest Lakes, September 29.

WHITETOP CLOVER

Trifolium variegatum
Pea Family (Fabaceae)

Height: To 16″.

Flowers: Red-purple with white to pinkish tips; pealike; to 5⁄16″ long; in roundish flower head of 2 to 10 or more flowers. Calyx lobes just below flower head are triangular; each has a sharp spine.

Leaves: Dark green, with 3 narrow, oblong leaflets with saw-toothed margins; leaflets to 2″ long.

Blooms: April–July.

Elevation: Not available. Photograph taken at 4,000′.

Habitat: Areas with moist soil and wet canyons.

Comments: More than twenty species of *Trifolium* in Arizona. Photograph taken north of Superior, April 20.

AMERICAN VETCH

Purple Vetch
Vicia americana
Pea Family (Fabaceae)

Height: A climber to 4′.

Flowers: Pinkish to bluish purple; pealike; to 1¼″ long, 4 to 9 in loose cluster on stalk in leaf axil; followed by pealike pod.

Leaves: Green, pinnately compound, with 8 to 12 elliptical leaflets, each to 1½″ long. Leaf shape is variable.

Blooms: May–September.

Elevation: 5,000 to 10,000′.

Habitat: Clearings in ponderosa and mixed conifer forests.

Comments: Perennial herb. Clings to other plants with its coiling tendrils. Four species of *Vicia* in Arizona. Photograph taken near Willow Springs Lake, June 11. A variety of **American Vetch** (*Vicia americana* var. *linearis*) has very narrow leaflets to 1¼″ long with the edges rolled under.

ARIZONA CENTAURY

Rosita
Centaurium calycosum
Gentian Family (Gentianaceae)

Height: To 2′.

Flowers: Pink, with 5 pointed lobes, bright yellow anthers; to 1″ wide; terminal and in forks of branches.

Leaves: Light green, opposite, lance-shaped, and succulent; clasping the stem; to 2½″ long.

Blooms: April–June. Year-round at lower elevations in protected areas.

Elevation: 150 to 6,000′.

Habitat: Moist meadows and along streams.

Comments: Three species of *Centaurium* in Arizona. Photograph taken at Saguaro National Park West, April 17.

CATCHFLY GENTIAN

Alkali Chalice
Eustoma exaltatum ssp. *exaltatum*
Gentian Family (Gentianaceae)

Height: To 2′.

Flowers: Deep bluish purple, darker toward center; 5-petaled; to 1¼″ wide, 1¼″ long.

Leaves: Grayish green, leathery, smooth; oblong, stalkless, paired with bases surrounding stem; to 4″ long.

Blooms: June–September.

Elevation: 500 to 2,500′.

Habitat: Along rivers, streams, and ditches.

Comments: Rare in Arizona. One species of *Eustoma* in Arizona. Photograph taken along Salt River north of Granite Reef Dam, August 20.

PARRY GENTIAN

Closed Gentian
Gentiana parryi
Gentian Family (Gentianaceae)

Height: To 16″.

Flowers: Deep blue to violet, whitish within and streaked with green, 5-lobed corolla with fringed appendages between lobes. Flower to 2″ long, ⅞″ wide, in terminal cluster of 1 to 5 flowers.

Leaves: Dark green, smooth, thick; leathery, opposite, to 1½″ long.

Blooms: August–September.

Elevation: 8,500 to 11,500′.

Habitat: Alpine and subalpine meadows.

Comments: Has erect, pinkish stem. Three species of *Gentiana* in Arizona. Photograph taken in mountain meadow above Greer, August 12.

PLEATED GENTIAN

Marsh Gentian
Gentiana rusbyi (Gentiana affinis)
Gentian Family (Gentianaceae)

Height: to 16".

Flowers: Violet to bluish purple, 5-lobed, erect; in leaflike floral bracts; flower to 1" long, in elongated, terminal cluster along upper third of stem.

Leaves: Dark green, opposite, narrowly lance-shaped; to 2" long, occurring all along stem.

Blooms: August–October.

Elevation: 7,000 to 9,500'.

Habitat: Mountain meadows.

Comments: Perennial herb. Stems are reddish brown and clustered. Three species of *Gentiana* in Arizona. Photograph taken in mountain meadow above Greer, September 12.

NORTHERN GENTIAN

Gentianella amarella ssp. *acuta*
Gentian Family (Gentianaceae)

Height: To 20".

Flowers: Pink to lavender, tubular, with 5 spreading lobes; fringe of whitish hairs around center opening; to ½" long, ⅜" wide, numerous; in leaf axils along main stem.

Leaves: Dark green, opposite, stalkless; lance-shaped and reddish tipped; at right angles to main stem; to 1¾" long.

Blooms: June–September.

Elevation: 7,000 to 11,000'.

Habitat: Moist coniferous forests.

Comments: Main stem is reddish. Five species of *Gentianella* in Arizona. Photograph taken in vicinity of Woods Canyon Lake, August 3.

AUTUMN DWARF GENTIAN

Gentianella amarella ssp. *heterosepala*
Gentian Family (Gentianaceae)

Height: To 16".

Flowers: Deep purple, tubular, with 5 spreading lobes; fringe of whitish hairs around center opening; to ¼" wide, numerous; in leaf axils along main stem.

Leaves: Light green, opposite, stalkless; elliptical, curled upright along main stem; to 1½" long.

Blooms: August–September.

Elevation: 7,000 to 11,500'.

Habitat: Moist mountain meadows.

Comments: Five species of *Gentianella* in Arizona. Photograph taken in mountains above Greer, August 10.

FELWORT

Star Swertia
Swertia perennis
Gentian Family (Gentianaceae)

Height: To 2'.

Flowers: Deep purple and starlike, with 5 or 6 pointed, petallike lobes; 2 fringed, hairy glands at base of each lobe; to 1" wide; on erect stems in leaf axils on main stem.

Leaves: Light green, sunken veins, elliptical base leaves on long slender stalk; to 8" long, 2¼" wide. Leaves on upper stem are opposite, smaller, and clasp stem.

Blooms: July–September.

Elevation: 9,000 to 10,000'.

Habitat: Mountain meadows and along shallow mountain streams in spruce-fir forests.

Comments: Perennial. One species of *Swertia* in Arizona. Photograph taken in mountains above Greer, August 12.

FILAREE
Heron-Bill
Erodium cicutarium
Geranium Family (Geraniaceae)

Height: To 15".

Flowers: Pinkish violet, with 5 petals; to ¼" wide; in loose cluster; followed by fruit with still, awl-shaped projection.

Leaves: Dark green, mainly fernlike; forming basal rosette; to 4" long.

Blooms: February–July.

Elevation: Below 7,000'.

Habitat: Desert flats, plains, mesas, and hillsides.

Comments: Annual. Native to southern Europe, but introduced to this country by early Spanish settlers and now naturalized in the Southwest. Ripening seed pod (to 2" long) resembles a heron's beak, but when mature and dry it twists into a spiral. When moisture is available again, pod untwists, pushing point of seed pod into soil. Seeds are stored by harvester ants; foliage serves as forage for livestock. Three species of *Erodium* in Arizona. Photograph taken in Superstition Mountains, February 4.

WILD GERANIUM
Cranesbill
Geranium caespitosum
Geranium Family (Geraniaceae)

Height: To 1½'.

Flowers: Deep reddish pink, with whitish streaks (occasionally all white); 5 petals flared backward; partly hairy; to ½" wide.

Leaves: Green, with 5 deeply cut lobes; both surfaces covered with soft hairs; to 1½" wide.

Blooms: May–September.

Elevation: 5,000 to 9,000'.

Habitat: Rich soil in pine forests.

Comments: Perennial herb. Good forage for sheep. Six species of *Geranium* in Arizona. Photograph taken near Greer, July 20.

CAROLINA GERANIUM
Geranium carolinianum
Geranium Family (Geraniaceae)

Height: To 16".

Flowers: Pale pink, with 5 petals barely extending beyond the sepals; to ¼" wide.

Leaves: Grayish green, with 5 or 7 segments; divisions are cleft into narrowly oblong lobes; to 1¼" wide.

Blooms: End of March–May.

Elevation: 2,000 to 5,500'.

Habitat: Desert washes, slopes, and mesas.

Comments: Pinkish stems. Six species of *Geranium* in Arizona. Photograph taken in Superstition Mountains, March 26.

RICHARDSON'S GERANIUM
Geranium richardsonii
Geranium Family (Geraniaceae)

Height: To 1½'.

Flowers: White to pale pink, with purplish veins; 5-petaled; to 1" wide.

Leaves: Dark green, thin, palmately cleft into 5 to 7 main segments; on long stalks; to 4" wide.

Blooms: April–October.

Elevation: 6,500 to 11,500'.

Habitat: Moist soil of ponderosa pine and mixed conifer forests.

Comments: Perennial herb. Often hybridizes with other species. Six species of *Geranium* in Arizona. Photograph taken in Greer area, June 20.

WAX CURRANT

Ribes cereum var. *cereum*
Currant Family (Grossulariaceae)

Height: To 6' tall and 8' wide.

Flowers: White or pink, tubular, and sticky; to ⅜" long; in groups of 1 to 4; followed by red to yellowish red, sticky fruits to ¼" in diameter.

Leaves: Light green with tiny white dots; sticky, 3- to 5-lobed, glandular-haired; to 1" wide.

Blooms: May–July.

Elevation: 5,500 to 9,000'.

Habitat: Clearings in pine forests.

Comments: Fruits lack spines. Browsed by elk and deer; fruits eaten by wildlife. Hopi Indians ate berries and used plant medicinally. Ten species of *Ribes* in Arizona. Photograph taken at Nelson Reservoir, August 3.

WHISKY CURRANT

Grosellero
Ribes cereum var. *inebrians (Ribes inebrians)*
Currant Family (Grossulariaceae)

Height: To 4'.

Flowers: White to pinkish, tubular, to ⅜" long; followed by round, red, ⅜" diameter fruit.

Leaves: Light green, fan-shaped, lobed, and glandular; to 1½" wide.

Blooms: May–August.

Elevation: 5,000 to 9,000'.

Habitat: Dry, sunny slopes and ridges.

Comments: Lacks spines. Currants eaten by birds and animals; shrubs browsed by deer. Ten species of *Ribes* in Arizona. Photograph taken at North Rim of Grand Canyon National Park, June 25.

SMALL-FLOWERED EUCRYPTA

Eucrypta micrantha
Waterleaf Family (Hydrophyllaceae)

Height: To 10".

Flowers: Bluish purple or white, bell-like, with 5 united petals; to ½" wide; in loose cluster on slender, prickly stem.

Leaves: Dark green, prickly, pinnately lobed; clasping stem; to 2" long.

Blooms: February–May.

Elevation: Below 4,000'.

Habitat: Shade of shrubs and other sheltered areas.

Comments: Annual. Has very weak, prickly stems. Two species of *Eucrypta* in Arizona. Photograph taken at Alamo Lake State Park, February 26.

PURPLE MAT

Nama demissum
Waterleaf Family (Hydrophyllaceae)

Height: To 3", with 8" stems in favorable years.

Flowers: Reddish pink, tubular, trumpetlike, with 5 rounded lobes; to ⅜" wide.

Leaves: Green, sticky, and narrow; hairy, spatula-shaped; to 1½" long.

Blooms: February–May.

Elevation: Below 3,500'.

Habitat: Desert flats and washes.

Comments: Annual. In years of little rainfall, plant produces only a few flowers on very short stems. With plentiful rains, masses of these plants carpet broad areas of desert. Eight species of *Nama* in Arizona. Photograph taken at Organ Pipe Cactus National Monument, February 26.

SCORPIONWEED
Caterpillar Weed
Phacelia ambigua (Phacelia crenulata)
Waterleaf Family (Hydrophyllaceae)

Height: To 18".

Flowers: Violet-purple, bell-shaped, with 5 rounded, united petals; to ¼" wide; in finely haired, terminal coils.

Leaves: Dark green, hairy, much divided, variable; to 5" long.

Blooms: February–May.

Elevation: Below 4,000'.

Habitat: Roadsides, sandy washes, and desert flats.

Comments: Annual, with sticky, bristly stems. Emits onion-like odor when foliage is crushed. The name "scorpionweed" refers to curling flower head that resembles a scorpion's erect tail. Forty-six species of *Phacelia* in Arizona. Photograph taken at Golden Shores, February 25. A closely related species, **Cleftleaf Wild Heliotrope** (*Phacelia corrugate*), is found at 5,000 to 7,000'.

WILD HELIOTROPE
Scorpionweed
Phacelia distans
Waterleaf Family (Hydrophyllaceae)

Height: To 30".

Flowers: Pale blue or violet and bell-shaped, with 5 rounded, united petals; to ¼" wide; in finely haired, terminal coils.

Leaves: Green, finely haired, once or twice pinnately divided; highly variable; to 3" long.

Blooms: February–May.

Elevation: 1,000 to 4,000'.

Habitat: Along washes and slopes.

Comments: Annual, with reddish, hairy, branching stems. Straggly, often growing in tangles among shrubs. Forty-six species of *Phacelia* in Arizona. Photograph taken at Saguaro National Park, March 31.

ROCKY MOUNTAIN IRIS

Western Blue Flag
Iris missouriensis
Iris Family (Iridaceae)

Height: Flower stem to 3′.

Flowers: Pale blue to violet, streaked with white; 3 narrow, erect petals, 3 wider sepals curved downward; to 4″ wide; at top of stout, leafless stem, followed by a thin-walled, oblong capsule with 6 ribs; to 1½″ long, ¾″ wide.

Leaves: Dark green, sword-shaped; to 20″ long, ½″ wide.

Blooms: May–September.

Elevation: 6,000 to 9,500′.

Habitat: Wet meadows and moist forest clearings.

Comments: Perennial. Grows in clumps. Rootstalks and leaves are poisonous if eaten. One species of *Iris* in Arizona. Photograph taken at Ashurst Lake area, June 1.

BLUE-EYED GRASS

Blue Star Grass
Sisyrinchium demissum
Iris Family (Iridaceae)

Height: To 1′.

Flowers: Dark purple or deep blue with yellow stamens; 6 segments with fine point at tips; striped on undersides; to ¾″ wide; terminal on long, flat stem.

Leaves: Dark bluish green, narrow, sword-shaped; mainly basal; to 6″ long.

Blooms: June–September.

Elevation: 5,000 to 9,500′.

Habitat: Moist meadows and along streams.

Comments: Perennial herb. Wiry stem. Six species of *Sisyrinchium* in Arizona. Photograph taken in mountain meadow above Greer, August 10.

RANGE RATANY

Chacate

Krameria erecta (Krameria parvifolia)

Ratany Family (Krameriaceae)

Height: Shrub to 1½'.

Flowers: Pinkish, with 5 petals: 3 orangish, upper petals, 2 petals like lobes or oil glands; to ¾" wide; followed by small, roundish, burlike seed pod with barbs along length of spines.

Leaves: Small, grayish, alternate, and narrow; to ½" long.

Blooms: April–October.

Elevation: Below 5,000'.

Habitat: Dry plains and mesas.

Comments: Perennial; sprawling, many-branched, low shrub. Partly parasitic to other plants, absorbing water through their roots. Certain species of bees scrape oil glands for oil to bring to their eggs as food for the future offspring. Native Americans used twigs medicinally and made dyes from roots. Three species of *Krameria* in Arizona. Photograph taken at Coronado National Memorial, April 18. Distinguished from **White Ratany** (*Krameria grayi*) (at right) by barbs along entire length of fruit spines.

WHITE RATANY

Chacate

Krameria grayi

Ratany Family (Krameriaceae)

Height: Shrub to 2'.

Flowers: Reddish magenta, very irregular, with 5 petals; 3 long, upper petals and 2 lower petals like lobes or oil glands; to ½" wide. Followed by small, round, burlike seed pod with 3 or 4 barbs at tip of each spine.

Leaves: Gray, finely haired, narrow; to ½" long.

Blooms: April–October.

Elevation: 500 to 4,000'.

Habitat: Dry desert foothills and mesas.

Comments: Partly parasitic on roots of such plants as creosote and bursage. More common than range ratany in the desert. Browsed by cattle and wildlife. Native Americans used powdered roots for treating sores. Three species of *Krameria* in Arizona. Photograph taken at Saguaro Lake, October 18. Unlike **Range Ratany** (*Krameria erecta*) (at left), its barbs occur only on the tips of spines of fruit.

TRAILING KRAMERIA

Krameria lanceolata
Ratany Family (Krameriaceae)

Height: Trailing stems to 2′ long.

Flowers: Dark, magenta or wine-colored red petals: 4 pointed upper petals, 1 larger, pointed lower petal curved forward. 3 greenish, fan-shaped stigma lobes with reddish tips; flower to ¾″ long, ¾″ wide; followed by a round, spiny fruit to 5⁄16″ in diameter.

Leaves: Grayish green, hairy, alternate; narrow and lance-shaped, elliptical or linear, to 1″ long.

Blooms: May–August.

Elevation: 2,500 to 5,000′.

Habitat: Sandy slopes and dry, open plains.

Comments: Perennial herb. Spiny fruit can penetrate bare skin if stepped on. Three species of *Krameria* in Arizona. Photograph taken in vicinity of Fort Bowie, May 8.

MOCK PENNYROYAL

Hedeoma hyssopifolium
Mint Family (Lamiaceae)

Height: To 20″.

Flowers: Lavender; gaping throat whitish with purplish blotches inside; upper lip projecting; to ½″ long; in pairs along stems.

Leaves: Green, narrow, numerous, opposite; to ¾″ long; occurring at intervals along stem.

Blooms: May–October.

Elevation: 5,000 to 9,500′.

Habitat: Ponderosa pine forests, hillsides, and canyons.

Comments: Seven species of *Hedeoma* in Arizona. Photograph taken near Willow Springs Lake, July 6.

OBLONG MOCK PENNYROYAL

Hedeoma oblongifolia
Mint Family (Lamiaceae)

Height: To 20".

Flowers: Purple to rose-purple, white throat with purple stripes; tubular; to ⅛" wide, ½" long; clustered in leaf axils.

Leaves: Bright green, oval, hairy, untoothed; to ½" long.

Blooms: March–September.

Elevation: 1,800 to 8,000'.

Habitat: Desert washes, along streams, and in woodlands.

Comments: Perennial herb, with stiff and square stem. Seven species of *Hedeoma* in Arizona. Photograph taken at Tortilla Flat, March 19.

DESERT LAVENDER

Hyptis emoryi (Hyptis albida)
Mint Family (Lamiaceae)

Height: Shrub to 15'.

Flowers: Violet-blue, to 1"; in clusters in leaf axils.

Leaves: Gray-green and oval, with irregularly serrated margins; covered with woolly hairs; to 2½" long.

Blooms: Any time of year.

Elevation: Below 5,000'.

Habitat: Desert washes and dry, rocky slopes.

Comments: Deciduous to semi-deciduous, with ash gray bark and lavender-scented foliage. Frost-sensitive. Browsed by livestock; a valuable bee plant. Seeds used as food by wildlife. Minty leaves used to flavor tea. One species of *Hyptis* in Arizona. Photograph taken at Usery Mountain Recreation Area, February 14.

FIELD MINT
Wild Mint
Mentha arvensis
Mint Family (Lamiaceae)

Height: To 2′.

Flowers: Pale pink to lavender, bell-shaped; to ¼″ long, ⅛″ wide; clustered in axils of foliage leaves at intervals along stem.

Leaves: Dark green, aromatic, slightly downy; ovate to lanceolate, tapering on both ends; to 3″ long.

Blooms: July–October.

Elevation: 5,000 to 9,500′.

Habitat: Moist woods and stream banks.

Comments: Strong-scented perennial. Produces runners to 2′ long. Native Americans used leaves for flavoring. Mints recognizable by opposite leaves, square stems, and irregular flowers. Four species of *Mentha* in Arizona. Photograph taken at Nelson Reservoir, August 3.

WILD BERGAMOT
Monarda fistulosa var. *menthifolia (Monarda menthaefolia)*
Mint Family (Lamiaceae)

Height: To 3′.

Flowers: Pinkish, tubular, 2-lipped; 1″ long; in single, dense, globular, terminal cluster above whorl of purplish green leaves.

Leaves: Yellowish green with purplish tinge in places; lance-shaped, opposite, toothed; to 2½″ long.

Blooms: July–August.

Elevation: 5,000 to 9,000′.

Habitat: Roadsides and moist pine and spruce-fir forests.

Comments: Perennial. Square stem tinged with purple. Leaves have strong mint odor when crushed. Four species of *Monarda* recorded for Arizona. Photograph taken south of Alpine, July 23.

PLAINS BEEBALM

Pony Mint
Monarda pectinata
Mint Family (Lamiaceae)

Height: To 1'.

Flowers: Light rose to white; to ½"; in clusters at intervals on a spike; each separated by a spinelike bract and a whorl of leaves.

Leaves: Dark green, oblong to lance-shaped, finely toothed; to 1½" long.

Blooms: August–September.

Elevation: 5,000 to 7,000'.

Habitat: Dry soil of plains and pastures.

Comments: Square stem. Four species of *Monarda* recorded for Arizona. Photograph taken south of Alpine, July 23.

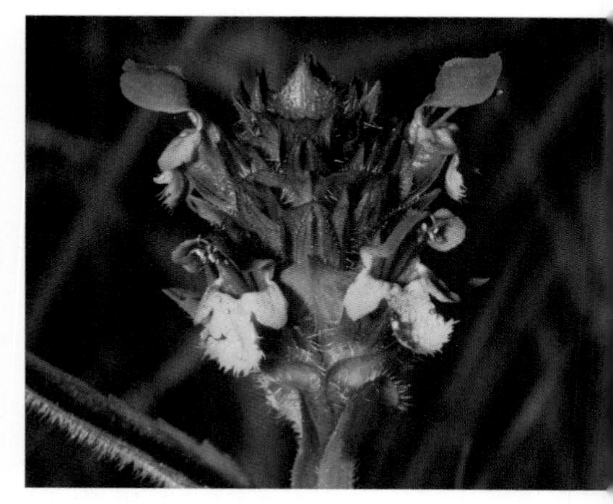

HEAL ALL

Self Heal
Prunella vulgaris ssp. *lanceolata*
Mint Family (Lamiaceae)

Height: To 12", sometimes prostrate.

Flowers: Lavender to purple; 2-lipped, with upper lip forming a hood, lower lip drooping; green bracts between flowers; to ½" long; in terminal spike.

Leaves: Dark green, oval to lance-shaped, toothed; to 4" long.

Blooms: June–September.

Elevation: 5,000 to 9,000'.

Habitat: Moist meadows along streams, lakeshores, and roadsides.

Comments: Perennial herb. Introduced from Europe; now naturalized. Square-stemmed. Flower spikes elongate after blooming. One species of *Prunella* in Arizona. Photograph taken south of Alpine, July 10.

MARSH SKULLCAP

Scutellaria galericulata
Mint Family (Lamiaceae)

Height: To 2′.

Flowers: Lavender, slipper-shaped, 2-lipped, tubular; lower lobe facing downward; to ¾″ long; 1 flower in each upper leaf axil.

Leaves: Light green, opposite, slightly scalloped on margins; lance-shaped, pinkish, with sunken veins; granular-feeling; clasping stem; to 1½″ long, ½″ wide.

Blooms: June–August.

Elevation: 6,000 to 9,500′.

Habitat: Moist ground of wet meadows and swampy areas.

Comments: Perennial herb. Square stems. Five species of *Scutellaria* in Arizona. Photograph taken in vicinity of McNary, July 7.

MARSH HEDGENETTLE

Stachys palustris
Mint Family (Lamiaceae)

Height: To 2′.

Flowers: Pale lavender with darker lavender markings; short upper lip projecting like a hood, 3-lobed lower lip long and bent downward; to 1″ long, ¼″ wide; in whorls of 6 at intervals on terminal spike to 10″ tall.

Leaves: Pale green, opposite, thin, spreading hairs, clasping stem, deeply veined, broadly lance-shaped, toothed margins, to 3″ long. 4-sided, hairy stem.

Blooms: July–August.

Elevation: 7,000 to 9,000′.

Habitat: Moist, shady areas.

Comments: Perennial herb. Tubers can be eaten. Five species of *Stachys* in Arizona. Photograph taken near Greer, August 15.

LILAC CHASTE-TREE

Monk's Pepper Tree
Vitex agnus-castus
Mint Family (Lamiaceae)

Height: To 20'.

Flowers: Lavender to lavender-blue, tubular, 5-lobed (lower lobe largest); 4 long stamens; to $\frac{3}{8}$" wide, $\frac{5}{16}$" long; in dense, terminal spike to 7" long, 1$\frac{1}{4}$" wide; followed by brownish, round fruit to $\frac{3}{16}$" in diameter.

Leaves: Dark green above, grayish beneath; long-stalked, opposite, palmately compound; to 5" long; 5 to 7 lance-shaped leaflets, center leaflet longest.

Blooms: May–October.

Elevation: Not available. Photograph taken at 1,300'.

Habitat: Undocumented in Arizona; this specimen and others were found growing in moist soil of a cove at north end of Saguaro Lake.

Comments: Deciduous; has reddish brown stems. Native of China and India. This species not previously recorded for Arizona; escapees from cultivation. Photograph taken at Saguaro Lake, October 18.

NODDING WILD ONION

Allium cernuum
Lily Family (Liliaceae)

Height: To 20".

Flowers: Pink to white with long stamens, bell-shaped; to $\frac{1}{4}$" long; in clustered, nodding umbel at top of leafless stem; to 1$\frac{3}{4}$" wide.

Leaves: Dark green, basal, grasslike; to 14" long.

Blooms: July–October.

Elevation: 5,000 to 8,500'.

Habitat: Cool, moist ponderosa forests.

Comments: No other species has a nodding umbel. Crushed leaves have onion smell. Thirteen species of *Allium* in Arizona. Photograph taken in vicinity of Ashurst Lake, September 5.

GEYER'S ONION

Wild Onion

Allium geyeri var. *geyeri*

Lily Family (Liliaceae)

Height: To 1'.

Flowers: Pink, ¼" long, in 1½" erect umbel at top of leafless stalk; thick rib on back of each petal.

Leaves: Grasslike, shorter than flower stem; to 8" long; have onion odor.

Blooms: June–August.

Elevation: 5,000 to 10,000'.

Habitat: Moist pine forests and clearings.

Comments: In certain sections of Arizona flowers replaced by small bulbils. Thirteen species of *Allium* in Arizona. Photograph taken in vicinity of Willow Springs Lake, August 5.

RED ONION

Allium geyeri var. *tenerum (Allium rubrum)*

Lily Family (Liliaceae)

Height: Flowers stems to 16".

Flowers: Pink (when present), to 5⁄16" long, in erect umbel; sterile; replaced by small pinkish bulbils.

Leaves: Dark green, few, onion-like; flat or curved; to 8" long.

Blooms: July–August.

Elevation: 7,000 to 9,500'.

Habitat: Moist mountain meadows.

Comments: Perennial herb. Solitary, fibrous-coated bulb. Thirteen species of *Allium* in Arizona. Photograph taken in mountains above Greer, July 8.

DESERT ONION
Arizona Onion
Allium macropetalum
Lily Family (Liliaceae)

Height: To 8".

Flowers: Pale pink, 6 petals; each petal with reddish brown, vertical stripe; to ¾" wide.

Leaves: Yellowish green, grasslike; to 4" long.

Blooms: March–June.

Elevation: 1,000 to 7,000'.

Habitat: Desert flats.

Comments: Crushed leaves smell like onion. Grows down from bulb to 12" below soil surface. Thirteen species of *Allium* in Arizona. Photograph taken south of Globe, April 20.

SEGO LILY
Mariposa Lily
Calochortus nuttallii
Lily Family (Liliaceae)

Height: To 20".

Flowers: Tuliplike, with 3 creamy white to lavender petals; yellow petal base is marked with a crescent-shaped purple band; dense, slender hairs near base of petals; to 2" wide; to 5 flowers per stalk.

Leaves: Grayish green, narrow, grasslike; margins rolled upward; to 4" long.

Blooms: May–July.

Elevation: 5,000 to 8,000'.

Habitat: Dry mesas, open pine forests, and hillsides.

Comments: State flower of Utah. Bulbous root once used for food by Native Americans and Mormon settlers. Six species of *Calochortus* in Arizona. Photograph taken near Willow Springs Lake, June 10.

BLUE DICKS

Desert Hyacinth
Dichelostemma capitatum (Dichelostemma pulchellum)
Lily Family (Liliaceae)

Height: To 30", but usually much less.

Flowers: Lavender, 6 segments, in terminal cluster on slender stem; flower to 1" wide.

Leaves: Dark green, few, grasslike; rising from bulb; to 15".

Blooms: February–May.

Elevation: Below 5,000'.

Habitat: Mesas, open slopes, and plains.

Comments: Perennial lily. Onion-like bulb used for food by pioneers and Native Americans. One species of *Dichelostemma* in Arizona. Photograph taken in Superstition Mountains, February 23.

DESERT FIVE SPOT

Lantern Flower
Eremalche rotundifolia
Mallow Family (Malvaceae)

Height: To 2'.

Flowers: Pink to lilac globes, with 5 petals; opening at top to reveal cream-colored center with 5 carmine spots; to 1¼" wide.

Leaves: Dark green to reddish, round to heart-shaped, with scalloped margins; to 2" wide.

Blooms: March–May.

Elevation: 100 to 1,500'.

Habitat: Dry, open desert and washes in sandy soil.

Comments: Annual herb. When light passes through globe of petals, flower resembles a lighted lantern. Two species of *Eremalche* in Arizona. Photograph taken north of Yuma, March 29.

ROCK HIBISCUS

Pale Face
Hibiscus denudatus
Mallow Family (Malvaceae)

Height: Straggly plant to 3′ long.

Flowers: Pale pink to lavender or white; deeper pink toward center; 5 broad, rounded petals; in center, dark pink to maroon, stamen column with pollen heads emerging from sides; to 1½″ wide; in upper leaf axils and tips of branches; followed by a 5-chambered capsule.

Leaves: Yellowish green, densely woolly haired, somewhat oval to elliptical, with toothed margins; to 1¼″ long.

Blooms: January–October.

Elevation: Below 2,000′.

Habitat: Rocky slopes, flats, and washes in desert.

Comments: Perennial herb with hairy stems; many-branched. Three species of *Hibiscus* in Arizona. Photograph taken at Saguaro Lake, February 6.

COMMON MALLOW

Malva neglecta
Mallow Family (Malvaceae)

Height: Spreading flat on ground, to 16″ long.

Flowers: Light pink striped with darker pink; 5 notched petals; cup-shaped; numerous stamens united into tube around style; to ¾″ wide; occurring singly or in clusters in leaf axils; followed by flat, disk-shaped fruit with up to 15 segments.

Leaves: Dark green, hairy, roundish, crinkly; alternate, shallowly lobed, with scalloped margins; to 2″ wide, 3″ long including stem.

Blooms: July–September.

Elevation: 3,500 to 7,500′.

Habitat: Roadsides and wastelands.

Comments: Introduced from Europe, now naturalized. Three species of *Malva* in Arizona. Photograph taken at Rainbow Lake, Lakeside, July 6.

CHEESEWEED

Little Mallow
Malva parviflora
Mallow Family (Malvaceae)

Height: To 4′.

Flowers: Pinkish, 5 petals; to ¼″ wide; in small clusters at base of leaf stalks; followed by rounded, disklike seed pods, to ½″ wide, containing 11 to 12 sections resembling cheese wedges.

Leaves: Dark green, soft, almost circular, with 5 to 7 toothed lobes; to 5″ wide, on stalks to 10″ long.

Blooms: March–September.

Elevation: 100 to 8,500′.

Habitat: Fields, lots, and roadsides.

Comments: Annual or biennial. Introduced from Europe. A common weed of fields and open lots. Boiled and eaten by Native Americans. Three species of *Malva* in Arizona. Photograph taken near Granite Reef Dam, March 1.

NEW MEXICAN CHECKERMALLOW

Prairie Mallow
Sidalcea neomexicana
Mallow Family (Malvaceae)

Height: To 3′.

Flowers: Deep pink to purple, with 5 petals; to 1½″ wide.

Leaves: Yellow green, lower leaves nearly round; rounded teeth; to 4″ wide; upper leaves are smaller, palmately divided.

Blooms: June–September.

Elevation: 5,000 to 9,500′.

Habitat: Along streams and in wet meadows.

Comments: Perennial. One species of *Sidalcea* in Arizona. Photograph taken near Greer, July 20.

DEVIL'S CLAW

Unicorn Plant
Proboscidea parviflora
Unicorn Plant Family (Martyniaceae)

Height: Sprawling stems to 3′ long.

Flowers: Reddish purple to pinkish with yellowish, striped throat; tubular, 5-lobed, hairy; to 1½″ long, ¾″ wide; followed by dark green, very sticky, hairy, fleshy fruit with curved tip; fruit to 12″ along the curve. When mature, each fruit sheds its fleshy skin and splits open, revealing a black, woody shell ending in 2 curved prongs.

Leaves: Dark green, very hairy and sticky, triangular; to 7″ long, 7″ wide; on pinkish stem to 1′ long.

Blooms: April–October.

Elevation: 1,000 to 5,000′.

Habitat: Roadsides, mesas, plains, and disturbed areas.

Comments: Annual herb. Entire plant is very sticky. Prongs on fruits hook on passing animals. Immature fruits are eaten as a vegetable. Black mature pods used by Native Americans in basketry. Two species of *Proboscidea* in Arizona. Photograph taken at Dead Horse Ranch State Park, October 3.

CHINABERRY

Umbrella Tree
Melia azedarach
Melia Family (Meliaceae)

Height: To 40′.

Trunk: To 30″ in diameter, but usually less.

Bark: Dark brown to reddish brown, furrowed.

Flowers: Purplish, fragrant; to ¾″ wide; in loose cluster to 8″ long; followed by round, yellowish, poisonous fruits to ½″ in diameter, hanging on tree through winter.

Leaves: Bright green, bipinnately compound, to 2′ long; with pointed, tooth-margined leaflets to 3″ long.

Blooms: March–April.

Elevation: Not available. Photograph taken at 2,600′.

Habitat: Desert soils where a moderate amount of water is available.

Comments: Deciduous. Native of Southeast Asia. Escapee from home plantings. Rapid grower, but short-lived. Leaves turn golden in the fall. One species of *Melia* in Arizona. Photograph taken at Catalina State Park, April 15.

SAND VERBENA

Abronia villosa
Four O'Clock Family (Nyctaginaceae)

Height: Trailing plant to 3' long; upright flower stalks to 12" high.

Flowers: Pink-purple, to 3"-wide heads of clusters of individual tubular flowers.

Leaves: Dark green, oval, sticky, and finely haired; to 1½" long.

Blooms: February–May.

Elevation: Below 1,500'.

Habitat: Roadsides, sandy flats, and dunes.

Comments: Annual. Flowers have delicate fragrance, principally at night. Soft hairs on leaves restrict water loss through evaporation. Five species of *Abronia* in Arizona. Photograph taken at Painted Rocks State Park, March 29.

TRAILING FOUR O'CLOCK

Desert Windmills
Allionia incarnata var. *villosa*
Four O'Clock Family (Nyctaginaceae)

Height: Trailing to 10' long.

Flowers: Pinkish-purple; 3 irregular flowers close together appear as 1 regular flower in cluster of 3; to 1" wide; in leaf axils (flower cluster only ¼" wide in variety *incarnata;* not shown).

Leaves: Dirty green above, silvery beneath; sticky, hairy, oval; to 2" long.

Blooms: March–October.

Elevation: Below 6,000'.

Habitat: Roadsides, open sandy plains, and mesas.

Comments: Perennial herb. Characterized by sticky stems. Two species of *Allionia* in Arizona. Photograph taken in Patagonia area, April 27.

RED SPIDERLING
Boerhavia coccinea
Four O'Clock Family (Nyctaginaceae)

Height: Trailing to 3' long.

Flowers: Purplish red, short tube, with 5 spreading lobes; to ⅛" wide; in tight clusters to ¼" wide, on stems at leaf axils.

Leaves: Dark green, some pink along margins; lighter green beneath; sticky, wavy-margined; oval to elliptical; to 2" long.

Blooms: April–November.

Elevation: Below 5,500'.

Habitat: Deserts, roadsides, and fields.

Comments: Perennial herb. Pinkish, with sticky stems. Twelve species of *Boerhavia* in Arizona. Photograph taken near Granite Reef Dam, May 14.

COULTER SPIDERLING
Boerhavia coulteri
Four O'Clock Family (Nyctaginaceae)

Height: To 3'.

Flowers: Pink with darker pink toward center, or white; tubular with 5 flaring lobes, each notched at tips; long stamens; to ⅜" wide, in raceme to 1¼" long.

Leaves: Dark green above, lighter green beneath, very hairy on both surfaces; triangular; to 3½" long. Leaves variable.

Blooms: August–November.

Elevation: 500 to 5,000'.

Habitat: Roadsides and dry plains.

Comments: Pinkish stems. Twelve species of *Boerhavia* in Arizona. Photograph taken in vicinity of Pinnacle Peak, Scottsdale, August 25.

TUFTED FOUR O'CLOCK

Mirabilis albida (Oxybaphus comatus)
Four O'Clock Family (Nyctaginaceae)

Height: To 2'.

Flowers: Purplish-red to rose, flaring tube; long stamens extend way beyond tube; to ¾" wide; in terminal clusters.

Leaves: Grayish green, triangular, finely haired; to 2¾" long, 1¾" wide at widest point.

Blooms: May–October.

Elevation: 3,500 to 9,000'.

Habitat: Meadows and clearings in ponderosa pine forests.

Comments: The plant's many branches spring from the ground. Thirteen species of *Mirabilis* in Arizona. Photograph taken at Lynx Lake, September 11.

DESERT WISHBONE BUSH

Mirabilis laevis var. *villosa (Mirabilis bigelovii)*
Four O'Clock Family (Nyctaginaceae)

Height: To 2'.

Flowers: White to pale pink, funnel-shaped, broad, petallike calyx (no true petals); to ¾" wide.

Leaves: Dark green, hairy, oval to kidney-shaped; to 1¼" long.

Blooms: March–October.

Elevation: Below 3,000'.

Habitat: Lower desert flats, slopes, and canyons.

Comments: Perennial herb. Weak-stemmed, straggling; with sticky stems. Branches form "wishbones." Thirteen species of *Mirabilis* in Arizona. Photograph taken in Superstition Mountains, March 26.

RIBBON FOUR O'CLOCK

Mirabilis linearis (Oxybaphus linearis)
Four O'Clock Family (Nyctaginaceae)

Height: To 3'.

Flowers: Reddish purple, 5-lobed, trumpet-shaped, with long stamens; flower to ¾" wide, 1" long.

Leaves: Grayish green, linear, to 3" long.

Blooms: April–September.

Elevation: 4,500 to 9,500'.

Habitat: Oak woodlands and pine forests.

Comments: Has sticky stem. Thirteen species of *Mirabilis* in Arizona. Photograph taken in Santa Rita Mountains, April 28.

DESERT FOUR O'CLOCK

Mirabilis multiflora
Four O'Clock Family (Nyctaginaceae)

Height: To 2'.

Flowers: Magenta-purple, funnel-shaped, petal-like calyx (no true petals), to 1" wide; in groups of 3 to 6 bell-shaped cups in leaf axils, followed by a smooth, dark brown, ½"-long fruit.

Leaves: Dark green, smooth, oval to heart-shaped; to 4" long.

Blooms: April–September.

Elevation: 2,500 to 6,500'.

Habitat: Roadsides, open sandy areas, and mesas.

Comments: Perennial. Forms a rounded clump, appearing almost shrublike. Flowers open in late afternoon, wither the next morning. The root is used for various remedies. Thirteen species of *Mirabilis* in Arizona. Photograph taken near Nutrioso, August 3.

SPREADING FOUR O'CLOCK

Mirabilis oxybaphoides
Four O'Clock Family (Nyctaginaceae)

Height: To 2'.

Flowers: Purplish red, funnel-shaped, with notched lobes; 3 flowers within the hairy bracts, 3 or 4 yellow-tipped stamens; flowers to ⅜" long, 3-flower cluster to ¾" wide.

Leaves: Dark green, wavy margins with sharp hairs, long-stalked, elliptical to oval but variable in shape; to 3½" long; few, opposite.

Blooms: August–September.

Elevation: 6,000 to 8,000'.

Habitat: Clearings and roadsides.

Comments: Spreading or erect plant, with many branches and hairy stems and buds. Thirteen species of *Mirabilis* in Arizona. Photograph taken at Nelson Reservoir, August 18.

FIREWEED

Chamerion angustifolia ssp. *circumvagum (Epilobium angustifolium)*
Evening Primrose Family (Onagraceae)

Height: To 6'.

Flowers: Deep rose-purple, with 4 widely spreading petals; to ¾" long, 1" wide; in long, terminal raceme; followed by slender pod to 3" long; each seed bearing a tuft of hairs.

Leaves: Green, lance-shaped, narrow; veins curved into scallops along margins; to 6" long.

Blooms: July–September.

Elevation: 7,000 to 11,500'.

Habitat: Roadsides, burned areas, and logged areas.

Comments: Perennial herb. Young shoots are potherbs. So named because it grows after forest fires. One species of *Chamerion* in Arizona. Photograph taken at Greer, July 21.

PARCHED FIREWEED

Epilobium brachycarpum (Epilobium paniculatum fa. adenocladon)
Evening Primrose Family (Onagraceae)

Height: To 4'.

Flowers: Rose to purplish, with 4 deeply cleft petals; to ½" wide, ¼" long; in widely branching cluster on flower stalk to ½" long; followed by 4-sided seed pod to 1" long.

Leaves: Dark green, linear, narrow, alternate; to 2" long. Clusters of shorter leaves in axils.

Blooms: August–October.

Elevation: 5,000 to 8,500'.

Habitat: Roadsides and dry, open, disturbed areas.

Comments: Thirteen species of *Epilobium* in Arizona. Photograph taken at McNary, August 10.

GLANDULAR WILLOWHERB

Epilobium halleanum
Evening Primrose Family (Onagraceae)

Height: To 12".

Flowers: Pink, striped with darker pin; with 4 deeply notched petals; to ⅛" wide; in cluster at upper leaf axil; followed by elongated, narrow, hairy, upright seed capsule to 1¾" long.

Leaves: Dark green, clasping stem, pointing upward close to stem; lance-shaped, toothed, opposite; to 1½" long.

Blooms: July–August.

Elevation: 8,000 to 10,000'.

Habitat: Wet mountain meadows and lakesides.

Comments: Has reddish, hairy stems. Thirteen species of *Epilobium* in Arizona. Photograph taken at Carnero Lake near Greer, July 11.

SCARLET BEEBLOSSOM

Gaura coccinea
Evening Primrose Family (Onagraceae)

Height: To 20".

Flowers: Reddish pink to white (lighter-colored in evening, becoming darker by mid-morning); irregular, shapeless; to ½" wide; on nodding spike blooming upward from base; followed by club-shaped, grooved seed capsule to ½".

Leaves: Light green, narrowly lanceolate; to 2½" long.

Blooms: April–September.

Elevation: 2,000 to 8,000'.

Habitat: Roadsides, fields, plains, and pine and juniper woodlands.

Comments: Perennial herb. Light-colored, newly opened flowers attract pollinating moths at night. By morning, flowers are pink, deepening to reddish as day progresses. Three species of *Gaura* in Arizona. Photograph taken in Nutrioso area, July 23.

LIZARD-TAIL

Smallflower Gaura
Gaura mollis (Gaura parviflora)
Evening Primrose Family (Onagraceae)

Height: To 6'.

Flowers: Pink, tiny, 4-petaled; to ⅜" long, $\frac{3}{16}$" wide; in dense, nodding, hairy, terminal spike to 18" long.

Leaves: Dark green, lance-shaped, slightly toothed; to 8" long in basal rosette. Leaves on stem are lance-shaped, velvety-haired, alternate; to 4" long.

Blooms: April–October.

Elevation: 100 to 6,800'.

Habitat: Roadsides, fields, disturbed ground, and desert washes.

Comments: Has hairy stems; 1 main stem with several smaller side branches. Three species of *Gaura* in Arizona. Photograph taken at Saguaro National Monument West, April 17.

CALYPSO
Fairy Slipper
Calypso bulbosa
Orchid Family (Orchidaceae)

Height: Flower stalk to 8".

Flowers: Rose-pink, orchidlike, bearded with yellow hairs; to 2" long, 1½" wide; single on leafless stalk.

Leaves: Dark green, single, oval to egg-shaped; to 2½" long; at base of tall, pinkish flower stalk.

Blooms: May–July.

Elevation: 8,000 to 10,000'.

Habitat: Cool, moist spruce-fir forests.

Comments: Flower hangs at tip of tall pink stem that grows from perennial bulbous root. One species of *Calypso* in Arizona. Photograph taken in vicinity of Hannagan Meadow, May 28.

BURRO WEED STRANGLER
Desert Broomrape
Orobanche cooperi
Broomrape Family (Orobanchaceae)

Height: To 15".

Flowers: Purplish, tubular, with 2-lobed upper lip and 3-lobed lower lip; to 1" long; many on erect, conelike spike.

Leaves: Purplish brown, scalelike bracts.

Blooms: February–September.

Elevation: 200 to 5,000'.

Habitat: Sandy deserts and washes.

Comments: Lacks chlorophyll. Parasitic herb usually on roots of bursage and other composites. Used by Native Americans for food and medicine. Six species of *Orobanche* in Arizona. Photograph taken at Cattail Cove State Park, February 23.

WOOD SORREL
Oxalis decaphylla
Oxalis Family (Oxalidaceae)

Height: To 5".

Flowers: Pink, with yellow throat, 5 petals; funnel-shaped; to ½" long, on leafless stem.

Leaves: Dark green, wedge-shaped, each is V-notched at tip; in umbrellalike arrangement; to 1¾" wide.

Blooms: June–August.

Elevation: 5,000 to 9,500'.

Habitat: Clearings in pine and mixed coniferous forests.

Comments: Perennial herb. Eight species of *Oxalis* in Arizona. Photograph taken near Greer, June 22.

BROAD-LEAVED GILIA
Aliciella latifolia var. *latifolia (Gilia latifolia)*
Phlox Family (Polemoniaceae)

Height: To 20".

Flowers: Pink, tubular, with 5 spreading lobes at right angles to narrow, funnel-shaped tube; to ½" long; terminal on stems.

Leaves: Green, mostly basal, oblong, many-lobed, to 8" long.

Blooms: February–April.

Elevation: Below 2,000'.

Habitat: Desert washes and slopes.

Comments: Annual. Attractive to butterflies and hummingbirds. Seven species of *Aliciella* in Arizona. Photograph taken at Usery Mountain Recreation Area, February 18.

MANY-FLOWERED GILIA

Ipomopsis multiflora (Gilia multiflora)
Phlox Family (Polemoniaceae)

Height: Sprawling to 2'.

Flowers: Violet-blue, tubular, with 5 flaring lobes, and 5 long stamens tipped with bright blue; to ½" wide, ½" long; in small, loose clusters along upper stems.

Leaves: Dark green, downy, alternate; threadlike; pinnately divided into 3 to 5 lobes, or undivided; to 1½" long, occurring along stems.

Blooms: July–October.

Elevation: 4,000 to 9,000'.

Habitat: Roadsides and dry slopes.

Comments: Many-stemmed; attractive to bees. Native Americans used plant medicinally. Fourteen species of *Ipomopsis* in Arizona. Photograph taken near Heber, August 4.

SLENDER PHLOX

Microsteris gracilis (Phlox gracilis)
Phlox Family (Polemoniaceae)

Height: To 8".

Flowers: Rose to lavender, tubular, with 5 lobes, each notched in center; yellowish tube with long, slender, bright yellow-tipped stamens; flower to ⅛" wide, ½" long, in terminal cluster.

Leaves: Dark green above, lighter green beneath; very hairy, lance-shaped, opposite; clasping stem; to ¾" long.

Blooms: February–May.

Elevation: 3,000 to 7,000'.

Habitat: Moist areas.

Comments: Annual. Pinkish and sticky, with hairy stems. One species of *Microsteris* in Arizona. Photograph taken northeast of Superior, April 3.

SPREADING PHLOX

Phlox austromontana (Phlox diffusa ssp. *subcarinata)*
Phlox Family (Polemoniaceae)

Height: To 4".

Flowers: Pure white to various shades of pink; upright trumpet with 5 broad, rounded lobes; to ⅝" wide; in large, colorful mats.

Leaves: Greenish, sharp, needlelike, and stiff, to ⅝" long; occurring all along stem; hairy clusters at leaf bases.

Blooms: May–August.

Elevation: 6,000 to 9,000'.

Habitat: Plateaus and canyon rims.

Comments: Woody, perennial herb. Fourteen species of *Phlox* in Arizona. Photograph taken at North Rim of Grand Canyon National Park, June 25. This low-growing species is distinguished by its broad, rounded petals and hairy clusters at leaf bases.

WOODHOUSE'S PHLOX

Phlox speciosa ssp. *woodhousei (Phlox woodhousei)*
Phlox Family (Polemoniaceae)

Height: To 6".

Flowers: Pink petal lobes (occasionally all white) above; white beneath; white center or "eye"; deeply notched; wedge-shaped petals; tubular with 5 or 6 petal lobes at right angles; to ¾" wide.

Leaves: Dark green, shiny, stiff, and thick; oblong, opposite, rough; to 2" long. Downy upper foliage.

Blooms: Spring and autumn.

Elevation: 3,500 to 8,000'.

Habitat: Open woods and pine forests.

Comments: Has woody base. Fourteen species of *Phlox* in Arizona. Photograph taken at Upper Lake Mary, June 1. Notched petal lobes and white "eye" differentiate this species.

TOWERING JACOB'S LADDER

Polemonium foliosissimum
Phlox Family (Polemoniaceae)

Height: To 3'.

Flowers: White, blue, or purple, with yellow center; shaped like a shallow bowl; to 1" wide; in terminal cluster.

Leaves: Dark green, pinnate, with up to 25 elliptical to lance-shaped leaflets to 2½" long.

Blooms: July–August.

Elevation: 8,000 to 9,000'.

Habitat: Moist soil along streams in mountains.

Comments: Perennial herb with hairy stem. Four species of *Polemonium* in Arizona. Photograph taken south of Alpine, August 2.

FLAT-TOP BUCKWHEAT

Eriogonum fasciculatum var. *polifolium*
Buckwheat Family (Polygonaceae)

Height: Shrub to 2'.

Flowers: White to pink, slightly fragrant; to ⅛" wide, in slightly flat, terminal clusters.

Leaves: Grayish green, finely haired, leathery, and narrow; to ⅜" long.

Blooms: February–June.

Elevation: 1,000 to 4,500'.

Habitat: Rocky slopes, flats, and along washes.

Comments: Attractive to bees; a source of honey. Over fifty species of *Eriogonum* in Arizona. Photograph taken at Usery Mountain Recreation Area, February 14.

WATER SMARTWEED

Knotweed

Persicaria amphibia (Polygonum amphibium)

Buckwheat Family (Polygonaceae)

Height: Flower stalk to 6".

Flowers: Deep pink; to 1⁄16" wide, 3⁄16" long; petalless and in a slender, spikelike cluster, to 2" long.

Leaves: Dark green, shiny, and leathery; alternate, mainly floating, lance-shaped to oval; to 6" long.

Blooms: July–September.

Elevation: 5,000 to 9,000'.

Habitat: Marshes, ponds, and lakes.

Comments: Six species of *Persicaria* in Arizona. Photograph taken at Nelson Reservoir, August 3.

LADY'S THUMB

Smartweed

Persicaria maculosa (Polygonum persicaria)

Buckwheat Family (Polygonaceae)

Height: To 32".

Flowers: Pinkish, 1⁄16" long; in dense spikes; to 2" long, some erect, others arched from weight of flowers.

Leaves: Dark green, lance-shaped; to 6" long; stems swollen and streaked with reddish brown at joint where leaf base clasps stem.

Blooms: July–September.

Elevation: 5,000 to 7,000'.

Habitat: Along ditches and in moist pastures and marshy areas.

Comments: Of European origin; now naturalized. Six species of *Persicaria* in Arizona. Photograph taken at Nelson Reservoir, August 3.

CANAIGRE DOCK

Desert Rhubarb
Rumex hymenosepalus
Buckwheat Family (Polygonaceae)

Height: To 2½'.

Flowers: Pinkish green; to ¾" wide; in erect, crowded, terminal raceme to 1' long; followed by clusters of pinkish, heart-shaped, 3-sided, winged seed capsules.

Leaves: Dark green, broadly lance-shaped, and thick; to 1' long.

Blooms: Mid-February–April.

Elevation: 1,000 to 6,000'.

Habitat: Fields, sandy washes, and roadsides.

Comments: A perennial herb. Tubers are a source of tannin. Fifteen species of *Rumex* in Arizona. Photograph taken at Alamo Lake, February 26.

RED MAIDS

Rock Purslane
Calandrinia ciliata
Purslane Family (Portulacaceae)

Height: Prostrate or semi-prostrate stems to 16" long.

Flowers: Reddish-pink or whitish, usually with 5 petals with rounded tips; hairy sepals; to ½" wide; on short stalks in leaf axils.

Leaves: Green, narrow, and thick; hairy, succulent; to 3" long.

Blooms: February–April.

Elevation: 1,500 to 4,000'.

Habitat: Washes, foothills, and plains.

Comments: One species of *Calandrinia* in Arizona. Photograph taken at Usery Mountain Recreation Area, February 14.

SOUTHWESTERN LEWISIA

Lewisia brachycalyx
Purslane Family (Portulacaceae)

Height: To 2½".

Flowers: Very pale pink, 5 to 9 petals arranged singly on short stem; to 2" wide.

Leaves: Green, flat, smooth; broadly linear and succulent; to 4" long, in basal rosette.

Blooms: April–June.

Elevation: 5,000 to 8,000'.

Habitat: Oak-juniper woodlands and ponderosa pine forests.

Comments: A perennial herb with thick, starchy roots. Four species of *Lewisia* in Arizona. Photograph taken near Willow Springs Lake, May 9.

DWARF LEWISIA

Lewisia pygmaea
Purslane Family (Portulacaceae)

Height: To 3".

Flowers: Deep pink to white, 5 to 9 petals with faint stripes radiating outward; to ¾" wide.

Leaves: Dark green, fleshy, rounded, and linear; to 3" long; in basal tuft.

Blooms: June–August.

Elevation: 8,000 to 9,000'.

Habitat: Moist mountain meadows.

Comments: Perennial herb. Four species of *Lewisia* in Arizona. Photograph taken at Crescent Lake, June 17.

PYGMY FLAMEFLOWER

Phemeranthus brevifolius (Talinum brevifolium)
Purslane Family (Portulacaceae)

Height: To 4" tall.

Flowers: Lavender or rose, with 5 widely separated petals; about 20 yellow stamens; to ¾" wide.

Leaves: Grayish green, succulent, semi-rounded; to ½" long, crowded all along stem.

Blooms: May–September.

Elevation: 5,000 to 8,000'.

Habitat: Clearings in coniferous forests.

Comments: Perennial herb. Sedumlike. Reddish, spreading stems. Seven species of *Phemeranthus* in Arizona. Photograph taken in vicinity of Luna Lake, June 29.

WESTERN SHOOTING STAR

Dodecatheon pulchellum var. *pulchellum (Dodecantheon radicatum)*
Primrose Family (Primulaceae)

Height: To 2'.

Flowers: Purplish pink with yellow at base; 4- or 5-lobed, with lobes sharply swept back; dark stamens forming cone-shaped tip, resembling a miniature rocket; nodding; to ¾" long; in terminal cluster on leafless stem.

Leaves: Dull green, broadly lance-shaped, but variable in shape; to 10" long.

Blooms: June–August.

Elevation: 6,500 to 9,500'.

Habitat: Moist meadows.

Comments: Perennial herb. Two species of *Dodecatheon* in Arizona. Photograph taken near Greer, June 17.

PARRY'S PRIMROSE
Alpine Primrose
Primula parryi
Primrose Family (Primulaceae)

Height: Flower stalk to 18".

Flowers: Magenta-pink, with 5 spreading lobes joined at base into narrow tube; yellow markings at throat of tube; to ¾" wide; in loose, rounded umbel or cluster at top of dark stalk.

Leaves: Dark green, oblong, basal; fleshy, usually erect; to 12" long.

Blooms: June–August.

Elevation: 10,000 to 12,000'.

Habitat: Moist rock crevices, wet meadows, and along mountain streams.

Comments: Flowers smell like carrion. Three species of *Primula* in Arizona. Photograph taken in Mount Baldy Wilderness, August 14.

PRAIRIE SMOKE
Old Man's Whiskers
Geum triflorum
Rose Family (Rosaceae)

Height: Flower stalk to 20".

Flowers: Reddish sepals hide pinkish to yellowish petals; bell-shaped, nodding; on long stalks; to ½" long; followed by reddish, seed-carrying, silky plumes, to 2" long.

Leaves: Green, hairy, fernlike, pinnately divided, mostly basal; to 7" long.

Blooms: May–August.

Elevation: 6,000 to 9,500'.

Habitat: Ponderosa pine and mixed conifer forests.

Comments: Perennial herb. Good forage for sheep. After fertilization, flowers turn upward. Four species of *Geum* in Arizona. Photograph taken near Mormon Lake, June 2.

ARIZONA ROSE

Wild Rose

Rosa woodsii var. *ultramontana (Rosa arizonica)*

Rose Family (Rosaceae)

Height: To 3'.

Flowers: Pink, with 5 wavy petals, which are uneven in size; yellow stamens; fragrant; to 1¾" wide; followed by berrylike fruit called a hip, which turns red when mature.

Leaves: Dark green above, lighter beneath; pinnate, 3- to 9-toothed; to 2½" long; oval to elliptical leaflets, each to ¾" long.

Blooms: May–July.

Elevation: 4,000 to 9,000'.

Habitat: Along streams and small clearings in ponderosa pine forests.

Comments: The most abundant rose in Arizona. Thorns, to ¼" long, are hooked; stems are grayish, brownish on twigs. Browsed by wildlife. Fruits, or hips, used for making wines, jams, and jellies. Three species of *Rosa* in Arizona. Photograph taken at Woods Canyon Lake, July 7. This species is shorter than the related **Fendler Rose** (*Rosa woodsii* var. *woodsii*) (at right), and has smaller flowers and larger, stout, curved thorns.

FENDLER ROSE

Rosa woodsii var. *woodsii (Rosa fendleri)*

Rose Family (Rosaceae)

Height: To 7'.

Flowers: Fragrant, with 5 broad, pink petals and yellow stamens; to 2¼" wide; followed by berrylike fruit called a hip, which turns red when mature.

Leaves: Dark green, pinnate, 3- to 7-toothed; oval to elliptical leaflets to 1" long; leaf to 3" long.

Blooms: June–August.

Elevation: 5,500 to 9,000'.

Habitat: Roadsides, slopes, and clearings in ponderosa forests.

Comments: Thorns are short, red, slender, and nearly straight; stems are reddish. Browsed by wildlife. Fruits, or hips, used in vitamin supplements and for making wine, jams, and jellies; also eaten by Native Americans as well as by birds and small mammals. Three species of *Rosa* in Arizona. Photograph taken near Prescott, May 26. This species is taller than the **Arizona Rose** (Rosa woodsii var. *ultramontana*), has larger flowers, and slender, nearly straight thorns.

HIMALAYA-BERRY

Rubus discolor (Rubus procerus)
Rose Family (Rosaceae)

Height: Creeping or clambering stems to 6′ long.

Flowers: Pinkish white, with 5 petals, numerous yellowish stamens; to 1¼″ wide; followed by a blackberry, which when ripened is thimble-shaped.

Leaves: Dark green above, white-felty beneath; pinnate, 3 to 5 toothed leaflets; to 9″ long with stem.

Blooms: June.

Elevation: Not available. Photograph taken at 6,000′.

Habitat: Mainly Oak Creek Canyon and Grand Canyon National Park.

Comments: An introduced species thriving in limited areas. Stem is grooved, very prickly, and white-spined. Six species of *Rubus* in Arizona. Photograph taken at Oak Creek Canyon, June 18.

HOUSTONIA

Houstonia rubra (Hedyotis rubra)
Madder Family (Rubiaceae)

Height: To 3″.

Flowers: Bright pink, tubular, 4 lobes spreading at right angles; to ½″ wide, 1″ long.

Leaves: Grayish green, linear, thick, and succulent; to 1″ long.

Blooms: April–August.

Elevation: 4,000 to 6,000′.

Habitat: Washes, mesas, and rocky hillsides.

Comments: Perennial. Dense, baseball-shaped mound of leaves and flowers: Three species of *Houstonia* in Arizona. Photograph taken in wash at Dead Horse Ranch State Park, May 28.

WRIGHT'S BLUETS

Houstonia wrightii (Hedyotis pygmaea)
Madder Family (Rubiaceae)

Height: To 8".

Flowers: Purplish, pinkish, or white; funnel-shaped; 4 flaring lobes; to ¼" wide; closely clustered.

Leaves: Dark green, thick, narrow, numerous; to ½" long.

Blooms: May–September.

Elevation: 5,000 to 9,000'.

Habitat: Slopes, dry mesas, and edges of coniferous forests.

Comments: Three species of *Houstonia* in Arizona. Photograph taken in Hannagan Meadow area, June 24.

TURPENTINE BROOM

Thamnosma montana
Rue Family (Rutaceae)

Height: To 4'.

Flowers: Deep blue-purple, cylinder-shaped, with 4 upright petals; to ½" long; followed by yellowish green, double, pea-sized, saclike fruit with gland-dotted skin.

Leaves: When present, yellowish green, linear to spatula-shaped, sparse, short-lived.

Blooms: February–April.

Elevation: Below 4,500'.

Habitat: Desert slopes and mesas.

Comment: Yellowish green, shrubby plant, which conducts photosynthesis in its stems. Aromatic glands on fruits yield an oil skin irritant. Crushed stems give off strong, citruslike odor. Native Americans used plant medicinally. Two species of *Thamnosma* in Arizona. Photograph taken in Superstition Mountains, February 4.

NARROWLEAF COTTONWOOD
Mountain Cottonwood
Populus angustifolia
Willow Family (Salicaceae)

Height: To 50′.

Trunk: To 1½′ in diameter.

Bark: Yellowish green and smooth when young, brownish to grayish brown and deeply furrowed on large trunks.

Flowers: Narrow, reddish pink catkins to 3″ long, with male and female on separate trees.

Leaves: Shiny, yellowish green above, paler beneath; lance-shaped, finely saw-toothed; leathery; yellowish midvein; tapering and long-pointed; to 5″ long, 1″ wide.

Blooms: Early spring.

Elevation: 5,000 to 7,000′.

Habitat: Along mountain streams.

Comments: Pointed, very resinous leaf buds. Wood used for fence posts and fuel. Eight species (including three natural hybrids) of *Populus* in Arizona. Photograph taken at Eager, August 18.

CORAL BELLS
Alum Root
Heuchera sanguinea
Saxifrage Family (Saxifragaceae)

Height: To 2′.

Flowers: Pinkish to coral-red, bell-shaped; with 5 petals; to ½″ long; in loose clusters on leafless stalk.

Leaves: Dark green, roundish, with pointed lobes; basal; on long stalk; to 3″ wide.

Blooms: March–October.

Elevation: 4,000 to 8,500′.

Habitat: Shaded hillsides and moist, rocky areas in shade.

Comments: Perennial herb with woolly stems. Used as an ornamental. Six species of *Heuchera* in Arizona. Photograph taken in Santa Rita Mountains, June 18.

FOOTHILL KITTENTAILS
Besseya plantaginea
Figwort Family (Scrophulariaceae)

Height: To 20".

Flowers: Pinkish, purplish, or white; 4-lobed (1 forming upper lip and others, lower lip); conspicuous bracts: to ½" long; in dense, terminal spike to 16" long.

Leaves: Green, often with reddish veins and tints of red; mostly basal; finely haired, finely scalloped, and oblong, to 8" long.

Blooms: June–August.

Elevation: 7,000 to 9,500'.

Habitat: Moist meadows and mixed coniferous forests.

Comments: Perennial herb. Two species of *Besseya* in Arizona. Photograph taken south of Alpine, June 30. A similar species **Arizona Coraldrops** (*Besseya arizonica*) has small, rounded leaves and a shorter flower stem.

OWL CLOVER
Escobita
Castilleja exserta ssp. *exserta (Orthocarpus purpurascens)*
Figwort Family (Scrophulariaceae)

Height: To 16".

Flowers: Rose-purple upper and lower lips, lower lip with white or yellow tip; surrounded by 5- to 7-lobed, rose-purple bracts, each to 1" long; flowers to 1¼" long; in dense, erect spike.

Leaves: Greenish to purplish and hairy; threadlike, with threadlike segments; to 2" long.

Blooms: March–May.

Elevation: 1,500 to 4,500'.

Habitat: Open mesas, slopes, and desert.

Comments: Annual. *Escobita* means "little broom" in Spanish. Can be partly parasitic on roots or other desert wildflowers. Seventeen species of *Castilleja* in Arizona. Photograph taken in Superstition Mountains, March 19.

CLUB-FLOWER
Purple Birdbeak
Cordylanthus parviflorus
Figwort Family (Scrophulariaceae)

Height: To 3′, sprawling to 4′ wide.

Flowers: Pink with yellow at tip; tubular with white, tonguelike projection along upper surface at tip; hairy; to ¾″ long, 5⁄16″ wide; terminal on branches. Flower is actually turned upside down with lower lip longer and uppermost.

Leaves: Dark green, linear, alternate; hairy and sticky; to ⅝″ long.

Blooms: August–October.

Elevation: 2,500 to 7,000′.

Habitat: Rocky slopes and mesas.

Comments: Annual. Many-branched. Partially root-parasitic. All parts of plant are glandular and very sticky. Four species of *Cordylanthus* in Arizona. Photograph taken at Oak Creek Canyon, October 1.

BIGELOW'S MONKEYFLOWER
Mimulus bigelovii var. *bigelovii*
Figwort Family (Scrophulariaceae)

Height: To 10″.

Flowers: Pink with darker pink in center; tubular; 5 crinkled lobes with white hairs; bright yellow stamens fused to lower petal; lower petal projecting slightly forward; hairy calyx, flower to 1″ long, 1″ wide.

Leaves: Dark green tinged with pink; hairy, broadly elliptical; to 2″ long.

Blooms: February–April.

Elevation: 500 to 2,500′.

Habitat: Sandy desert washes and open, sandy plains.

Comments: Hairy stems. Fourteen species of *Mimulus* in Arizona. Photograph taken at Cattail Cove State Park, March 8.

BLUE TOADFLAX

Texas Toadflax
Nuttallanthus texanus (Linaria texana)
Figwort Family (Scrophulariaceae)

Height: To 32".

Flowers: Blue-violet, with 2-lipped, long, slender spur projecting backward; to ⅜" long; loosely grouped on slender stems.

Leaves: Dark green, shiny, linear; to 1¼" long.

Blooms: February–May.

Elevation: 1,500 to 5,000'.

Habitat: Roadsides, plains, and mesas.

Comments: Annual. Two species of *Nuttallanthus* in Arizona. Photograph taken near Sells, March 30.

TWOTONE OWL'S CLOVER

Purple-White Owl's Clover
Orthocarpus purpureo-albus
Figwort Family (Scrophulariaceae)

Height: To 12".

Flowers: White and pinkish purple, with lower lip greatly inflated; small, hooked beak above; to ¾" long; spaced loosely along stem.

Leaves: Dark green, threadlike, 3-lobed; to 1¼" long, growing all along stem.

Blooms: July–September.

Elevation: 5,500 to 9,000'.

Habitat: Pinyon-juniper woodlands and ponderosa forests.

Comments: Annual. Six species of *Orthocarpus* in Arizona. Photograph taken near Ashurst Lake, September 5.

WOOD BETONY
Juniper Lousewort
Pedicularis centranthera
Figwort Family (Scrophulariaceae)

Height: To 4½".

Flowers: White; 2-lipped with purplish tips on upper and lower lips; 4 stamens curled under upper lip, with tips of anthers projecting like tiny upper teeth; flower to 1¾" long, ¼" wide; in short, dense, terminal cluster of very hairy bracts.

Leaves: Grayish green with reddish midrib; pinnate, finely dissected into toothed lobes, fernlike; to 4" long; mainly in basal rosette.

Blooms: April–June.

Elevation: 5,000 to 8,000'.

Habitat: Ponderosa pine forests.

Comments: Perennial herb. *Pediculus* means "louse" in Latin; in Roman times seeds were used to kill lice. Eight species of *Pedicularis* in Arizona. Photograph taken in vicinity of Willow Springs Lake, April 22.

PHOTO LICENSED BY SHUTTERSTOCK

ELEPHANTHEAD LOUSEWORT
Pedicularis groenlandica
Figwort Family (Scrophulariaceae)

Height: To 3'.

Flowers: Pink or reddish purple, shaped like a miniature elephant's head; to ½" long without "trunk"; twisted, long upper lip forms "trunk," shorter lower side lobes form "ears"; flowers arranged along upper half of erect stem.

Leaves: Green, long, narrow, alternate; pinnately divided into sharply toothed segments; to 10" long.

Blooms: August.

Elevation: 8,000 to 10,000'.

Habitat: Wet meadows and cold streams.

Comments: Perennial herb. *Pediculus* means "louse" in Latin; in Roman times seeds were used to kill lice. Eight species of *Pedicularis* in Arizona.

BUSH PENSTEMON

Gilia Beardtongue
Penstemon ambiguus
Figwort Family (Scrophulariaceae)

Height: To 3′.

Flowers: Pale pink, tubular, with 5 united petals; tube is curved with 2 large, upper petal lobes bent backward and 3 lower lobes projecting forward; to ¾″ wide, ½″ long.

Leaves: Green, linear, grasslike, opposite; to 1″ long; occurring up along stem.

Blooms: June–July.

Elevation: 4,500 to 6,500′.

Habitat: Sandy mesas and grasslands.

Comments: Branches freely forming a large, rounded, colorful bush when in bloom. More than three dozen species of *Penstemon* in Arizona. Photograph taken near Page, June 26. Note the oblique angle of the lobes to the tubular section of the flower, thus distinguishing this species from all other penstemons.

SUNSET CRATER BEARDTONGUE

Penstemon clutei
Figwort Family (Scrophulariaceae)

Height: To 20″.

Flowers: Deep pink to rose-purple, very hairy, tubular; tube widens gradually; 2 small, earlike upper lobes; 3 larger, rounded lower lobes; to 1″ long, ¾″ wide; in elongated cluster.

Leaves: Grayish green, lance-shaped, opposite; crinkled margins, toothed; to 2″ long; clasping stem.

Blooms: June–July.

Elevation: About 7,000′.

Habitat: Volcanic cinders.

Comments: In Arizona found only at Sunset Crater National Monument. More than three dozen species of *Penstemon* in Arizona. Photograph taken at Sunset Crater National Monument, June 5.

NARROWLEAF PENSTEMON

Toadflax Penstemon
Penstemon linarioides
Figwort Family (Scrophulariaceae)

Height: To 18".

Flowers: Lavender to bluish purple, tubular; 2 upper lobes flaring upward, 3 lower lobes flaring downward; whitish throat with yellow hairs; to ¾" long, ½" wide; all facing in one direction on long, narrow flower stalk.

Leaves: Grayish green, linear, very narrow, pointing upward; to ¾" long; occurring all along stem.

Blooms: June–August.

Elevation: 4,500 to 9,000'.

Habitat: Dry slopes and clearings in woodlands and pine forests.

Comments: Shrubby growth; has branching root system. More than three dozen species of *Penstemon* in Arizona. Photograph taken at Oak Creek Canyon, June 18. The subspecies *sileri* forms a small mound to 6" tall, and is woody at base.

PALMER'S PENSTEMON

Penstemon palmeri
Figwort Family (Scrophulariaceae)

Height: Flower stem to 5'.

Flowers: Pale pink, fragrant, tubular; 2 upper lobes flared backward, 3 lower lobes flared downward; flower tube swollen; to 1⅛" wide, 1½" long; in long, narrow cluster often bending under the weight of buds and flowers.

Leaves: Grayish green, with waxy, blue coating; leathery, toothed, wavy; upper leaves paired together surrounding stem; to 5" long on each side, to 10" long for pair.

Blooms: March–September.

Elevation: 3,500 to 6,500'.

Habitat: Roadsides, washes, and mountain slopes.

Comments: A spectacular *Penstemon*. More than three dozen species of *Penstemon* in Arizona. Photograph taken on Mingus Mountain, May 28.

PARRY'S PENSTEMON
Beardtongue
Penstemon parryi
Figwort Family (Scrophulariaceae)

Height: To 4'.

Flowers: Pinkish to lavender, broadly funnel-shaped; petal lobes short and round, lower petal lobes project forward; to ¾" long; in long, open, terminal cluster.

Leaves: Bluish green, fleshy, without stalks; smooth, narrowly triangular; to 5" long.

Blooms: March.

Elevation: 1,500 to 5,000'.

Habitat: Well-drained slopes, mountain canyons, and roadsides.

Comments: Perennial herb. Well-scattered, does not grow in clumps. Flowers attract hummingbirds, bees, and other insects. More than three dozen species of *Penstemon* in Arizona. Photograph taken at Horseshoe Lake Dam, March 21.

DESERT PENSTEMON
Penstemon pseudospectabilis
Figwort Family (Scrophulariaceae)

Height: To 4'.

Flowers: Bright lavender-red; wide throat with lower, one-sided bulge; to ¾" long, in clusters along stem.

Leaves: Gray, triangular, leathery, and toothed; bases joined together around stem; to 3" long.

Blooms: February–May.

Elevation: 2,000 to 7,000'.

Habitat: Roadsides, hillsides, and canyons.

Comments: More than three dozen species of *Penstemon* in Arizona. Photograph taken in Superstition Mountains, March 15.

ROCKY MOUNTAIN PENSTEMON

Penstemon strictus
Figwort Family (Scrophulariaceae)

Height: To 32".

Flowers: Purple to violet-blue, tubular; 2 upper petals project forward like roof of a porch, lower petals slope downward; stamen heads with white, twisted hairs; flower to 1" long, spaced along stem in loose raceme.

Leaves: Dark green, narrow to lance-shaped, to 3" long.

Blooms: June–July.

Elevation: 7,000 to 8,000'.

Habitat: Roadsides and dry, gravelly slopes.

Comments: A variable species. More than three dozen species of *Penstemon* in Arizona. Photograph taken near Nutrioso, June 23.

WANDBLOOM PENSTEMON

Penstemon virgatus
Figwort Family (Scrophulariaceae)

Height: To 30".

Flowers: Pale pinkish violet, light purplish lines within; funnel-shaped, 2-lipped, 2 lobes above, 3 lobes below; to ¾" wide, 1" long; along only one side of stem.

Leaves: Dark green, narrow, linear; to 4" long.

Blooms: June–September.

Elevation: 5,000 to 11,000'.

Habitat: Pine forests and mountain meadows.

Comments: Very variable species. More than three dozen species of *Penstemon* in Arizona. Photograph taken near Willow Springs Lake, August 5.

WHIPPLE'S PENSTEMON

Penstemon whippleanus
Figwort Family (Scrophulariaceae)

Height: To 2'.

Flowers: Whitish to lavender to deep purple; sticky-haired; yellow-orange in throat; strongly 2-lipped, with 2 lobes of upper lip flaring forward and upward, 3 lower lobes spread apart and projecting forward more than upper lobes; to 1" long; in several downward-facing clusters on stem.

Leaves: Dark green; basal leaves elliptical, to 4" long; stem leaves opposite and lance-shaped, to 1" long.

Blooms: July–August.

Elevation: 6,500 to 12,000'.

Habitat: Moist meadows and rocky slopes.

Comments: Plants in Arizona are in the purple color range. More than three dozen species of *Penstemon* in Arizona. Photograph taken in mountains above Greer, July 3. This species of *Penstemon* identifiable by its longer lower lip and by downward-facing arrangement of flower clusters on stem.

NUTTALL'S SNAPDRAGON

Sairocarpus nuttallianus (Antirrhinum nuttallianum)
Figwort Family (Scrophulariaceae)

Height: To 3'.

Flowers: Violet with a few cream-colored markings; snapdragon-like; 2 upper lobes (erect, sharply pointed, earlike, darker violet), 3 lower lobes (flaring outward, ruffled, lighter violet); flower stalks as long or longer than calyx; flower to ⅜" wide, ⅜" long; in long spike with side branches.

Leaves: Dark green, hairy, and sticky; lance-shaped, opposite and alternate; to 1" long.

Blooms: March–May.

Elevation: Below 4,000'.

Habitat: Canyons.

Comments: Very hairy, sticky plant. Two species of *Sairocarpus* in Arizona. Photograph taken at Saguaro National Park West, April 17.

AMERICAN BROOKLIME

American Speedwell
Veronica americana
Figwort Family (Scrophulariaceae)

Height: To 2′.

Flowers: Blue streaked with lavender; 4-lobed with lower lobe narrowest; to ½″ wide; in loose clusters arising from leaf axils.

Leaves: Dark green, shiny, broadly lance-shaped to oval; toothed margins variable; short-stalked; to 3″ long.

Blooms: May–August.

Elevation: 1,500 to 9,500′.

Habitat: In and around springs and along streamsides.

Comments: Semi-aquatic succulent with creeping, reddish stems. Spreads easily, forming dense areas. Named for Saint Veronica. Eighteen species of *Veronica* in Arizona. Photograph taken in Hannagan Meadow area, June 30.

THYMELEAF SPEEDWELL

Veronica serpyllifolia
Figwort Family (Scrophulariaceae)

Height: To 6″.

Flowers: Pale blue or violet; 4-lobed with lower lobe narrowest; to ¼″ wide, arising from leaf axils.

Leaves: Dark green, oval to oblong, hairy, and toothed; to 1″ long.

Blooms: June–August.

Elevation: 8,000 to 10,000′.

Habitat: Moist areas in coniferous forests.

Comments: Creeping, hairy stems. Introduced from Eurasia; now naturalized in U.S. Named for Saint Veronica. Eighteen species of *Veronica* in Arizona. Photograph taken at Lee Valley Reservoir area, June 30.

ANDERSON'S THORNBUSH

Tomatillo
Lycium andersonii
Nightshade Family (Solanaceae)

Height: To 9', but usually less.

Flowers: Pale lavender or whitish, narrow, and tubular; usually 5-lobed (sometimes 4), with hairy calyx, stigma doesn't protrude beyond lobes; stamens level with or protruding beyond lobes; flower to ½" long; followed by reddish orange, fleshy, juicy, egg-shaped, many-seeded fruit to ⅜" long.

Leaves: Dark green, very finely haired, succulent, and leathery; rather thick, spatula-shaped; to 1¾" long.

Blooms: February–March.

Elevation: Below 5,500'.

Habitat: Flats and along desert washes.

Comments: Has spine-tipped branches. Older branches are grayish; newer growth is brownish. Birds and small mammals feed on fruit. Flowers attract bees and other insects. Eleven species of *Lycium* in Arizona. There are several varieties of this species, differing in flower length and in leaf thickness and size. Photograph taken in Superstition Mountains, February 4.

BERLANDIER'S WOLFBERRY

Lycium berlandieri
Nightshade Family (Solanaceae)

Height: To 7'.

Flowers: Bluish to lavender, funnel-shaped (flaring at top); 4- or 5-lobed, with protruding stamens; flower to ⅜" long; solitary or in clusters; followed by roundish berry, to 5⁄16" in diameter, red at maturity.

Leaves: Green, minutely hairy, linear to spatula-shaped; to 1¼" long.

Blooms: March–November.

Elevation: Below 3,000'.

Habitat: Plans and rocky hillsides.

Comments: Shrub with spines on branches and branch tips. Favorite of butterflies and bees. Eleven species of *Lycium* in Arizona. Photograph taken in Tucson area, November 12.

FREMONT'S THORNBUSH

Wolfberry
Lycium fremontii
Nightshade Family (Solanaceae)

Height: To 9'.

Flowers: Purplish to lavender, tubular, erect, 5-lobed; to ⅜" long; followed by orangish red, fleshy, juicy fruit to ⅜" long.

Leaves: Light green, spatula-shaped, succulent, glandular-hairy; to 1" long, ¼" wide; occurring all along branches.

Blooms: Throughout year, but primarily January–March.

Elevation: Below 2,500'.

Habitat: Desert.

Comments: Has sharp-pointed branches. Fruits are eaten by desert Native Americans. Eleven species of *Lycium* in Arizona. Photograph in flower and in fruit taken at Organ Pipe Cactus National Monument, March 7.

PURPLE GROUND CHERRY

Husk Tomato
Physalis lobata (Quincula lobata)
Nightshade Family (Solanaceae)

Height: Usually prostrate, to 16" high.

Flowers: Purple, saucer-shaped; to 1" wide; on slender flower stalks in leaf axils; bright yellow, knobby anthers; followed by ¼" berry.

Leaves: Dark green, lance-shaped, pinnately lobed; to 4" long.

Blooms: March–September.

Elevation: 1,000 to 5,000'.

Habitat: Desert roadsides, flats, and mesas.

Comments: Fourteen species of *Physalis* in Arizona. Photograph taken near Pisinimo, March 30.

SILVERLEAF NIGHTSHADE

Horse Nettle
Solanum elaeagnifolium
Nightshade Family (Solanaceae)

Height: To 3′.

Flowers: Violet to bluish violet; starlike with 5 points; bright yellow anthers in center; to 1½″ wide; followed by yellow, ½″-diameter berry.

Leaves: Silvery, oblong to lance-shaped, wavy-margined, spines on underside; to 4″ long.

Blooms: May–October.

Elevation: 1,000 to 5,500′.

Habitat: Roadsides and fields.

Comments: Perennial herb, and a poisonous weed. Spines on stem. Plant produces a protein-digesting enzyme; in cheese-making, Native Americans added crushed berries to curdle milk. Fifteen species of *Solanum* in Arizona. Photograph taken at Hassayampa River Preserve, Wickenburg, May 7.

PURPLE NIGHTSHADE

Solanum xanti
Nightshade Family (Solanaceae)

Height: To 3′.

Flowers: Deep violet to dark lavender, starlike, 5 united petals, crinkled; to 1″ wide; 5 bright yellow stamens attached in center; in small cluster, followed by round, green berry to ½″ in diameter.

Leaves: Dark green, very hairy, alternate; oval to elliptical, occasionally lobed; to 3″ long.

Blooms: April–November.

Elevation: 3,500 to 5,500′.

Habitat: Rocky slopes, mostly in chaparral.

Comments: Perennial herb. A bushy plant, woody at base. Fifteen species of *Solanum* in Arizona. Photograph taken in vicinity of Sedona, June 9.

SALT CEDAR

Five-Stamen Tamarisk
Tamarix chinensis (Tamarix pentandra)
Tamarix Family (Tamaricaceae)

Height: Shrub, or small tree to 15'.

Flowers: Deep pink to nearly white; 1⁄16" long; crowded in narrow, 1- to 2"-long racemes grouped together in drooping, terminal clusters.

Leaves: Bluish green, scalelike; narrow, pointed, and wiry; to 1⁄16" long.

Blooms: March–August.

Elevation: Below 5,000'.

Habitat: Along streams, irrigation ditches, and other moist areas.

Comments: Deciduous. Forms extensive thickets along rivers and lakes. Introduced from Eurasia; now naturalized in U.S. Robs native plants of water. Foliage salty to taste. Visited by bees. Two species of *Tamarix* in Arizona. Photograph taken at Roper Lake State Park, April 20.

ARIZONA VALERIAN

Tobacco-Root
Valeriana arizonica
Valerian Family (Valerianaceae)

Height: To 14".

Flowers: Pinkish white to lavender, tubular, 5-lobed; white stamens and anthers extend beyond lobes; to ¼" wide, 1" long (including stamens); in rounded cluster to 2" wide.

Leaves: Dark green, thin, basal, broadly elliptical; to 2" long. Stem leaves are clasping, arrow-shaped, divided into a pair or pairs of narrow lobes, with end lobe largest and pointed. Leaf to 2½" long; leaf and stalk to 6½" long.

Blooms: April–July.

Elevation: 4,500 to 8,000'.

Habitat: Moist coniferous forests.

Comments: Perennial. Has smooth, hollow stem. When dried, plants give off unpleasant odor. Some species used as food by Native Americans. Twelve species of *Valeriana* in Arizona. Photograph taken at Sharp Creek, northeast of Christopher Creek, April 22.

DAKOTA VERBENA

Vervain
Glandularia bipinnatifida (Verbena bipinnatifida, V. ciliata)
Vervain Family (Verbenaceae)

Height: To 18".

Flowers: Pink to purple, fragrant, and tubular, with 5 abruptly flaring lobes; each with a notch; to ½" wide; in somewhat flat, terminal cluster.

Leaves: Dark green, long-haired, with edges curled under; sunken veins; much-divided or pinnately cleft into linear lobes; to 1½" long.

Blooms: April–September.

Elevation: 2,000 to 10,000'.

Habitat: Roadsides and clearings from grasslands to coniferous forests.

Comments: Forms a mounded bush. Stems are very hairy. Frequented by bees and butterflies. Seven species of *Glandularia* in Arizona. Photograph taken north of Springerville, August 5.

GOODDING'S VERBENA

Southwestern Vervain
Glandularia gooddingii (Verbena gooddingii)
Vervain Family (Verbenaceae)

Height: To 12".

Flowers: Pink to lavender, 5-notched, with joined petals; to ½" wide, in headlike cluster to 1¼" wide.

Leaves: Dark green, very hairy above and beneath; to 1½" long; cleft into 3 main lobes, which are many toothed or cleft.

Blooms: February–October.

Elevation: Below 5,000'.

Habitat: Dry slopes, mesas, and roadsides.

Comments: Perennial. Square stems. A favorite of butterflies and moths. *Glandularia* species are very difficult to identify. Seven species in Arizona. Photograph taken at Desert Botanical Garden, Phoenix, March 22.

PROSTRATE VERVAIN
Verbena bracteata
Vervain Family (Verbenaceae)

Height: To 5", stems spread out on ground.

Flowers: Pale lavender to purple, small; to ⅛" wide, on a broad spike.

Leaves: Dark green, hairy, 3-lobed, jagged-toothed; to 3" long.

Blooms: May–September.

Elevation: 1,000 to 7,500'.

Habitat: Roadsides, disturbed ground, and dry river bottoms.

Comments: Plant is quite hairy. Seventeen species of *Verbena* in Arizona. Photograph taken near Woods Canyon Lake, August 5.

NEW MEXICAN VERVAIN
Spike Verbena
Verbena macdougalii
Vervain Family (Verbenaceae)

Height: To 3'.

Flowers: Lavender to purple; 5-lobed, with 3 lobes bent downward, 2 bent upward; to ¼" wide; on long, erect spike. Flowers open first at bottom of stalk and progress upward.

Leaves: Dark green, lance-shaped, prominently veined, irregularly toothed; to 4" long.

Blooms: June–September.

Elevation: 6,000 to 7,500'.

Habitat: Open flats, valleys, and roadsides.

Comments: Square-stemmed like a mint. Native Americans used plant for medicinal and ceremonial purposes. Seventeen species of *Verbena* in Arizona. Photograph taken near Nutrioso, July 23.

HILLSIDE VERVAIN

Verbena neomexicana
Vervain Family (Verbenaceae)

Height: To 30".

Flowers: Lavender, 2-lipped, 5-lobed; to ½" long, ⅜" wide; in long, slender, interrupted spike with flowers opening from base upward.

Leaves: Grayish green, hairy, opposite; pinnately lobed to sharply toothed; to 3½" long.

Blooms: March–October.

Elevation: 2,000 to 6,000'.

Habitat: Canyons and foothills.

Comments: Natural hybrids are common in this genus. Seventeen species of *Verbena* in Arizona. Photograph taken at Saguaro National Park West, April 17.

WESTERN DOG VIOLET

Hookspur Violet
Viola adunca
Violet Family (Violaceae)

Height: Flower stem to 4".

Flowers: Bluish violet, with 5 petals; 2 pointed upper petals; 2 side petals cupped forward with bases bearing tufts of hairs; lower petal whitish with purple, vertical veins at base; prominent backward spur extending beyond upper petals. Flower to 1" long, ½" wide.

Leaves: Dark green, spreading outward from stems, thick, round to oval; finely scalloped on margins; saw-toothed margins on lance-shaped stipules at bases of leaf stalks; to 1" wide, 1" long; leaf with stem to 2½" long.

Blooms: June–July.

Elevation: 7,000 to 9,800'.

Habitat: Moist spruce-fir forests and borders of mountain meadows in shaded areas.

Comments: A variable species. Has hairy stem. Ten species of *Viola* in Arizona. Photograph taken in mountains above Greer, July 11. Note small size, presence of main stem with side stems, and long spur.

CANADA VIOLET

Viola canadensis
Violet Family (Violaceae)

Height: To 1′, occasionally taller.

Flowers: White, growing from axils of upper leaves; bases of petals are yellow with purplish veins; fading to pinkish; broadly triangular lip petal; petals tinged with purple on back; flower to ¾″ wide.

Leaves: Dark green, broadly heart-shaped, with toothed margins; to 3″ long.

Blooms: April–September.

Elevation: 6,000 to 11,500′.

Habitat: Rich, moist soil in coniferous forests.

Comments: Perennial. Ten species of *Viola* in Arizona. Photograph taken at Greer, June 18.

MEADOW VIOLET

Northern Bog Violet
Viola nephrophylla (Viola sonoria ssp. *affinis)*
Violet Family (Violaceae)

Height: To 6″.

Flowers: Deep bluish violet, with 5 petals, 3 lower petals with darker purple veins on white areas in throat and bearing tufts of hair at base; short-spurred flower to ¾″ long, ¾″ wide.

Leaves: Dark green, kidney-shaped to heart-shaped; broad, with rounded teeth; to 3½″ long (including stem), 3″ wide.

Blooms: April–July.

Elevation: 5,000 to 9,500′.

Habitat: Moist meadows and moist mountain slopes in coniferous forests.

Comments: Ten species of *Viola* in Arizona. Photograph taken in mountains above Greer, July 8. Recognizable by absence of main stem; all leaves arise from base of plant.

FAGONIA

Fagonia laevis
Caltrop Family (Zygophyllaceae)

Height: To 18".

Flowers: Pinkish to lavender, 5 petals; to ½" wide; in leaf axils and branch ends.

Leaves: Dark green, narrow, each divided into 3 lance-shaped leaflets, each to ½" long.

Blooms: January–April.

Elevation: Below 2,500'.

Habitat: Dry, rocky slopes and mesas.

Comments: Grows in low rounded mounds. Stems are angular and sticky. Flowers open early in day and close by early afternoon. Two species of *Fagonia* in Arizona. Photograph taken at Cattail Cove State Park, February 24.

ANGIOSPERMS: BLUE FLOWERS

MANY-FLOWERED STICKSEED

Hackelia floribunda
Forget-me-not Family (Boraginaceae)

Height: To 3'.

Flowers: Pale blue, tubular; yellow center of tiny teeth surrounding tube opening; 5-lobed; to ¼" wide; in cluster on curving flower stalk and on stems at leaf axils; followed by nutlet with barbed prickles on margins.

Leaves: Grayish green, velvet-haired, with sunken midvein; alternate, narrowly lance-shaped, gradually becoming smaller up the stem; to 5" long.

Blooms: July–August.

Elevation: Above 7,000'.

Habitat: Moist meadows, stream banks, and clearings in coniferous forests.

Comments: Biennial, with hairy stem. Hooks on nutlets adhere to fabric and fur. Four species of *Hackelia* in Arizona. Photograph taken near Green's Peak, Greer area, August 9. Barbed prickles on margins of nutlets help identify this species. **Livermore Stickseed** (*Hackelia pinetorum*) has wider leaves and prickles all over nutlets.

FRANCISCAN BLUEBELLS

Lungwort
Mertensia franciscana
Forget-me-not Family (Boraginaceae)

Height: To 3'.

Flowers: Dark to pale blue (can be pinkish or white); tubular, with hairy sepal margins; pendent, in loose clusters; to ⅝" long.

Leaves: Dark green, lance-shaped, narrowing to a point; with short, flattened hairs on upper surface; to 6" long.

Blooms: June–September.

Elevation: Above 7,000'.

Habitat: Moist, shaded areas in ponderosa and spruce-fir forests.

Comments: Five species of *Mertensia* in Arizona. Photograph taken south of Alpine, August 2. The very similar species **Macdougall's Bluebells** *(Mertensia macdougalii)* has hairless sepal margins and hairless leaves.

WESTERN DAYFLOWER

Commelina dianthifolia
Spiderwort Family (Commelinaceae)

Height: To 15".

Flowers: Blue, to 1" wide; 3 petals, lower petal a bit smaller; yellow stamens; hairless; boat-shaped bract beneath flower cluster. Flower opens early in the day, wilts by midday.

Leaves: Green, narrow, to 6" long, ¼" wide.

Blooms: August–September.

Elevation: 3,500 to 9,500'.

Habitat: Pine woods and mixed conifer forests.

Comments: Two species of *Commelina* in Arizona. Photograph taken in Woods Canyon Lake area, August 3. A similar species, **Whitemouth Dayflower** (*Commelina erecta*), has 2 blue petals instead of 3; the third or lower petal is very small and white.

ARIZONA BLUE EYES

Evolvulus arizonicus
Morning Glory Family (Convolvulaceae)

Height: To 1'.

Flowers: Sky blue, like those of a flattened morning glory; to ¾" wide; on slender stalk; in upper leaf axil.

Leaves: Grayish green, lance-shaped, to 1" long.

Blooms: April–October.

Elevation: 3,500 to 5,000'.

Habitat: Deserts, grasslands, and pinyon-juniper woodlands.

Comments: Perennial herb. Four species of *Evolvulus* in Arizona. Photograph taken at Catalina State Park, April 15.

WOOLLY MORNING GLORY

Ipomoea purpurea (Ipomoea hirsutula)
Morning Glory Family (Convolvulaceae)

Height: Twining vine.

Flowers: Purplish blue, tubular, with 5 pointed, hairy sepals; to 1¼" long, 1" wide.

Leaves: Dark green, hairy above and beneath; lacking lobes or with 3 pointed lobes; to 3" wide, to 3" long; on stems up to 5" long.

Blooms: July–November.

Elevation: 1,000 to 5,500'.

Habitat: Roadsides and fields.

Comments: Often becoming a weed. Has hairy stems. Eighteen species of Ipomoea in Arizona. Photograph taken at Clear Creek near Camp Verde, September 30.

DESERT CANTERBURY BELLS

Desert Bluebell
Phacelia campanularia
Waterleaf Family (Hydrophyllaceae)

Height: To 2'.

Flowers: Dark blue to purplish blue, bell-shaped, tubular; 5 rounded lobes, each with white marking; purple filaments tipped with yellow-white anthers; hairy sepals; to 1" wide, 1½" long; in loose flower cluster.

Leaves: Dark green above, edged in deep red; lighter green beneath; scalloped, ruffled, velvety-haired on both surfaces, oval to heart-shaped; to 3½" long (including stem), 1½" wide.

Blooms: February–April.

Elevation: Not available. Photograph taken at 2,700'.

Habitat: Washes and sandy areas.

Comments: Annual. Hairy, glandular plant with reddish stems. Branches freely. Forty-six species of *Phacelia* in Arizona. It may be native to far western Arizona, but it is apparently becoming naturalized throughout the low deserts of the state. Photograph taken in a remote wash of the Santa Catalina Mountains, April 14. We have also seen the plant growing in the Santa Rita Mountains, May 11.

CHIA

Salvia columbariae var. *columbariae*
Mint Family (Lamiaceae)

Height: To 20".

Flowers: Deep blue, prominent upper and lower lips; to ½" long; in dense, rounded clusters on terminal spike.

Leaves: Green, mostly basal, oblong, much-divided; to 4" long.

Blooms: March–May.

Elevation: Below 3,500'.

Habitat: Sandy washes and desert slopes.

Comments: Annual; smells skunky. Like most members of the mint family, chia has square stems. When placed in water, seeds form sticky, mucilaginous mass, believed to aid in germination. Native Americans used seeds for food and to make mucilaginous poultices and certain beverages. Fifteen species of *Salvia* in Arizona. Photograph taken in Superstition Mountains, March 26.

WESTERN BLUE FLAX

Linum lewisii
Flax Family (Linaceae)

Height: To 3'.

Flowers: Sky blue, often nearly white, with 5 petals; to 2" wide, in loose, terminal cluster.

Leaves: Grayish green, long, narrow, sharp-pointed; to 1" long; crowded along stem.

Blooms: April–September.

Elevation: 2,000 to 9,500'.

Habitat: Roadsides, clearings in ponderosa pine forests, and open mesas.

Comments: Perennial herb (an uncommon winter/spring ephemeral at lower elevations). Petals usually drop by noon. Fiber is used by Native Americans for cord, fishing nets, mats, and baskets. Cultivated flax is used to make linen; seeds are crushed to produce linseed oil. Nine species of *Linum* in Arizona. Photograph taken on Mount Lemmon, April 30.

MINIATURE WOOL STAR

Starflower
Eriastrum diffusum
Phlox Family (Polemoniaceae)

Height: To 4½".

Flowers: Pale blue to whitish, tubular (long tube for tiny flower); to ½" long; in terminal clusters on bristle-tipped, woolly heads.

Leaves: Grayish green, very narrow, threadlike; to ¾" long.

Blooms: March–June.

Elevation: 1,000 to 5,500'.

Habitat: Sandy areas, deserts, and mesas.

Comments: Annual. Stems are reddish brown. Two species of *Eriastrum* in Arizona. Photograph taken in Superstition Mountains, March 26.

BLUE GILIA

Ipomopsis longiflora (Gilia longiflora)
Phlox Family (Polemoniaceae)

Height: To 2'.

Flowers: Pale blue to nearly white, with 5 spreading lobes at right angles to long, narrow tube; to 2" long; solitary or in pairs; on leafy cluster.

Leaves: Green, threadlike, to 1" long; lower leaves with narrow segments; to 2" long.

Blooms: May–November.

Elevation: 1,000 to 8,000'.

Habitat: Dry mesas, plains, and roadsides.

Comments: Often grows on limestone soil. Many-branched herb. Fourteen species of *Ipomopsis* in Arizona. Photograph taken in Tucson area, November 10.

COLUMBIA MONKSHOOD

Blue-Weed
Aconitum columbianum ssp. *columbianum*
Buttercup Family (Ranunculaceae)

Height: To 7', but averaging to 3'.

Flowers: Dark blue to blue-violet, occasionally white; to 1¼" long; in raceme; 5 sepals resembling petals, upper sepal forming hood, 2 oval side sepals, 2 narrow bottom sepals; petals concealed under hood.

Leaves: Dark green, palmately lobed, jagged-margined; to 8" wide.

Blooms: June–September.

Elevation: 5,000 to 9,500'.

Habitat: Along mountain streams and in meadows and rich, moist forests.

Comments: Very poisonous to livestock and humans if ingested. Two species of *Aconitum* in Arizona. Photograph taken at Greer, July 21.

ROCKY MOUNTAIN COLUMBINE

Blue Columbine
Aquilegia coerulea var. *pinetorum*
Buttercup Family (Ranunculaceae)

Height: To 30".

Flowers: To 3" wide; 5 petallike, pale blue sepals; 5 paler blue to white petals with backward-pointing spurs to 2" long.

Leaves: Bluish green, mostly basal, divided into several lobes; leaflets to 1¼" long, about as wide.

Blooms: June–July.

Elevation: 8,000 to 11,000'.

Habitat: Coniferous forests in rich, moist soil.

Comments: State flower of Colorado. Seven species of *Aquilegia* in Arizona. Photograph taken Colorado, July 16. In Arizona, plants of variety *pinetorum* are paler in color and have longer, more slender spurs than the typical phase in the photograph.

MOGOLLON LARKSPUR

Towering Larkspur

Delphinium geraniifolium (Delphinium tenuisectum)

Buttercup Family (Ranunculaceae)

Height: To 3'.

Flowers: 4 dark blue to bluish purple petals; 5 colored, petallike sepals, 1 forming a long, backward-facing spur; flower to 1" long, 1" wide, in long, densely flowered spike.

Leaves: Dark green, smooth, finely dissected into wedge-shaped lobes that are again cleft into narrow lobes; leaf to 6" long, 4" wide.

Blooms: July–August.

Elevation: 8,500 to 9,000'.

Habitat: Mountain meadows and forest clearings.

Comments: Perennial herb, with very leafy stem and no hair. Plant contains a poisonous juice. Ten species of *Delphinium* in Arizona. Photograph taken above Greer, August 10.

NELSON'S LARKSPUR

Delphinium nuttallianum (Delphinium nelsonii)

Buttercup Family (Ranunculaceae)

Height: To 15".

Flowers: Deep blue to bluish purple; upper petals in center of flower are white, notched, and with faint blue lines; 4 petals, 5 sepals, backward-projecting spur; flower including spur to 1¼" long, clustered on spikelike raceme; followed by 3-sectioned seed capsule.

Leaves: Grayish green, finely haired, succulent, and palmate; 3- to 4-lobed, each lobe generally 3-cleft with well-rounded tips; basal; to 1¼" wide.

Blooms: June.

Elevation: 6,000 to 8,500'.

Habitat: Pine forests.

Comments: Perennial herb. Contains delphinine and other toxic alkaloids. Ten species of *Delphinium* in Arizona. Photograph taken at Upper Lake Mary, June 1.

PALEFACE LARKSPUR

Delphinium parishii ssp. *parishii (Delphinium amabile)*
Buttercup Family (Ranunculaceae)

Height: To 4'.

Flowers: Pale blue to lavender, with 4 petals in unequal pairs and backward-pointing spur; to 1" long; in elongated cluster.

Leaves: Green, cleft into narrow lobes with sharp points; on long stalks, mainly at or near base of stem; to 3" long. Leaves often wither at blooming time.

Blooms: February–May.

Elevation: Below 5,000'.

Habitat: Desert mesas and washes.

Comments: Perennial herb. Ten species of *Delphinium* in Arizona. Photograph taken at Boyce Thompson Southwestern Arboretum, Superior, April 12.

BARESTEM LARKSPUR

Delphinium scaposum
Buttercup Family (Ranunculaceae)

Height: To 30".

Flowers: Royal blue with whitish center; backward-projecting spur, 4 petals, and 5 sepals; to 1" wide; clustered on spikelike raceme, followed by 3-sectioned seed capsule.

Leaves: Dark green, basal, palmately divided into lobes with rounded tips; to 2½" wide.

Blooms: March–May.

Elevation: Below 5,000'.

Habitat: Gravelly mesas, hillsides, and open desert.

Comments: Perennial herb with leafless stem. Contains delphinine and other toxins poisonous to humans and livestock. Hopi Indians use ground flowers in religious ceremonies. Ten species of *Delphinium* in Arizona. Photograph taken at Saguaro National Park, March 31.

ANGIOSPERMS: GREEN TO BROWN FLOWERS

ROCKY MOUNTAIN MAPLE

Acer glabrum
Maple Family (Aceraceae)

Height: Shrub, or small tree to 25'.

Trunk: To 1' in diameter.

Bark: Light brown to gray; smooth, thin.

Flowers: Greenish yellow, small, in cluster; followed by paired, winged seed cases or "keys" to 1" long, set in narrow V. Male and female flowers usually on different trees.

Leaves: Shiny, dark green above, lighter beneath; 3- to 5-lobed, toothed margins, red leaf stems; to 5" wide.

Blooms: May–June.

Elevation: 5,000 to 9,000'.

Habitat: Moist, rich soil along streams in ponderosa pine and spruce-fir forests.

Comments: Foliage turns yellow and red in fall. Seeds eaten by rodents and foliage browsed by deer, elk, and cattle. Three species of *Acer* in Arizona. Photograph taken at Woods Canyon Lake, August 3. Pointed teeth on leaf margins identify this maple.

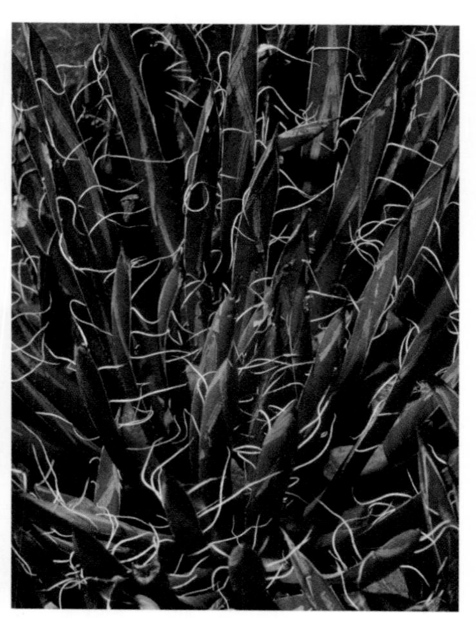

FAIRY RING AGAVE

Agave toumeyana var. *bella*
Agave Family (Agavaceae)

Height: Flower stalk to 6'.

Flowers: Green to pale yellow, sac-shaped perianth with large anthers extended; flower to 1" long; in elongated terminal cluster on long spike, followed by oblong, thin-walled capsule to ½" long.

Leaves: Light green with grayish white slashings; whitish threads along leaf margins; reddish tips; linear, concave on upper surface; to 8" long, 1" wide; in compact, circular rosette of 100 or more leaves at maturity.

Blooms: May–July.

Elevation: 4,000 to 5,000' in south-central Arizona.

Habitat: Rocky slopes in chaparral.

Comments: Native to higher elevations in south-central Arizona. Some authorities rank this variety as a subspecies. Twelve species of *Agave* in Arizona. Photograph taken at the Arizona-Sonora Desert Museum, November 12.

SOTOL

Desert Spoon

Dasylirion wheeleri

Agave Family (Agavaceae)

Height: To 3'; flower stem to 15'.

Flowers: Greenish white with 6 petallike segments; tiny, thousands in narrow cluster to 8' long. Male and female flowers on different plants.

Leaves: Green, ribbonlike; margins with forward-facing teeth, split ends; to 3' long, 1" wide; is rounded, basal cluster.

Blooms: May–August.

Elevation: 4,000 to 6,000'.

Habitat: Rocky slopes in desert grassland and oak woodlands.

Comments: Produces a flower stalk yearly. Spaniards called plant "desert spoon" because when pulled from the plant the dried leaf and its base resemble a *cuchara* or "spoon." Bighorn sheep browse sotols. Heads contain a sugary sap that, when fermented, is used to produce a potent beverage called *sotol*. One species of *Dasylirion* in Arizona. Photograph taken at Madera Canyon, June 10. Unlike sharply pointed yucca leaves, sotol leaves have split ends; its flowers, unlike the large bell-shaped flowers of the yucca, are small.

POISON IVY

Toxicodendron rydbergii (Rhus radicans var. *rydbergii)*

Cashew Family (Anacardiaceae)

Height: An erect shrub to 2', or an ascending vine with aerial roots.

Flowers: Greenish white; ¼" wide; in loose clusters to 3" long at lower leaf axils; followed by cluster of yellowish white, ¼" berrylike fruits.

Leaves: Dark green, compound, divided into 3 leaflets; usually shiny; oblong to lance-shaped to oval; notched or toothed; to 4" long. Immature leaves have red tinge.

Blooms: April–September.

Elevation: 3,000 to 8,000'.

Habitat: Rich soils in canyons, ravines, and disturbed areas.

Comments: Leaves turn reddish orange in fall. All parts of plant contain an oil that can cause skin eruptions. Remember: "Leaflets three, let it be." Two species of *Toxicodendron* in Arizona. Photograph taken at Oak Creek Canyon, June 9.

SWEET CICELY
Sweet Root
Osmorhiza depauperata
Carrot Family (Apiaceae)

Height: To 30".

Flowers: Tiny, greenish white, with 5 petals; to 1⁄16" wide, 3⁄16" long; in clusters at ends of 2 to 5 long-spreading stalks; followed by narrow, hairy, cylindrical fruit with club-shaped tip, to 3⁄8" long, on slender stem to 3⁄8" long.

Leaves: Dark green, hairy, several times thrice-divided; toothed, lobed; to 4" long, 6" wide.

Blooms: May–October.

Elevation: 7,000 to 10,000'.

Habitat: Moist, shady coniferous forests.

Comments: Perennial. Roots have aniselike flavor. Three species of *Osmorhiza* in Arizona. Photograph taken in vicinity of Mormon Lake, June 1.

ANTELOPE HORNS
Asclepias asperula
Milkweed Family (Asclepiadaceae)

Height: To 1'.

Flowers: Greenish yellow with maroon tinges; fragrant; to ½" wide, in cluster to 3" wide; followed by stout, 6"-long, green pod with olive-green and pink streaks.

Leaves: Dark green and narrow, with grayish stripe down midvein; to 6" long.

Blooms: April–August.

Elevation: 3,000 to 9,000'.

Habitat: Dry plains, slopes, and clearings in oak woodlands and pine forests.

Comments: Perennial herb; used medicinally. Twenty-nine species of *Asclepias* in Arizona. Photograph taken at Chiricahua National Monument, April 24.

CANYON RAGWEED

Ambrosia ambrosioides
Sunflower Family (Asteraceae)

Height: To 40".

Flowers: Yellowish green, to ⅜" wide, in terminal spike; followed by clusters of cocklebur-like fruit with slender spines; to 1" long.

Leaves: Green above and below, hairy; elongated to lance-shaped, saw-toothed; to 5" long, 1" wide.

Blooms: February–May.

Elevation: Below 4,500'.

Habitat: Sandy washes and canyons.

Comments: Branches are reddish brown with long white hairs. Fourteen species of *Ambrosia* in Arizona. Photograph taken in Superstition Mountains, March 26. A similar species, **Giant Ragweed** (*Ambrosia trifida*), is found in fields and along roadsides. Its abundant airborne pollen is dreaded by those who suffer with hay fever.

TRIANGLE-LEAF BURSAGE

Ambrosia deltoidea
Sunflower Family (Asteraceae)

Height: To 2'.

Flowers: Greenish, petalless, to ¼" wide; in terminal spike, followed by ¼"-long, burlike fruits with hook-tipped spines.

Leaves: Grayish green above, white and woolly beneath; triangular-shaped, finely toothed; to 1¼" long.

Blooms: December–April.

Elevation: 1,000 to 3,000'.

Habitat: Washes, gravelly slopes, and desert flats.

Comments: Perennial shrub. Causes severe hay fever in spring. Stabilizes soil. A "nurse" shrub for young cacti growing in its shade. Fourteen species of *Ambrosia* in Arizona. Photograph taken at Usery Mountain Recreation Area, February 14.

WHITE BURSAGE

Ambrosia dumosa
Sunflower Family (Asteraceae)

Height: To 2′.

Flowers: Greenish, petalless, to ⅛″ wide, in terminal spike; followed by ¼″-long spiny fruits (not hooked).

Leaves: Smoky-colored due to dense, short hairs; pinnately divided several times into small divisions; to ¾″ long.

Blooms: Late March–November.

Elevation: Below 3,000′.

Habitat: Dry plains and mesas.

Comments: Rounded, many-branched shrub with short, dense, white hairs on branches. Eaten by burros, horses, and sheep. Fourteen species of *Ambrosia* in Arizona. Photograph taken at Salome, March 28.

© MAX LICHER

COMMON COCKLEBUR

Xanthium strumarium (Xanthium saccharatum)
Sunflower Family (Asteraceae)

Height: To 4′.

Flowers: Male: tiny, in clusters at top of stem; female: spiny, oval, brown burs to 1½″ long in leaf axils. Bur encloses 2 female flowers covered with hundreds of stiff, hooked prickles.

Leaves: Yellowish green, heart-shaped, coarsely toothed; rough, glandular, long-stalked; to 14″ long, 8″ wide.

Blooms: June–October.

Elevation: 100 to 6,000′.

Habitat: Waste areas, roadsides, and cultivated fields.

Comments: This bushy annual is exotic to Arizona. Spiny burs get tangled in fur and fabric; 2 seeds inside each bur. Seeds and young plants are poisonous to livestock, especially chickens and hogs. Two species of *Xanthium* in Arizona. Photograph taken north of Payson, September 2.

THINLEAF ALDER

Black Alder
Alnus incana ssp. *teniufolia (Alnus tenuifolia)*
Birch Family (Betulaceae)

Height: Large shrub, or small tree to 25′.

Trunk: Several trunks to 6″ in diameter.

Bark: Grayish, reddish brown, and scaly with age.

Flowers: Minute, male and female flowers in separate catkins on same tree; female flowers followed by small cones ⅜″ to ½″ long.

Leaves: Dark green above, yellowish and slightly hairy beneath; egg-shaped to oblong, rounded at base; doubly toothed margins, bent stem; to 4″ long, to 2½″ wide.

Blooms: Early spring.

Elevation: 7,000 to 9,000′.

Habitat: Along streams and in moist meadows in ponderosa pine and spruce-fir forests.

Comments: Rounded crown. Thick and woody alder cones remain on trees. The Navajo used powdered bark to make red dye for wool. Deer, rabbits, and beaver eat alder bark; birds eat the seeds. Two species of *Alnus* in Arizona. Photograph taken south of Alpine, August 8. This alder recognizable by its oval to oblong leaves, which are rounded or heart-shaped at base.

ARIZONA ALDER

Mexican Alder
Alnus oblongifolia
Birch Family (Betulaceae)

Height: To 60′.

Trunk: To 3′ in diameter.

Bark: Dark gray, thin, and smooth; scaly with age.

Flowers: Minute, male and female flowers on separate catkins on same tree; female flowers followed by small cones to ½″ long.

Leaves: Dark green above, paler and slightly hairy beneath; elliptical, narrowed at base, with doubly toothed margins; to 3″ long, 2″ wide.

Blooms: March.

Elevation: 4,500 to 7,500′.

Habitat: Along streams and canyons in ponderosa pine forests and oak woodlands.

Comments: Rounded crown. Alders add nitrogen to soil thus improving it. Two species of *Alnus* in Arizona. Photograph taken at Lynx Creek, Prescott, September 11. This alder recognizable by more elliptical leaves, which gradually narrow at base.

GREEN-FLOWERED MACROMERIA

Macromeria viridiflora
Forget-me-not Family (Boraginaceae)

Height: To 4′.

Flowers: Yellowish to greenish white, narrow, 5-lobed; very hairy, trumpet-shaped, and drooping; to 2½″ long, in clusters at ends of branches.

Leaves: Grayish green, prominently veined, very hairy, lance-shaped to 7″ long.

Blooms: July–September.

Elevation: 6,000 to 9,000′.

Habitat: Roadsides, rocky slopes, and valleys in coniferous forests.

Comments: Dried flowers and leaves mixed with wild tobacco are used by Hopi Indians during rainmaking ceremonies. One species of *Macromeria* in Arizona. Photograph taken south of Alpine, June 30.

SHE-OAK

Beefwood
Casuarina equisetifolia
Casuarina Family (Casuarinaceae)

Height: To 100′.

Trunk: To 30″ in diameter.

Bark: Reddish brown, deeply furrowed on mature trees.

Flowers: Tiny, male and female on separate trees. Female flowers followed by brownish, hard, warty, conelike fruit which grow to ½″ in diameter.

Leaves: Grayish green, scalelike; in whorls on drooping stems.

Blooms: Spring.

Habitat: Floodplains and moist areas.

Comments: Native to Australia. Fast-growing and long-lived. Branchlets are wirelike and jointed. Although not listed in *A Catalogue of the Flora of Arizona*, *Casuarina* is included here because of its massiveness and frequency in certain areas of Arizona. Photograph taken at Catalina State Park, April 15.

CANOTIA

Crucifixion-Thorn
Canotia holacantha
Bittersweet Family (Celastraceae)

Height: Spiny shrub, or small tree to 18′.

Trunk: To 8″ in diameter.

Bark: Yellowish green; smooth when young, gray and rough with age.

Flowers: Greenish white, 5 petals, 5 stamens; to ¼″ long; 3 to 5 in small cluster along branches; followed by reddish brown, egg-shaped, long-pointed, 5-valved capsule, to ½″ long, later splitting along 10 lines.

Leaves: Greenish, scalelike, short-lived, deciduous.

Blooms: May–August.

Elevation: 2,000 to 4,500′.

Habitat: Dry slopes and mesas in chaparral and desert.

Comments: The most common of the crucifixion-thorns. Twigs are spine-tipped and form masses; very flexible when young, becoming rigid with age. Twigs and branches take the place of leaves in food manufacture. Fruit capsule persists until the following spring. One species of *Canotia* in Arizona. Photograph taken at Sedona, June 18.

MOUNTAIN LOVER

Oregon Boxwood
Paxistima myrsinites (Pachystima myrsinites)
Bittersweet Family (Celastraceae)

Height: Spreading shrub to 2′, usually shorter.

Flowers: Reddish brown, with 4 petals, 4 pointed sepals; to ⅛″ wide; on ¼″-long stem; 2 or 3 in cluster in leaf axils; followed by green, oval fruit to ⅛″ long.

Leaves: Dark green, shiny, thick; oval to elliptical, tooted; to 1″ long.

Blooms: May–July.

Elevation: 6,000 to 10,000′.

Habitat: Coniferous forests in moist, shady locations.

Comments: Prostrate evergreen ground cover. Woody at base, with brownish stems. Browsed by deer. One species of *Paxistima* in Arizona. Photograph taken south of Alpine, June 30.

BLACK GREASEWOOD

Sarcobatus vermiculatus
Goosefoot Family (Chenopodiaceae)

Height: To 8'.

Flowers: Inconspicuous female flowers are borne in axils of leaves, followed by a green to tan, winglike, membranous disk surrounding the seeds. Male flowers are in conelike spikes at branch ends.

Leaves: Grayish green, fleshy, narrow; to 1¼" long, ¹⁄₁₆" wide; dense along branches.

Blooms: June–September.

Elevation: 1,000 to 6,000'.

Habitat: Alkaline soil.

Comments: Creosote bush is often incorrectly called greasewood, but this is actually the only true greasewood. It has white or tannish bark, many rigid branches, and salty tasting leaves. Browsed by cattle and sheep, but overeating causes bloating; also a favorite of jackrabbits. Native Americans used wood to make digging sticks for planting corn, and for dice and knitting needles. One species of *Sarcobatus* in Arizona. Photograph taken at Canyon de Chelly National Monument, June 27.

ROUNDLEAF BUFFALO BERRY

Lead Bush
Shepherdia rotundifolia
Oleaster Family (Elaeagnaceae)

Height: To 5'.

Flowers: Grayish green, petalless, tiny, inconspicuous; coated with grayish green scales; in leaf axils; followed by silvery, scaly, soft, juicy, egg-shaped berry to ⅜" long. Male and female flowers on separate plants.

Leaves: Silvery gray and covered with silvery scales above, white- or yellow-woolly and scaly beneath; curved under; thick, oval to roundish; to 1¼" long.

Blooms: May–June.

Elevation: 5,000 to 8,000'.

Habitat: Steep slopes in northern Arizona.

Comments: Evergreen shrub. Ripened fruit contains sweet, watery, yellowish juice. Berries eaten by birds, bears, and small mammals; also used for making jelly. Salve made from plant to treat sheep eye irritation. Three species of *Shepherdia* in Arizona. Photograph taken at North Rim of Grand Canyon National Park, June 25.

WOODLAND PINEDROPS

Pterospora andromedea
Heather Family (Ericaceae)

Height: To 40'.

Flowers: Yellowish brown, urn-shaped, nodding; to ¼" long; loosely scattered along erect airy, sticky, reddish brown, leafless stem.

Leaves: Brown and scalelike.

Blooms: June–September.

Elevation: 6,000 to 9,500'.

Habitat: Rich soil in coniferous forests.

Comments: Lacks chlorophyll. A saprophytic herb feeding on decaying plant matter in soil. Roots are like matted hair. Plant feels sticky when touched. Some taxonomists choose to separate this species into the family Monotropaceae. One species of *Pterospora* in Arizona. Photograph taken at Greer, September 11.

DESERT POINSETTIA

Beetle Spurge
Euphorbia eriantha
Spurge Family (Euphorbiaceae)

Height: To 15".

Flowers: Greenish, tiny, clustered in center of floral leaves.

Leaves: Bronze-green, slender; to 3" long; issue from stem. Threadlike floral leaves emerge from flower cluster.

Blooms: February–October.

Elevation: 300 to 3,500'.

Habitat: Dry desert areas in the shelter of small shrubs.

Comments: Produces a milky sap. More than three dozen species of *Euphorbia* in Arizona. Photograph taken at Usery Mountain Recreation Area, February 14.

ARIZONA WHITE OAK

Roble
Quercus arizonica
Beech Family (Fagaceae)

Height: To 60′.

Trunk: To 3′ in diameter.

Bark: Light gray, fissured into thick plates.

Flowers: Tiny; male and female on same tree. Male: drooping catkin; female in short spike, followed by light brown, to 1″-long acorn, with lower half partly covered by hairy-scaled, shallow, bowl-like cup.

Leaves: Dull blue green, with sunken veins above; lighter color, densely haired with brownish yellow fuzz and raised veins beneath. Broadly oval, thick, and stiff; margins slightly wavy and toothed toward apex; to 3″ long.

Blooms: Spring.

Elevation: 5,000 to 7,500′.

Habitat: Oak woodlands, foothills, mountains, and canyons.

Comments: Evergreen; among the largest southwestern oaks. New leaves emerge when old leaves drop in May. Rounded crown. Found to occasionally hybridize with **Gambel Oak** (*Quercus gambelii*) (page 367) and **Gray Oak** (*Quercus grisea*). More than a dozen species of *Quercus* in Arizona. Photograph taken at Harshaw, April 27.

EMORY OAK

Bellota
Quercus emoryi
Beech Family (Fagaceae)

Height: Shrub, small or large tree to 50′.

Trunk: To 2½′ in diameter.

Bark: Black to gray and thick, in rectangular plates.

Flowers: Tiny; male and female on same tree. Male in drooping catkin, female in short spike; followed by brownish, oblong, ¾″-long acorn, with lower third covered by a hairy cup.

Leaves: Yellowish green on both sides; shiny, thick, stiff, and leathery; broadly lance-shaped, with spiny tip, some spiny teeth on margins; base of leaf blade is fuzzy where it joins stem on over half-mature leaves; to 2½″ long.

Blooms: Spring.

Elevation: 4,000 to 7,000′.

Habitat: Moist canyons and dry foothills.

Comments: Evergreen; found from southeastern to central Arizona. An important source of firewood in southern Arizona. Acorns (*bellota* in Spanish) mature in first year. Sweet and tasty, they are eaten by wildlife as well as by people, in particular Native Americans and Hispanics of the region. More than a dozen species of *Quercus* in Arizona. Photograph taken at Harshaw, April 27. The shiny yellow-green, hairless, spiny-toothed leaves are distinguishing features of this oak.

GAMBEL OAK

Quercus gambelii
Beech Family (Fagaceae)

Height: Shrub to 6 ½′, tree to 50′.

Trunk: To 2½′ in diameter.

Bark: Gray, thick, deeply furrowed or scaly.

Flowers: Very tiny; male and female on same tree. Male in drooping catkin, female in short spike; followed by brownish, broadly oval, ¾″-long acorn, with lower half covered by hairy scales, bowl-like cup.

Leaves: Dark green and smooth above, lighter green and with soft hairs beneath; oblong, with 7 to 11 rounded lobes; to 6″ long.

Blooms: Spring.

Elevation: 5,000 to 8,000′.

Habitat: Mountains and plateaus in ponderosa forests.

Comments: Deciduous; with rounded crown. Turns yellow and reddish in autumn before leaves fall. Known to hybridize with **Shrub Live Oak** (*Quercus turbinella*) (page 369); see also **Wavyleaf Oak** (page 370). Also found to occasionally hybridize with **Arizona White Oak** (*Quercus arizonica*) (page 366) and **Havard Oak** (*Quercus havardii*). More than a dozen species of *Quercus* in Arizona. Photograph taken near Ashurst Lake, September 5. Has a distinctive, lobed leaf for an Arizona oak.

SILVERLEAF OAK

Quercus hypoleucoides
Beech Family (Fagaceae)

Height: Tree to 60′.

Trunk: To 2½′ in diameter.

Bark: Blackish, deeply furrowed into plates and ridges.

Flowers: Tiny; males in drooping clusters, females in short spikes; followed by egg-shaped acorn with thick, scaly cup; to ⅝″ long.

Leaves: Yellow-green and shiny above, silvery white and woolly beneath; lance-shaped, leathery, with edges rolled under; evergreen; to 4″ long, 1″ wide.

Blooms: April–May.

Elevation: 5,000 to 7,000′.

Habitat: Mountain slopes and canyons in oak woodlands.

Comments: Acorns mature in 1 or 2 years. Found in southern Arizona. More than a dozen species of *Quercus* in Arizona. Photograph taken at Chiricahua National Monument, April 25. An easily recognizable oak by its narrow, yellow-green leaves with silvery undersides.

MEXICAN BLUE OAK

Quercus oblongifolia
Beech Family (Fagaceae)

Height: To 30′.

Trunk: To 18″ in diameter.

Bark: Gray, fissured in square plates.

Flowers: Tiny; males in drooping clusters, females in short spikes; followed by egg-shaped acorn, ¾″ long, ⅓ enclosed by scaly cup; 1 to a stalk (sometimes stalkless).

Leaves: Blue-green above, paler beneath; evergreen, stiff, hairless, and toothless; rounded at both ends; short-stalked; to 2″ long.

Blooms: April–May.

Elevation: 4,000 to 6,000′.

Habitat: Oak woodlands at upper edge of desert; foothills and canyons of southern Arizona.

Comments: Has spreading, rounded crown. Acorns mature first year. Leaves fall in spring when new leaves appear. Browsed by deer. More than a dozen species of *Quercus* in Arizona. Photograph taken in vicinity of Mount Lemmon, April 30. Identifiable by blue-green, rounded, toothless, hairless leaves.

PALMER OAK

Quercus palmeri (Quercus dunnii & Quercus chrysolepis var. *palmeri)*
Beech Family (Fagaceae)

Height: Shrub, or small tree to 20′.

Trunk: To 4″ in diameter.

Bark: Grayish or brown, smooth or scaly.

Flowers: Tiny; males in drooping clusters, females in short spikes; followed by a narrow or broad acorn to 1½″ long with an oversized cup, coated with golden hairs, variable.

Leaves: Yellowish green; shiny above, yellowish beneath; elliptical to oval; stiff, leathery, spiny-toothed (hollylike) or untoothed and short-pointed; to 1½″ long.

Blooms: April–May.

Elevation: 3,500 to 7,000′.

Habitat: Mountainsides, canyons, and oak woodlands.

Comments: Has broad crown. Acorns take 2 years to mature. More than a dozen species of *Quercus* in Arizona. Photograph taken along East Verde River north of Payson, September 2.

NETLEAF OAK

Quercus rugosa (Quercus reticulata)
Beech Family (Fagaceae)

Height: Tree to 40′.

Trunk: To 1′ in diameter.

Bark: Gray, fissured, scaly.

Flowers: Tiny; males in narrow, drooping clusters, females in leaf axils; followed by two or three ¾″-long oblong acorns, ¼ of each enclosed in scaly cup on long, slender stalk.

Leaves: Dark green; finely haired, with sunken veins above, yellow-haired with raised veins beneath; stiff, thick, broad, variable in size and shape; a few spiny teeth toward rounded tip; notched at base; to 2¾″ long.

Blooms: April–May.

Elevation: 4,000 to 8,000′.

Habitat: Canyons, mountain slopes, and oak woodlands.

Comments: Has broad, rounded crown. Acorns mature first year; are eaten by wildlife. More than a dozen species of *Quercus* in Arizona. Photograph taken at Chiricahua National Monument, April 25. Recognizable by broadness of leaf at tip end and by prominent network of leaf veins.

SHRUB LIVE OAK

Scrub Oak
Quercus turbinella
Beech Family (Fagaceae)

Height: Many-branched shrub or small tree to 13′.

Bark: Gray, scaly, fissured.

Flowers: Tiny; male and female on same tree. Male in drooping catkin, female in short spike; followed by brownish; narrow, pointed ¾″-long acorn in shallow, scaly cup.

Leaves: Bluish green with a bloom (delicate powdery coating); nearly hairless above, yellowish green and finely haired beneath; elliptical to oblong, thick, stiff; margins have small, spinelike teeth; evergreen; to 1¼″ long.

Blooms: Spring.

Elevation: 4,500 to 8,000′.

Habitat: Lower ponderosa forests, hillsides, and chaparral.

Comments: Characteristic tree of Arizona chaparral. Often forms dense thickets. Retards soil erosion. Provides browse for livestock, acorns for wildlife. Known to hybridize with **Gambel Oak** (*Quercus gambelii*) (page 367); see **Wavyleaf Oak** (page 370). More than a dozen species of *Quercus* in Arizona. Photograph taken at Oak Creek Canyon, September 9. Identifiable by spiny-toothed, thick, stiff leaves with whitish bloom.

WAVYLEAF OAK
Shin Oak
Quercus × pauciloba (Quercus undulata)
Beech Family (Fagaceae)

Height: Shrub to 6', tree to 15'.

Trunk: To 4" in diameter.

Bark: Light gray and scaly.

Flowers: Tiny; male in drooping cluster, female in short spike, followed by oblong acorn to ⅞" long with deep cup.

Leaves: Grayish green, pinkish velvety cast when young; evergreen, wavy-margined; variable in shape, elliptical or oblong; wavy-lobed or toothed; to 2½" long.

Blooms: April–May.

Elevation: 6,000 to 9,000'.

Habitat: Mountainsides and canyons.

Comments: Often found in burned areas. A natural hybrid between **Gambel Oak** (*Quercus gambelii*) (page 367) and **Shrub Live Oak** (*Quercus turbinella*) (page 369). Over a dozen species of *Quercus* in Arizona. Photograph taken in Santa Catalina Mountains, April 30. Recognizable by shallow, wavy lobes of leaf.

QUININE BUSH
Silktassel Bush
Garrya flavescens
Silktassel Family (Garryaceae)

Height: To 6'.

Flowers: Small, petalless, abundant; in a dense, drooping, grayish green tassel of bell-like, hairy bracts; tassel to 4" long; followed by a grapelike cluster of berrylike fruits. Each fruit is round to oval, downy-covered, pointed at tip; to ½" long, 5⁄16" wide; cluster is up to 3" long. Male and female flowers are on separate plants.

Leaves: Dull grayish green; leathery, thick; woolly hair is silky to the touch; untoothed, opposite, oval to elliptical; evergreen; to 4" long, to 2" wide.

Blooms: January–April.

Elevation: 2,500 to 7,000'.

Habitat: Dry mountain slopes and canyons in woodlands and chaparral.

Comments: Garryin, an alkaloid derived from several *Garrya* species, has medicinal uses. The leaves on *G. flavescens* are larger, covered with silky hairs, and appear much grayer than those of **Wright Silktassel** (*Garrya wrightii*) (page 371). Two species of *Garrya* in Arizona. Photograph taken at Oak Creek Canyon, Sedona, June 8.

WRIGHT SILKTASSEL

Garrya wrightii
Silktassel Family (Garryaceae)

Height: To 15′.

Flowers: Small, petalless, few in number, in a loose, hanging, grayish green tassel; followed by roundish fruit to ⅓″ in diameter, dark bluish purple when mature.

Leaves: Light greenish gray, elliptical, rough, evergreen; untoothed, leathery, opposite and alternate; to 2″ long.

Blooms: March–August.

Elevation: 3,000 to 8,000′.

Habitat: Dry slopes.

Comments: A small amount of rubber can be extracted from this plant. Browsed by deer. Two species of *Garrya* in Arizona. Photograph taken on Mount Graham, May 3.

UTAH SWERTIA

Frasera paniculata (Swertia utahensis)
Gentian Family (Gentianaceae)

Height: To 3′.

Flowers: Yellowish green with purple streaks and dots; greenish at tip of lobes; 4 lobes, each with an elongated, basal gland lined with yellowish hairs; to 1″ wide; in loose, terminal raceme.

Leaves: Dark green with white margins; narrow-leaved, basal, wavy margined; to 3″ long.

Blooms: June–September.

Elevation: 4,000 to 7,500′.

Habitat: Sandy washes and dunes.

Comments: Perennial herb. Has erect stem. Four species of *Frasera* in Arizona. Photograph taken at Dead Horse Ranch State Park, May 28.

DEERS EARS

Elkweed
Frasera speciosa (Swertia radiata)
Gentian Family (Gentianaceae)

Height: To 6'.

Flowers: Greenish white dotted with purple; star-shaped; 4-lobed with 2 fringed glands on each lobe; in leaf axils on stem; to 1½" wide, in elongated cluster.

Leaves: Light green, in whorls of 4 to 6; linear to lance-shaped; to 12" at base, diminishing in size upward on stem.

Blooms: May–August.

Elevation: 5,000 to 10,000'.

Habitat: Rich soil in open pine forests and mixed aspen-conifer forests.

Comments: Perennial herb. Four species of *Frasera* in Arizona. Photograph taken near Payson, June 10.

WHITE-STEM GOOSEBERRY

Ribes inerme
Currant Family (Grossulariaceae)

Height: To 5'.

Flowers: Greenish to purplish sepals, with hidden petals; bell-shaped; up to 3 flowers on stems in leaf axils; followed by round, striped, smooth berry, to ¼" in diameter; turning purplish red when mature.

Leaves: Dark green above and beneath, 3- to 5-lobed, toothed; to 1½" wide.

Blooms: April–May.

Elevation: 7,000 to 8,500'.

Habitat: Clearings in pine forests.

Comments: Deciduous plant, with whitish stems; greenish spines on stems to ½" long. Ten species of *Ribes* in Arizona. Photograph taken at North Rim of Grand Canyon National Park, June 25.

ARIZONA WALNUT

Nogal
Juglans major var. *major*
Walnut Family (Juglandaceae)

Height: To 50', but usually less.

Trunk: To 3' in diameter.

Bark: Grayish brown, furrowed on mature trees.

Flowers: Greenish, male and female separate on same tree, male flowers are hanging catkin; female flowers in erect cluster; no petals; followed by round, brown-haired husks, to 1½" in diameter, each encasing a deeply grooved nutshell.

Leaves: Yellowish green, pinnately compound; 9 to 13 coarsely toothed, lance-shaped leaflets, each to 4" long, 1½" wide; leaf to 14" long.

Blooms: Before or during leaf development.

Elevation: 3,500 to 7,000'.

Habitat: Along streams and in canyons in upper desert, grasslands, and oak woodlands.

Comments: Deciduous. Rounded crown of widely spreading branches. Favorite of squirrels. Only species of *Juglans* in Arizona. Photograph taken at Mormon Lake, September 3.

WHITE MULBERRY

Russian Mulberry
Morus alba
Mulberry Family (Moraceae)

Height: To 40'.

Trunk: To 1' in diameter.

Bark: Tan, smooth; on older trees furrowed into scaly ridges.

Flowers: Green, tiny, in short clusters; followed by pinkish white fruits, to ¾" long, maturing to dark purple.

Leaves: Dark green, broadly oval, toothed, long-stalked; often 3- to 5-lobed; to 7" long, 5" wide.

Blooms: Spring.

Elevation: Not available. Photograph taken at 2,100'.

Habitat: Cities and lots.

Comments: This species of *Morus* was introduced from China; now naturalized throughout eastern and western U.S. Birds relish the fruits; also a main food source for silkworms. The **Texas Mulberry** (page 374) is the only native mulberry found in Arizona. Photograph taken at Hassayampa River Preserve, Wickenburg, May 7.

TEXAS MULBERRY

Morus microphylla
Mulberry Family (Moraceae)

Height: Shrub, or small tree to 20′.

Trunk: To 8″ in diameter.

Bark: Pale gray, smooth, becoming scaly and fissured.

Flowers: Green, tiny, in dense catkins to ¾″ long; male and female on separate trees; followed by red to black, ½″-long cylindrical fruits.

Leaves: Dark green above, paler and hairy beneath; variable in shape from oval to 3- to 5-lobed; coarsely saw-toothed; to 2½″ long.

Blooms: March–April.

Elevation: 2,000 to 6,000′.

Habitat: Moist areas in upper desert; grasslands, and woodlands among streams, washes, and in canyons.

Comments: Fruits are juicy, acidic, and edible; eaten by humans as well as by wildlife. The Texas mulberry is the only native species of *Morus* in Arizona. Photograph taken at Catalina State Park, November 9.

NEW MEXICAN OLIVE

Wild Privet
Forestiera pubescens var. *pubescens (Forestiera neomexicana)*
Olive Family (Oleaceae)

Height: Sprawling shrub to 6′, or small tree to 10′.

Bark: Gray to light tan blotched with gray; reddish brown twigs.

Flowers: Tiny and petalless; followed by dark bluish green, oval fruits, to ¼″ long, 3⁄16″ wide, in clusters on branches. Male and female on separate plants.

Leaves: Green to grayish green, opposite, alternate and in clusters; oval to elliptical but highly variable; minutely toothed or untoothed; to 1½″ long.

Blooms: March–May.

Elevation: 2,000 to 7,000′.

Habitat: Hillsides, mesas, and lakeshores.

Comments: Very hard wood. Navajo Indians used wood for prayer sticks; the Hopi used wood as digging sticks. Two species of *Forestiera* in Arizona. Photograph taken at Wupatki National Monument, June 5.

DESERT OLIVE
Forestiera shrevei
Olive Family (Oleaceae)

Height: Shrub to 12′.

Bark: Gray or blackish, smooth.

Flowers: Greenish, dark purple anthers; to ¼″ long; in small clusters; followed by brownish, egg-shaped, 1-seeded, fleshy fruit, to ⅜″ long. Male and female on separate plants.

Leaves: Green, finely haired on both surfaces, untoothed; opposite, margins rolled under, oblong or lance-shaped or reverse lance-shaped; to 1″ long, ¼″ wide.

Blooms: December–March.

Elevation: 2,500 to 4,500′.

Habitat: Dry, rocky slopes and desert canyons.

Comments: Evergreen or nearly so. Often forms dense thickets. Named for Charles Le Forestier, a French naturalist and physician. Two species of *Forestiera* in Arizona. Photograph taken in Kofa Mountains, February 22.

LOWELL ASH
Fraxinus anomala var. *lowellii (Fraxinus lowellii)*
Olive Family (Oleaceae)

Height: Shrub, or small tree to 25′.

Trunk: To 6″ in diameter.

Bark: Grayish brown, deeply furrowed.

Flowers: Greenish, ⅛″ long, in clusters to 1½″ long; followed by light brown, long-winged, flattened "keys" (dry, 1-seeded, winged fruit); in clusters to 1½″ long. Male and female on separate trees.

Leaves: Dark green, paired, pinnately compound; to 7″ long: 3, 5, or 7 leaflets; oval but variable, with saw-toothed margins; slightly leathery; leaflets to 3″ long.

Blooms: March–May.

Elevation: 3,200 to 6,500′.

Habitat: Along streams and in moist canyon soils in oak woodlands and upper desert areas in central Arizona.

Comments: Named for Percival Lowell, the famous astronomer, who first found this ash in Oak Creek Canyon. Seven species of *Fraxinus* in Arizona. Photograph taken at Lowell Observatory in Flagstaff, June 22.

VELVET ASH

Fresno

Fraxinus velutina (Fraxinus pennsylvanica ssp. *velutina)*

Olive Family (Oleaceae)

Height: To 30′.

Trunk: To 1′ in diameter.

Bark: Gray, and deeply furrowed into broad ridges.

Flowers: Small; male and female on different trees. Male flower is yellow; female is greenish, in clusters, followed by elliptical, long-winged "keys" or samaras to 1" long.

Leaves: Green, pinnately compound; with 5 to 9 elliptical to lance-shaped leaflets, each to 1½" long; margins with or without teeth; leaf to 6" long. Leaves vary greatly in all characteristics. Young leaves feel velvety, but velvet soon disappears.

Blooms: March–April.

Elevation: 2,000 to 7,000′.

Habitat: Along streams, in moist canyons, and along moist washes.

Comments: Deciduous; the most common ash of the Southwest; spreading branches and rounded crown. Flowers appear in spring before leaves. Birds and other animals eat seeds. Seven species of *Fraxinus* in Arizona. Photograph taken at Oak Creek Canyon, September 9.

SATYR ORCHID

Frog Orchid

Coeloglossum viride var. *virescens (Habenaria viridis* var. *bracteata)*

Orchid Family (Orchidaceae)

Height: To 2′.

Flowers: Greenish; flowers borne in narrow, elongated cluster. Upper sepal and 2 erect, pointed, lateral petals forming hood; rectangular-shaped, greenish yellow, 2- or 3-lobed lip pointing downward to ½" long; flower to ½" long; floral bract much longer than flower.

Leaves: Dark green, lance-shaped; alternate; clasping stem; to 6" long, 1" wide.

Blooms: July–August.

Elevation: Not available. Photograph taken at 8,700′.

Habitat: Moist spruce-fir forests.

Comments: One species of *coeloglossum* in Arizona. Photograph taken in mountains above Greer, July 8. This species recognizable by its very long flower bracts and rectangular lip with 2 or 3 lobes.

SPOTTED CORAL ROOT

Large Coral Root
Corallorhiza maculata
Orchid Family (Orchidaceae)

Height: To 20".

Flowers: Brown to purplish red, orchidlike flowers along a purplish, leafless flower stem; lip cream-colored to white with purple spots; to ¾" long.

Leaves: Scalelike, tubular sheaths on stem.

Blooms: June–July.

Elevation: 6,000 to 10,000'.

Habitat: Coniferous forests.

Comments: Saprophytic orchid lacking chlorophyll. Has coral-like underground stem. Receives nourishment from a fungus that decomposes dead plant material. Three species of *Corallorhiza* in Arizona. Photograph taken near Greer, June 16.

SPRING CORAL ROOT

Corallorhiza wisteriana
Orchid Family (Orchidaceae)

Height: To 1'.

Flowers: Brown upper lobes and backward pointing spur; white lower lip petal with faint pink markings; unlobed lower lip projects noticeably forward and downward; flower to ⅛" wide, ½" long, in slender raceme.

Leaves: Leafless; reduced to scales.

Blooms: April–May.

Elevation: 6,000 to 8,000'.

Habitat: Hillsides and clearings in ponderosa pine forests.

Comments: A saprophyte lacking chlorophyll. Pinkish brown stem. Three species of *Corallorhiza* in Arizona. Photograph taken in Sharp Creek area northeast of Christopher Creek, April 22.

ADDER'S MOUTH

Malaxis macrostachya (Malaxis soulei)
Orchid Family (Orchidaceae)

Height: To 6".

Flowers: Yellowish green, less than ⅛" long, clustered on single flower spike to 4" long.

Leaves: Dark green, single, oval; clasping stem; to 3" long.

Blooms: July–September.

Elevation: 6,000 to 9,500'.

Habitat: Mixed conifer forests.

Comments: Grows from a corm. Five species of *Malaxis* in Arizona. Photograph taken south of Alpine, August 2.

SPARSELY-FLOWERED BOG ORCHID

Northern Rein Orchid
Platanthera sparsiflora (Habenaria sparsiflora)
Orchid Family (Orchidaceae)

Height: To 30".

Flowers: Greenish to yellowish, slender spur as long as or longer than lip petal; to ¼" wide, ¾" long, in elongated and loosely flowered flower spike.

Leaves: Light green, fleshy, clasping stem; strap-like to lance-shaped; alternate; to 7½" long, at intervals on flower stalk.

Blooms: June–October.

Elevation: 5,000 to 9,000'.

Habitat: Wet, boggy areas and moist ledges in coniferous forests.

Comments: Six species of *Plantanthera* in Arizona. Photograph taken in Oak Creek Canyon, Sedona, June 8.

SLENDER BOG ORCHID

Rein Orchid
Platanthera stricta (Habenaria saccata)
Orchid Family (Orchidaceae)

Height: To 3'.

Flowers: Light green, with 3 petals and 3 sepals; orchidlike, lower petal (called the "lip") is long and curved upward; flower to ⅜" long, in long, narrow, loosely flowered spike.

Leaves: Dark green, sunken midvein; alternate, lance-shaped; clasping stem; to 10" long, 1½" wide.

Blooms: July–September.

Elevation: 8,500 to 9,500'.

Habitat: Moist spruce-fir forests and along mountain streams in coniferous forests.

Comments: Six species of *Platanthera* in Arizona. Photograph taken in mountains above Greer, August 9.

POKEBERRY

Pokeweed
Phytolacca americana var. *americana*
Pokeberry Family (Phytolaccaceae)

Height: To 10', but usually to 6'.

Flowers: Greenish white to pinkish, petals absent, 5 sepals; to ¼" wide, in terminal raceme, followed at maturity by dark purplish to black ¼" berries in drooping cluster.

Leaves: Dark green, oval to lance-shaped, smooth; to 9" long.

Blooms: August.

Elevation: 4,000 to 6,000' in Chiricahua Mountains.

Habitat: Clearings, roadsides, and open woods.

Comments: A branched, perennial herb. Likely introduced to Arizona from eastern U.S. Root is used medicinally. Stems become reddish in fall. Berries are poisonous, and leave a purple stain when overly ripened. Three species of *Phytolacca* in Arizona. Photograph taken in northeastern U.S. where this same species is found.

© MAX LICHER

BUCKHORN PLANTAIN

Ribwort

Plantago lanceolata

Plantain Family (Plantaginaceae)

Height: To 24".

Flowers: Greenish white, to ⅛" long; spirally arranged in a dense, terminal spike, to 1½" long, on tall, leafless stalk; stamens projecting from hairlike stalks with cream-colored anthers at ends; followed by tiny, brown, 2-seeded capsule.

Leaves: Dark green; lancelike, in basal cluster; to 16" long.

Blooms: April–September.

Elevation: Widely distributed.

Habitat: Meadows, fields, and leaves.

Comments: Considered a weed. Native to Europe; now naturalized in U.S. Twenty-one species of *Plantago* in Arizona.

ARIZONA SYCAMORE

Platanus wrightii

Plane Tree Family (Platanaceae)

Height: To 80'.

Trunk: To 4' or more in diameter.

Bark: On branches, whitish, smooth, and thin; on trunk and very large branches, whitish and peeling in brownish flakes; on very large, old trunks, dark gray with odd-shaped, thick plates that hang loose.

Flowers: Inconspicuous; male and female flowers in separate, dense, round clusters of 2 to 4. The female cluster matures into light brown, hairy, globular, seed head; to 1" in diameter; with 2 to 4 along a raceme to 8" long.

Leaves: Light green above, paler beneath, with small hairs; palmately lobed, divided into 3, 5, or 7 narrow, pointed lobes; to 10" long and wide.

Blooms: March–April.

Elevation: 2,000 to 6,000'.

Habitat: Along streams and in rocky canyons.

Comments: Deciduous. Tree has large, spreading branches and broad, open crown. Its roots help slow down soil erosion along stream banks. Small owls and other birds nest in the hollows of old branches. One species of *Platanus* in Arizona. Photograph taken north of Payson, June 3.

CURLY DOCK

Rumex crispus
Buckwheat Family (Polygonaceae)

Height: To 4′.

Flowers: Small, yellowish green, with 6 sepals in 2 circles (inner 3 enlarge to become the fruit), later becoming rosy-colored; in long, loose clusters on upper branches, followed by small, reddish brown fruits to ¼″ long.

Leaves: Dark green to bluish green, lance-shaped, wavy-curled margins; alternate on stem, mostly basal; to 1′ long.

Blooms: May–October.

Elevation: 100 to 8,000′.

Habitat: Moist soil along streams, roadsides, ditches, and in pastures.

Comments: A perennial herb and a weed, occasionally used as a potherb. Native to Eurasia, it is now naturalized. Has deep taproot and reddish stems. Fifteen species of *Rumex* in Arizona. Photograph taken in vicinity of Prescott, June 7.

FENDLER'S MEADOW RUE

Thalictrum fendleri
Buttercup Family (Ranunculaceae)

Height: To 3′.

Flowers: Petalless; greenish to yellowish stamens on purplish, threadlike stalks; to ⅜″ wide. Male and female flowers on separate plants; male flowers resemble miniature tassels, female flowers are tiny clusters.

Leaves: Green, fernlike, delicate; thin, stalked, compound; divided several times into leaflets wider than long; to 1½″ long.

Blooms: May–August.

Elevation: 5,000 to 9,500′.

Habitat: Pine and spruce-fir forests.

Comments: Perennial herb. Two species of *Thalictrum* in Arizona. Photograph taken at Greer, July 5. A similar species, **Waxyleaf Meadow-Rue** (*Thalictrum revolutum*) has thick, rigid leaflets that are more long than wide, and white stamens.

CALIFORNIA SNAKE BUSH
Snakewood
Colubrina californica
Buckthorn Family (Rhamnaceae)

Height: To 10′.

Flowers: Greenish or yellowish, inconspicuous; to ⅛″ wide; solitary or in small cluster; followed by round, light brown, woody, 3-celled, drupelike seed capsule to ¼″ in diameter.

Leaves: Light green; finely haired, especially on margins; prominent sunken veins on upper surface; untoothed; oval, elliptical to oblong; to ½″ long, ⅜″ wide.

Blooms: June–August.

Elevation: 2,000 to 3,000′.

Habitat: Along desert washes and on dry, rocky slopes.

Comments: Young twigs are pinkish, with short hairs; older branches are light gray. Has spines at tips of branches. One species of *Colubrina* in Arizona. Photograph taken in Kofa Mountains, March 6.

BITTER CONDALIA
Condalia globosa var. *pubescens*
Buckthorn Family (Rhamnaceae)

Height: Shrub, or small tree to 20′.

Trunk: To 1′ in diameter.

Bark: Brownish gray, thin, fissured, shreddy.

Flowers: 5 yellowish green, pointed sepals; cup-shaped, petalless, fragrant; less than ⅛″ wide; in leaf axils, followed by dark blue to blackish, juicy, bitter, berrylike fruit to ¼″ in diameter in spring, when maturing.

Leaves: Yellowish green, usually finely haired, toothless; spoon-shaped; to ½″ long, ¼″ wide; occurring singly or in small clusters on branches.

Blooms: March or in the fall.

Elevation: 1,000 to 2,500′.

Habitat: Dry, sandy plains, along desert washes, and on rocky slopes.

Comments: Spine-tipped, many spreading branches. Six species of *Condalia* in Arizona. Photograph taken at Desert Botanical Garden, Phoenix, March 23. Native to southwestern Arizona.

WARNOCK CONDALIA

Condalia warnockii var. *kearneyana*
Buckthorn Family (Rhamnaceae)

Height: To 5′.

Flowers: Tiny and petalless; solitary or in cluster; 5 stamens; followed by roundish, dark red fruit, to ¼″ in diameter.

Leaves: Dark green, alternate, spatula-shaped or elliptical; to 3⁄16″ long, 1⁄16″ wide; crowded together on branches.

Blooms: Spring.

Elevation: 2,500 to 4,500′.

Habitat: Sandy or gravelly slopes and mesas.

Comments: Has thorn-tipped branches. Four species of *Condalia* in Arizona. Photograph taken at Catalina State Park, April 2.

BIRCHLEAF BUCKTHORN

Coffeeberry
Frangula betulifolia ssp. *betulifolia (Rhamnus betulaefolia)*
Buckthorn Family (Rhamnaceae)

Height: To 8′.

Trunk: Shrub to 4″ in diameter.

Bark: Smooth, gray.

Flowers: Greenish, with 5 joined, pointed lobes; slightly hairy; to ⅛″ wide; in clusters in leaf axils; followed by shiny, blackish purple fruits (in the fall); to ⅜″ in diameter.

Leaves: Deciduous; bright green and shiny above, lighter green and very finely haired beneath; broadly oblong to egg-shaped; prominent veins beneath are reddish pink, blunt or short-pointed at tip; very finely toothed or not toothed; thin, edges not rolled under; to 4″ long.

Blooms: May–June.

Elevation: 3,500 to 7,500′.

Habitat: Canyons, along streams in oak woodlands, and ponderosa pine forests.

Comments: Non-thorny, with reddish brown stems. (Older stems are dark red.) Leaves resemble birch leaves. Native Americans chewed inner bark as a medicine and ate the fruits. Fruits eaten by wildlife; foliage and twigs browsed by deer and other wildlife. Three species of *Frangula* in Arizona. Photograph taken at Oak Creek Canyon, May 29. A similar species, **California Buckthorn** (*Frangula californica*) (page 384), has thicker, narrower, evergreen leaves, which are slightly rolled under on margins.

CALIFORNIA BUCKTHORN

Coffeeberry
Frangula californica (Rhamnus californica)
Buckthorn Family (Rhamnaceae)

Height: To 20′, but commonly to 10′.

Trunk: To 6″ in diameter, generally less.

Bark: Pinkish when young, gray and smooth with age.

Flowers: Greenish, with 5 joined, pointed lobes; slightly hairy; to ⅛″ wide; in clusters in leaf axils; followed by shiny, juicy berries to ⅜″ in diameter; in small cluster; changing from green to red to black in fall.

Leaves: Dull green above, paler green and hairy beneath; elliptical to oval, short-pointed at tips; evergreen, very finely toothed; thick and leathery; prominent veins beneath; edges slightly rolled under; to 3″ long; on pinkish stem.

Blooms: May–June.

Elevation: 3,500 to 6,500′.

Habitat: Canyons, ponderosa pine forests, and mountainsides.

Comments: Birds, bears, and deer feed on fruits. Three species of *Frangula* in Arizona. Photograph taken near Christopher Creek, September 27. A similar species, **Birchleaf Buckthorn** (*Frangula betulifolia*) (page 383), has thinner, wider, deciduous leaves, which are not rolled under on margins.

HOLLYLEAF BUCKTHORN

Redberry Buckthorn
Rhamnus crocea
Buckthorn Family (Rhamnaceae)

Height: Shrub, or small tree to 15′.

Bark: Dark gray, rough, fissured.

Flowers: Yellowish green, with 5 joined, pointed lobes; to ⅛″ wide; in clusters in leaf axils, followed by bright red, ¼″ diameter, juicy fruits in fall. Male and female on different plants.

Leaves: Shiny, yellow-green above, paler beneath; hollylike, oval to nearly round; spiny-toothed, leathery; to 1½″ long.

Blooms: March–May.

Elevation: 3,000 to 7,000′.

Habitat: Chaparral and lower elevation ponderosa pine forests.

Comments: Evergreen. A slow grower. Browsed by bighorn sheep and deer. Native Americans consumed fruits. Four species of *Rhamnus* in Arizona. Photograph taken in Superstition Mountains, April 6.

GRAY THORN

Lotebush
Ziziphus obtusifolia
Buckthorn Family (Rhamnaceae)

Height: Shrub to 10'.

Flowers: Whitish green, tiny, less than ⅛" long, in a stalked cluster, followed by round to elliptical fruits, to ¼" long, maturing to blue-black.

Leaves: Dark green, oblong, finely haired; to ¾" long, ⅜" wide.

Blooms: May–September.

Elevation: 1,000 to 5,000'.

Habitat: Desert, grassland, and mesas.

Comments: Spiny-branched with gray bark. Fruits eaten by birds, especially white-winged doves and Gambel's quail. Native Americans used parts of plant for medicinal purposes. Solution made from roots used as soap substitute. One species of *Ziziphus* in Arizona. Photograph taken at Tortilla flat, May 7.

CURL-LEAF MOUNTAIN MAHOGANY

Cercocarpus ledifolius
Rose Family (Rosaceae)

Height: Shrub or small tree to 25'.

Trunk: To 1' in diameter.

Bark: Reddish brown, deeply furrowed, and scaly.

Flowers: Yellowish green, petalless, funnel-shaped; 5-lobed, stalkless; to ⅜" wide; growing at leaf bases; followed by reddish brown, narrow, ¼"-long fruit with twisted, 3"-long, hairy tail.

Leaves: Evergreen. Dark green above, hairy beneath; shiny, elliptical, thick, leathery; edges rolled under; almost stalkless; aromatic; grooved midvein; to 1¼" long, usually in clusters.

Blooms: April–June.

Elevation: 5,000 to 9,000'.

Habitat: Dry, rocky mountain slopes.

Comments: Evergreen, with very hard wood. Unrelated to true mahogany. Browsed by elk, deer, and livestock. Native Americans concocted a red dye from roots. Three species of *Cercocarpus* in Arizona. Photograph taken at Madera Canyon, April 29.

ALDERLEAF MOUNTAIN MAHOGANY

Cercocarpus montanus
Rose Family (Rosaceae)

Height: To 10'.

Flowers: Greenish, petalless, with green sepals forming a tube with pinkish, flared lobes; followed by ½"-long seed attached to fuzzy, spirally twisted tail to 3" long.

Leaves: Grayish green, paler green beneath; wedge-shaped, toothed on upper margins; deeply veined; to 1" long, ½" wide.

Blooms: Spring.

Elevation: 4,500 to 7,000'.

Habitat: Canyons and hillsides in pinyon-juniper and pine belts.

Comments: Hardwood shrub. Browsed by livestock, bighorns, and deer. Navajo Indians used shrub to make a red dye for wool. Three species of *Cercocarpus* in Arizona. Photograph taken at Canyon de Chelly National Monument, June 27.

BIRCHLEAF MOUNTAIN MAHOGANY

Hardtack
Cercocarpus montanus var. *glaber (Cercocarpus betuloides)*
Rose Family (Rosaceae)

Height: Shrub to 8', or small tree to 20'.

Trunk: To 6" in diameter.

Bark: Gray or brown, smooth, becoming scaly.

Flowers: Yellowish green, petalless, funnel-shaped, 5-lobed; nearly stalkless; to ⅜" wide; 1 to 3 at leaf base, followed by narrow ⅜" long fruit, with hairy, twisted tail, to 3¼" long.

Leaves: Evergreen; dark green above, paler and hairy beneath; elliptical, finely pointed, tapering toward base, toothed beyond middle; prominent sunken veins; short-stalked; to 1¼" long.

Blooms: March–July.

Elevation: 3,500 to 6,500'.

Habitat: Dry, rocky mountain slopes in oak and chaparral areas.

Comments: Crossbreeds with **Hairy Mountain Mahogany** (*Cercocarpus montanus* var. *paucidentatus*) (page 387). Browsed by livestock, pronghorn, elk, and deer. Sharp end and twisted tail of hard fruit aid in penetration into soil. Three species of *Cercocarpus* in Arizona. Photograph taken north of Superior, April 20.

HAIRY MOUNTAIN MAHOGANY

Cercocarpus montanus var. *paucidentatus (Cercocarpus breviflorus)*
Rose Family (Rosaceae)

Height: Shrub or small tree to 15'.

Trunk: To 6" in diameter.

Bark: Gray to reddish brown and smooth, becoming fissured and scaly with age.

Flowers: Yellowish green, funnel-shaped, and petalless; 5-lobed; to ¼" wide, ½" long; 1 to 3 in leaf axils; followed by reddish brown, hairy, ¼"-long fruit with twisted tail of whitish hairs, to 1½" long.

Leaves: Evergreen, dark green and slightly hairy above, paler green and hairy beneath; elliptical, tapering to base; edges turned under, rounded teeth near tip; to 1" long, ½" wide.

Blooms: March–November.

Elevation: 5,000 to 8,000'.

Habitat: Dry slopes in chaparral and oak woodlands.

Comments: Has open crown. Browsed by deer and livestock. Three species of *Cercocarpus* in Arizona. Photograph taken at Lynx Lake, May 27.

BLACKBRUSH

Coleogyne ramosissima
Rose Family (Rosaceae)

Height: To 5'.

Flowers: Four yellowish green sepals, petalless, numerous stamens; to ⅜" long; short-lived; solitary, at ends of short branchlets; followed by small, brown fruit capsule, to ½" long, which, when dried, reveals white hairs in center.

Leaves: Grayish green, hairy, narrow; somewhat thick, club-shaped; to ⅜" long, 1⁄16" wide; in clusters all along branchlets.

Blooms: March–May.

Elevation: 3,000 to 6,500'.

Habitat: Dry, gravelly or sandy open plains and mesas.

Comments: A rigid, many-branched shrub, often with spiny-tipped ends. Its dark gray bark turns blackish with age and very black when wet. Gives a blackish appearance to the landscape when found in dense stands devoid of other shrubs. Browsed mainly by sheep, goats, and deer. Wrongly called burro-brush. One species of *Coleogyne* in Arizona. Photograph taken in Page area, June 26.

FREMONT COTTONWOOD
Alamo
Populus fremontii
Willow Family (Salicaceae)

Height: To 100′, but usually less.

Trunk: To 4′ in diameter, but usually less.

Bark: Branches on young trees are gray-brown, thin, and smooth; on old trees, dark reddish brown, thick, and deeply furrowed.

Flowers: Greenish yellow, tiny, male and female on separate trees (female forms long, slender catkins); to 4″ long; followed by fuzzy-white, cottony seeds.

Leaves: Yellow-green, shiny, broadly triangular and pointed; nearly straight across base; coarsely toothed margins; to 2½″ long, 3″ wide.

Blooms: Early spring before leaves form.

Elevation: 150 to 6,000′.

Habitat: Along streams and moist areas.

Comments: In autumn, leaves turn golden yellow before falling. Wood used as fuel. Hopi Indians use roots for kachina dolls, wood for drums. A favorite of beavers for food and for dam building. Eight species of *Populus* in Arizona. Photograph taken at Saguaro Lake, October 18.

QUAKING ASPEN
Golden Aspen
Populus tremuloides
Willow Family (Salicaceae)

Height: Usually to 40′, rarely to 80′.

Trunk: Usually to 1′ in diameter, rarely to 30″.

Bark: Whitish or yellowish, waxy-appearing, smooth, thin. Older trees are dark gray, thick, furrowed.

Flowers: Tiny, inconspicuous, in catkins; male and female on separate trees; followed by cottony seeds on female trees.

Leaves: Almost round, short-pointed, finely saw-toothed; shiny green above, dull green beneath; leaf stalk longer than leaf blade and flattened lengthwise at right angles to leaf; to 3″ long.

Blooms: Early spring before leaves form.

Elevation: 6,500 to 10,000′.

Habitat: Ponderosa pine and spruce-fir forests.

Comments: Every breeze causes slender, flattened leaf stalks to tremble. In autumn, leaves turn golden yellow or orange before falling. Browsed by livestock, deer, and elk; a favorite of beavers. Bark used medicinally by pioneers and Native Americans. Early growth in burned or logged areas, later replaced by conifers. Wood used mainly for paper pulp. Eight species of *Populus* in Arizona. Photograph taken near Willow Springs Lake, September 13.

BEBB WILLOW

Sauz

Salix bebbiana

Willow Family (Salicaceae)

Height: Usually much-branched shrub or small tree to 15'.

Trunk: To 6" in diameter.

Bark: Grayish, with reddish orange twigs. Smooth when young; rough and furrowed with age.

Flowers: Catkins on short, leafy twigs; to 1½" long.

Leaves: Dull green and finely haired above; whitish, finely haired and strongly net-veined beneath. Untoothed; elliptical or oblong; to 3½" long, 1" wide.

Blooms: Spring.

Elevation: 7,000 to 11,000'.

Habitat: Coniferous forests along stream, springs, and lakes.

Comments: Often forming clusters. Broad, rounded crown. Browsed by elk and deer, and used by beavers for food and dam building. Twigs used in making baskets and furniture. Extract from bark used medicinally. Nearly twenty species of *Salix* in Arizona. Photograph taken in Mormon Lake area, June 2.

BONPLAND WILLOW

Red Willow

Salix bonplandiana

Willow Family (Salicaceae)

Height: To 50', usually to 25'.

Trunk: To 2' in diameter.

Bark: Dark gray to black; ridged and fissured. Twigs are reddish purple.

Flowers: Tiny, in catkins, to 1½" long.

Leaves: Yellowish green above, shiny, whitish beneath; yellowish midrib; narrowly lance-shaped, long-pointed, broadest in middle; very finely toothed; to 6" long, ¾" wide.

Blooms: April.

Elevation: 3,000 to 5,000'.

Habitat: Wet soils along lakes and streams in southeastern and central Arizona.

Comments: Has broad, rounded crown. Branches droop at ends. Semi-evergreen; leaves shed irregularly during winter, rather than in fall like other willows. Browsed by deer and livestock; bark is eaten by beavers, rabbits, and small rodents. Nearly twenty species of *Salix* in Arizona. Photograph taken at Patagonia Lake State Park, April 27.

COYOTE WILLOW
Basket Willow
Salix exigua
Willow Family (Salicaceae)

Height: Shrub to 15′.

Trunk: To 5″ in diameter.

Bark: Greenish; smooth when young, grayish brown with age.

Flowers: In catkins; hairy, with yellow scales; to 2½″ long.

Leaves: Silvery-hairy on unfolding, later becoming dull grayish green and less hairy; very long and narrow, untoothed or with a few teeth; leaf to 4″ long, to ¼″ wide.

Blooms: Spring.

Elevation: To 9,500′.

Habitat: Along streams, silt flats, and riverbanks.

Comments: Rarely treelike. Often forms thickets of clustered stems. Prevents erosion. Browsed by livestock and wildlife. Twigs and bark used by Native Americans for basket making. Nearly twenty species of *Salix* in Arizona. Photograph taken at Lynx Creek, Prescott, May 27.

GOODDING WILLOW
Salix gooddingii
Willow Family (Salicaceae)

Height: To 45′.

Trunk: To 30″ in diameter.

Bark: Gray, thick, rough, deeply furrowed with narrow ridges. Twigs are yellow.

Flowers: Tiny; in catkins to 3½″ long; followed by cottony seeds.

Leaves: Shiny green or yellowish green, narrowly lance-shaped, long-pointed; slightly curved to one side, finely toothed; to 5″ long, ¾″ wide.

Blooms: March.

Elevation: Below 7,000′.

Habitat: Along streams.

Comments: Largest willow in Arizona. Has broad, rounded crown. Prevents stream erosion with its deep root system. Nearly twenty species of *Salix* in Arizona. Photograph taken near Granite Reef Dam, March 1.

SCOULER WILLOW

Salix scouleriana
Willow Family (Salicaceae)

Height: Large shrub to small tree, rarely a tree to 30′ tall.

Trunk: To 4″ in diameter.

Bark: Grayish, smooth, and thin when young; dark brown and fissured with age.

Flowers: In catkins; nearly stalkless; stout; catkin to 2″ long.

Leaves: Dark green, shiny above, whitish with grayish to reddish hairs beneath; midvein yellow; untoothed; elliptical but variable in shape; to 4″ long, 1½″ wide.

Blooms: April–May.

Elevation: 7,000 to 10,000′.

Habitat: Coniferous forests and clearings in moist or dry conditions.

Comments: Form clusters with rounded crown. Among the first trees to appear after a fire, stopping erosion and providing protection for developing conifers. Browsed by livestock and deer. Nearly twenty species of *Salix* in Arizona. Photograph taken at River Reservoir, Greer, July 5.

YEW-LEAF WILLOW

Salix taxifolia
Willow Family (Salicaceae)

Height: Shrub, or tree to 40′, but smaller in Arizona.

Trunk: To 18″ in diameter.

Bark: Grayish brown, rough, and fissured.

Flowers: In catkins with yellowish, hairy scales; catkin to ¾″ long.

Leaves: Grayish green, linear, needlelike; crowded along stems; to 2″ long, ⅛″ wide.

Blooms: March and occasionally again in fall.

Elevation: 3,500 to 6,000′.

Habitat: In oak woodlands and along streams and washes in mountains and foothills.

Comments: Compact, rounded crown. Browsed by livestock. Slow grower. Branches droop at tips. A soil binder. Nearly twenty species of *Salix* in Arizona. Photograph taken at Catalina State Park, November 10.

ALUM-ROOT
Heuchera eastwoodiae
Saxifrage Family (Saxifragaceae)

Height: Flower stalk to 20".

Flowers: Yellowish green, without petals; pointed sepals, 6 short, yellow stamens; to 3⁄16" wide; in loose, terminal raceme on weak, leafless stalk.

Leaves: Dark green, roundish, finely haired, and scalloped; to 3½" wide; basal, on long leaf stalks.

Blooms: May–August.

Elevation: 5,000 to 8,000'.

Habitat: Moist slopes in ponderosa pine forests and canyons.

Comments: Perennial herb; found in central Arizona. Six species of *Heuchera* in Arizona. Photograph taken at Black Canyon Lake, June 4.

TREE OF HEAVEN
Ailanthus altissima
Simarouba Family (Simaroubaceae)

Height: To 80'.

Trunk: To 2' in diameter.

Bark: Dark gray, thin, rough.

Flowers: Yellowish green; to ¼" long; in large, dense, terminal cluster, to 10" long; followed by a reddish brown, twisted, winged fruit, to 1½" long; in very large, dense cluster. Male and female flowers on separate trees.

Leaves: Dark green, pinnately compound; to 2' long, with 6 to 12 pairs of broadly lance-shaped, pointed leaflets, to 5" long, 2" wide, toothed at base.

Blooms: Spring.

Elevation: Not available. Photograph taken at 3,300'.

Habitat: Roadsides and wastelands.

Comments: Deciduous; native of China. Escapees from cultivation in the Southwest. Spreads from seeds and root suckers. Clusters of seeds hang on tree most of winter. Flowers of male trees have objectionable odor. One species of *Ailanthus* in Arizona. Photograph taken at Dead Horse Ranch State Park, September 9.

CRUCIFIXION THORN

Corona-De-Cristo
Castela emoryi (Holacantha emoryi)
Simarouba Family (Simaroubaceae)

Height: Small tree, or large shrub to 12'.

Flowers: Yellowish green, hairy, with 4 to 8 petals; yellow stamens, pinkish buds; to ⅜" wide; occurring singly or in small cluster. Male and female flowers on separate plants; female flowers followed by a ring composed of 5 to 10 flattened, 1-seeded segments, to ¼" long, persisting for years on plant.

Leaves: Usually leafless; when present, very small or scalelike.

Blooms: June–July.

Elevation: 500 to 2,000'.

Habitat: Desert plains.

Comments: Grayish green, smooth, rigid twigs, up to 8" long and about ¼" in diameter, are tipped with sharp spines. Two species of *Castela* in Arizona. Photograph in flower taken at Desert Botanical Garden, Phoenix, May 27. Plant native mostly to the southwestern part of Arizona and elsewhere in state.

JOJOBA

Goatnut
Simmondsia chinensis
Jojoba Family (Simmondsiaceae)

Height: Shrub to 7'.

Flowers: Greenish yellow, tiny; male and female flowers on separate plants; male (staminate) in dense cluster producing much pollen; female (pistillate) to ½" long; followed by a green, hard-shelled, acornlike capsule, to 1" long, turning tan at maturity.

Leaves: Grayish green, leathery, thick, and elliptical, to 1½" long.

Blooms: December–July; extremely variable.

Elevation: 1,000 to 5,000'.

Habitat: Along washes, on alluvial fans, and dry, rocky slopes.

Comments: Evergreen. Browsed by deer and bighorn sheep; rodents eat seeds. Native Americans and pioneers used seeds as food and as substitute for coffee; their bitter taste improved with roasting. Waxy oil from seeds used commercially in medicines and cosmetics. One species of *Simmondsia* in Arizona. Photograph taken in Mesa on January 31.

RABBIT THORN

Pale Wolfberry
Lycium pallidum var. *pallidum*
Nightshade Family (Solanaceae)

Height: To 6′.

Flowers: Greenish yellow, funnel-shaped, and 5-lobed, with lobes flaring outward; 5 stamens extending beyond corolla tube; to ¾″ long, ⅝″ wide; followed by round, orange to red, juicy fruit, to ¼″ in diameter.

Leaves: Bluish white to bluish green, spatula-shaped to elliptical or oval; leathery, covered with bloom; to 3″ long, ½″ wide; in clusters on branches.

Blooms: April–June.

Elevation: 3,500 to 7,000′.

Habitat: Dry plains and slopes.

Comments: Has spines along branches. Browsed by livestock. The bitter fruits are eaten by birds, small animals, and people. Eleven species of *Lycium* in Arizona. Photograph taken in vicinity of Fort Bowie, May 8.

NARROWLEAF CATTAIL

Typha angustifolia
Cattail Family (Typhaceae)

Height: To 6′.

Flowers: Male: topmost, yellowish and minute, grow on clublike spike; to 5″ long (after pollen sheds, spike is bare). Female: beneath male flowers, minute, in brownish cylinder; grow to 4″ long. A definite bare space between male (staminate) and female (pistillate) flowers in this species.

Leaves: Dark green, erect, strap-like; to 6′ long, ½″ wide.

Blooms: June–September.

Elevation: 1,000 to 5,500′.

Habitat: Marshy areas in shallow water and at marshy areas at lake edges.

Comments: Perennial herb. Prefers alkaline water. Three species of *Typha* in Arizona. Photograph taken at Lynx Lake, September 11.

COMMON CATTAIL

Typha latifolia
Cattail Family (Typhaceae)

Height: To 9′.

Flowers: Male: topmost; yellowish and minute, grow on clublike spike; to 6″ long (after pollen sheds, spike is bare). Female: beneath male flowers, minute, in brownish cylinder; grow to 6″ long. No space between male (staminate) and female (pistillate) flowers in this species.

Leaves: Dark green, flat, reedlike, erect, strap-like; to 9′ long, 2″ wide.

Blooms: June–August.

Elevation: 3,500 to 7,500′.

Habitat: Marshy areas in shallow water, ponds, and edges of lakes.

Comments: Perennial herb. Leaves are used to weave mats. Pioneers used cattail fluff for bedding. Native Americans used rootstocks for food. Muskrats feed on rootstocks. Red-winged blackbirds, marsh wrens, and other birds make their nests among cattails. Three species of *Typha* in Arizona. Photograph taken at McNary, August 10.

NETLEAF HACKBERRY

Canyon Hackberry
Celtis reticulata
Elm Family (Ulmaceae)

Height: Large shrub, or small tree to 30′.

Trunk: To 1′ in diameter.

Bark: Gray, smooth; fissured and warty in older trees.

Flowers: Greenish, protrude from base of immature leaf, to ⅛″ wide; followed by orange-red, sweet berry to ⅜″ in diameter, on long stem at each leaf axil.

Leaves: Dark green and rough above, yellow-green and slightly hairy with prominent veins beneath; growing in 2 rows; margins can be coarsely saw-toothed; thick, lopsided, usually oval, but variable; to 2½″ long, 1½″ wide.

Blooms: March–April.

Elevation: 1,500 to 6,000′.

Habitat: Moist soils along streams, in canyons, and on hillsides from upper desert to oak woodlands.

Comments: Deciduous; in autumn, leaves turn yellow before falling. Mites and fungi often cause deformed, bushy growths called "witches'-brooms" in branches. At times plant lice (psyllids) cause galls to form on leaves. Two species of *Celtis* in Arizona. Photograph taken at Patagonia Lake State Park, April 26.

SIBERIAN ELM

Ulmus pumila
Elm Family (Ulmaceae)

Height: To 60', but smaller in Arizona.

Trunk: To 1½' in diameter; less in Arizona.

Bark: Grayish or brown, rough, and furrowed.

Flowers: Greenish, petalless, to ⅛" wide; in clusters, appearing before leaves unfold; followed by a roundish, flat, wafer-like, notched fruit, to ½" wide, in small clusters on branches.

Leaves: Dark green with toothed margins; lopsided at base, narrowly elliptical; to 2½" long.

Blooms: Early spring.

Elevation: 1,000 to 5,000'.

Habitat: Dry or moist soils in woodlands and parks.

Comments: Introduced from northern China and eastern Siberia. Grown for shade and windbreaks, they seed prolifically and have naturalized in some areas. Browsed by deer. No native species of *Ulmus* grow in Arizona. Photograph taken at Lynx Creek, near Prescott, May 27.

TALL WHITE NETTLE

Urtica dioica ssp. *gracilis (Urtica gracilis)*
Nettle Family (Urticaceae)

Height: To 5'.

Flowers: Greenish white, tiny, petalless, to ⅛" wide; in threadlike, drooping clusters in leaf axils.

Leaves: Dark green, opposite, lance-shaped, with prominent, spiny veins; triangular-toothed; to 7" long.

Blooms: July–August.

Elevation: To 9,000'.

Habitat: Along streams and springs.

Comments: Has spiny stems. Three species of *Urtica* in Arizona. Photograph taken at Greer, July 5.

DWARF MISTLETOE

Arceuthobium microcarpum
Mistletoe Family (Viscaceae)

Dimension: Aerial parasitic green to purple shrub to 10" across.

Flowers: Male and female on separate plants. Flowers tiny and greenish brown. Oblong fruit ⅛" long, bicolored. Eaten and distributed primarily by birds or dispersed explosively (to over 40').

Leaves: Reduced to minute scales.

Blooms: August–September.

Elevation: 7900 to 10,400'.

Habitat: Mixed conifer forests.

Comments: This evergreen is found at higher elevations and is parasitic on **Spruce** (*Picea* species) and **Bristlecone Pine** (*Pinus aristata*) (page 423). Nine species of *Arceuthobium* in Arizona. Photograph taken in mountains above Greer, August 7.

DESERT MISTLETOE

Phoradendron californicum
Mistletoe Family (Viscaceae)

Dimension: A mass in tree branches to 2' in diameter.

Flowers: Yellowish green, inconspicuous but sweetly fragrant; male and female on separate plants. Fruits, pinkish berries, grow to ⅛" in diameter.

Leaves: Scalelike, with brownish yellow stems; densely clustered on branches of host tree.

Blooms: Spring.

Elevation: Below 4,000'.

Habitat: Deserts and foothills.

Comments: Evergreen. A partial parasite, mainly on leguminous trees and shrubs such as paloverde, mesquite, ironwood, and acacia, although also observed on creosote bush, jojoba, and other nonlegumes. Over time, often kills host plant by invading bark and sap with its roots, draining moisture and nutrients from host. Five species of *Phoradendron* in Arizona. Photograph taken in Superstition Mountains, February 4. **American Mistletoe** (*Phoradendron serotinum* ssp. *macrophyllum*), a mistletoe of higher elevations, has bright green, 2"-long, rounded leaves, and white berries, and resembles the "kissing" mistletoe of Christmas season. It parasitizes cottonwood, sycamore, oak, and willow trees.

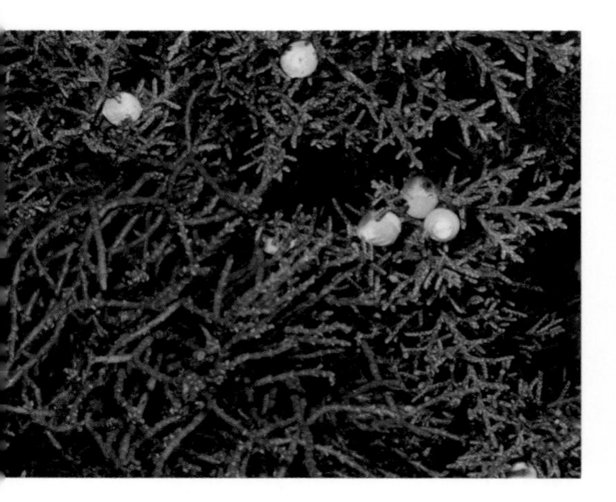

JUNIPER MISTLETOE
Phoradendron juniperinum
Mistletoe Family (Viscaceae)

Dimension: A mass in juniper branches growing to 2′ in diameter.

Flowers: Greenish, inconspicuous; male and female on separate plants; white to pinkish berries.

Leaves: Triangular and scalelike with yellowish green stems; jointed, shiny, growing in dense clusters on branches of juniper trees.

Blooms: July–August.

Elevation: 4,000 to 7,000′.

Habitat: Juniper woodlands.

Comments: Evergreen. A parasite on several species of juniper, sapping nutrients from host plant with its modified roots, sometimes killing host plant. Birds feed on berries, carrying sticky seeds from tree to tree, thus spreading this parasite. This species used medicinally by Hopi Indians. Five species of *Phoradendron* in Arizona. Photograph taken east of Camp Verde, September 30.

VIRGINIA CREEPER
Woodbine
Parthenocissus quinquefolia (Parthenocissus inserta)
Grape Family (Vitaceae)

Height: A woody vine climbing tree trunks and over walls.

Flowers: Greenish, inconspicuous, in clusters opposite the leaves; followed by bunches of bluish black berries to ¼″ wide.

Leaves: Dark green, shiny; to 4″ long; palmately compound, with 5 to 7 leaflets.

Blooms: May–September.

Elevation: 3,000 to 7,000′.

Habitat: Moist canyons and roadsides.

Comments: Deciduous; leaves turn red in fall. No aerial roots. Fruits eaten by birds and small mammals. Two species of *Parthenocissus* in Arizona. Photograph taken at Walnut Canyon National Monument, September 7.

CANYON GRAPE

Arizona Grape
Vitis arizonica
Grape Family (Vitaceae)

Height: A sprawling, scrambling, woody vine, with tendrils, often covering entire trees.

Flowers: Greenish white, small; followed by juicy, purple-black clusters of grapes. Male and female flowers on separate plants.

Leaves: Dark green, broadly heart-shaped, coarsely toothed; to 6" long, 4¾" wide.

Blooms: April–July.

Elevation: 2,000 to 7,500'.

Habitat: Along streams and in canyons.

Comments: Grapes used for making jelly, wine, and juice. Attractive to birds. One species of *Vitis* in Arizona. Photograph taken at Lynx Lake area, September 11.

CACTI

SAGUARO

Giant Cactus
Carnegiea gigantea
Cactus Family (Cactaceae)

Height: To 50', usually to 30' high; 2½' in diameter; and weighing 9 tons.

Flowers: Waxy-white, funnel-shaped, to 3" wide; in crown-like, terminal clusters on arms or main trunk; opening at night and remaining open part of the following day; egg-shaped, fleshy, green fruit (tinged with red), to 3" long, to 1¾" in diameter.

Stems: Green, columnar, treelike trunk, 12 to 30 prominent ribs; branching well up on main trunk when older, branches to 20" in diameter.

Spines: White-gray, sometimes tinged with pink; 15 to 30 per areole; to 3" long, the longest pointed downward.

Blooms: May–June.

Elevation: 600 to 3,600'.

Habitat: Rocky slopes and well-drained flats in the Sonoran Desert.

Comments: The largest cactus in the U.S., saguaros grow very slowly, but live for 150 to 200 years. In their first year, they grow only about ½". They reach about 1' in 15 years; 10' in 40 to 50 years; and 12 to 20' in 75 to 100 years, when arm buds usually appear.

Young saguaros cannot survive either the heat of the desert sun or trampling; they usually grow under a "nurse" plant, such as a creosote bush or a paloverde tree. Spines on the stems of older plants provide shade to the trunk.

Once saguaros reach a height of about 8', the first flower buds appear. Flowers smell like ripe melon and attract bees and other insects during the daytime, and bats and moths at night. Full-grown saguaros produce at least 100 fruits in a season. When mature in July, the fruits split open, revealing their scarlet linings, deep red pulp, and up to 2,000 tiny, jet-black seeds.

One species of *Carnegiea* in Arizona. Photograph taken at Usery Mountain Recreation Area, April 6.

BEEHIVE CACTUS

Coryphantha vivipara
Cactus Family (Cactaceae)

Height: To 8".

Flowers: Pink, 2½" wide, day-blooming; followed by green, oval fruit to 1" long, ⅝" in diameter.

Stems: Green and globular, to 3" in diameter; forming dense clumps.

Spines: Central spines: reddish to brown at tips, white at base, to ¾" long. Radial spines: white, to ⅝" long.

Blooms: May–July.

Elevation: 4,700 to 7,200'.

Habitat: Rocky, sandy ground in juniper woodlands and ponderosa pine forests.

Comments: Nine species of *Coryphantha* in Arizona. Photograph taken in vicinity of Kohls Ranch, May 23.

BUCKHORN CHOLLA

Cylindropuntia acanthocarpa (Opuntia acanthocarpa)
Cactus Family (Cactaceae)

Height: Shrubby to treelike, to 6'.

Flowers: Variable, from yellow to orange to red; to 2¼" wide, followed by oval, dry fruit with long spines; to 1½" long. When seeds mature, fruit falls from plant before winter.

Joints: Green, cylindrical, to 20" long, 1¼" in diameter, with elongated tubercles or knoblike projections.

Spines: Straw-colored, stout, 12 or more in a cluster; to 1½" long.

Blooms: Mid-April–late May.

Elevation: 500 to 3,500'.

Habitat: Sandy soils of slopes and washes.

Comments: Native Americans ate flower buds after steaming them. Seventeen species of *Cylindropuntia* in Arizona. Photograph taken at Usery Mountain Recreation Area, April 6.

PENCIL CHOLLA

Cylindropuntia arbuscula (Opuntia arbuscula)
Cactus Family (Cactaceae)

Height: Shrubby to treelike, to 9', but usually much less.

Flowers: Yellow, greenish or brownish red; to 1" long, ⅝" wide; followed by green fruit tinged with purple or red; smooth, spineless, and elongated, to 1½" long, ⅞" in diameter, lasting through winter.

Joints: Green, nearly smooth; to 6" long, ½" in diameter.

Spines: Reddish or tan, straight, up to 4 per cluster; largest spine facing downward; to 1½" long.

Blooms: May–June.

Elevation: 1,000 to 3,000'.

Habitat: Gravelly and sandy plains, valleys, and washes.

Comments: Seventeen species of *Cylindropuntia* in Arizona. Photograph taken at Saguaro National Park West, March 31.

TEDDY BEAR CHOLLA

Jumping Cholla
Cylindropuntia bigelovii var. *bigelovii (Opuntia bigelovii)*
Cactus Family (Cactaceae)

Height: Shrubby to treelike, to 5'; rarely to 9'.

Flowers: Greenish or yellowish streaked with lavender, to 1½" wide; near end of joint; followed by yellowish, egg-shaped, knobby fruit to ¾" long, ⅜" wide.

Joints: Light green to bluish green, cylindrical; to 10" long, 2½" in diameter; form arms at top of main stem.

Spines: Silvery to golden when young, black when old; dense, backward-facing barbs; to 1" long.

Blooms: February–May.

Elevation: 100 to 3,000'.

Habitat: Desert to rocky hillsides.

Comments: Joints detach very easily when brushed against, causing people to believe joints "jump off" plant. Detached joints root quickly in desert soil, creating dense thickets. Pack rats carry spiny joints to nest sites, often creating a huge pile to ward off enemies. Despite spines, cactus wrens favor this species and chain fruit cholla as nest sites. Seventeen species of *Cylindropuntia* in Arizona. Photograph taken at Usery Mountain Recreation Area, April 22.

SILVER CHOLLA

Golden Cholla

Cylindropuntia echinocarpa (Opuntia echinocarpa)

Cactus Family (Cactaceae)

Height: To 4'.

Flowers: Greenish yellow, outer parts streaked with red; to 2½" wide; followed by green, egg-shaped fruit with spines on upper half; to 1¼" long, ¾" wide; turning light tan at maturity.

Joints: Green, cylindrical, with conspicuous tubercles to ⅜" long; joints generally to 6" long, 1½" wide.

Spines: Silvery or golden, very dense, 3 to 12 per areole; straight, pointing in all directions, narrow; tapering from base; to 1½" long.

Blooms: April.

Elevation: 1,000 to 4,000'.

Habitat: Flats, slopes, and washes.

Comments: Many-branched. Seventeen species of *Cylindropuntia* in Arizona. Photograph taken in Alamo Lake area, February 26.

CHAIN-FRUIT CHOLLA

Jumping Cholla

Cylindropuntia fulgida var. *fulgida (Opuntia fulgida)*

Cactus Family (Cactaceae)

Height: Shrubby to treelike, to 15', usually less.

Flowers: Deep pink to lavender; to ¾" wide; forming on previous year's pendent fruit, thus length of fruit chain increases each year. Fruit is green, oval, to 1½" long, 1" in diameter.

Joints: Light green, cylindrical; to 6" long, detach easily.

Spines: Grayish to yellowish, barbed, 2 to 12 per areole; to 1¼" long. Variety *mamillata* has sparse, thin spines, giving the plant a much greener appearance.

Blooms: May–August.

Elevation: 1,000 to 3,000'.

Habitat: Deserts and hillsides.

Comments: Joints seem to "jump" from plant when only slightly touched. Seeds are rarely fertile. Plant reproduces when fruit or segment falls to ground and forms roots, starting new plant. Fruits eaten by cattle; plant is pollinated by bees. Cactus wrens and curve-billed thrashers nest in this cactus. Seventeen species of *Cylindropuntia* in Arizona. Photograph taken at Usery Mountain Recreation Area on January 24.

DESERT CHRISTMAS CACTUS

Tesajo

Cylindropuntia leptocaulis (Opuntia leptocaulis)

Cactus Family (Cactaceae)

Height: To 3′ when growing in the open; taller and more vinelike when growing among desert trees.

Flowers: Greenish yellow; to 1″ wide; followed by bright red fruit, fleshy at maturity; to ½″ long, 7⁄16″ in diameter.

Joints: Dark green, cylindrical, branched; to 16″ long, ½″ in diameter.

Spines: Grayish, straight, to 2″ long.

Blooms: May–June.

Elevation: 1,000 to 4,000′.

Habitat: Washes, slopes, and flat areas.

Comments: Red fruits remain on stems much of winter. Seventeen species of *Cylindropuntia* in Arizona. Photograph taken at Tortilla Flat, November 1.

DIAMOND CHOLLA

Cylindropuntia ramosissima (Opuntia ramosissima)

Cactus Family (Cactaceae)

Height: To 5′, but usually less.

Flowers: Apricot to brown with some lavender or red; to ½″ wide; followed by brownish, spiny, elliptical fruit to ¾″ long, ½″ in diameter.

Stems: Grayish green, slender, flattened, and platelike; diamond-shaped, grooved or notched tubercles (projections on joints); to 4″ long.

Spines: Yellow to tan part-way, grayish toward stem; barbed, thin, straight, set in a groove; to 2¼″ long.

Blooms: May–September.

Elevation: 100 to 2,000′; at times to 3,000′.

Habitat: Washes and desert flats in sandy soil.

Comments: Bushy, matted, many-branched. Seventeen species of *Cylindropuntia* in Arizona. Photograph taken in Kofa Mountains, February 22. Diamond-shaped, notched, or grooved tubercles are unique to this cholla.

CANE CHOLLA

Walkingstick Cholla
Cylindropuntia spinosior (Opuntia spinosior)
Cactus Family (Cactaceae)

Height: Shrubby to treelike, to 8′.

Flowers: Variable in color, white or yellowish or red or purple; to 3″ wide, followed by yellow, egg-shaped fruit to 1¾″ long, to 1″ in diameter; fruits persist through winter.

Joints: Light green, to 12″ long, 1″ in diameter; with numerous, elongated tubercles or projections.

Spines: Gray to pinkish, up to 20 per areole, straight, widely radiating, barbed, to ⅜″ long.

Blooms: May–June.

Elevation: 1,000 to 5,000′.

Habitat: Desert grasslands and desert mountainsides.

Comments: Fruits eaten by cattle; Native Americans ate fruits raw or cooked. Cactus wrens nest in branches. Seventeen species of *Cylindropuntia* in Arizona. Photograph taken at Roper Lake, May 2.

STAGHORN CHOLLA

Cylindropuntia versicolor (Opuntia versicolor)
Cactus Family (Cactaceae)

Height: Shrubby to treelike, to 15′, usually less.

Flowers: Orange, brown, yellow, or red (very variable); to 2¼″ wide; followed by green (tinged with purple to red), pear-shaped, fleshy, spineless fruit to 1¾″ long, which remain attached during winter.

Joints: Green, elongated, to 14″ long, 1″ in diameter, with long, knoblike projections.

Spines: Gray or purplish, 7 to 10 in a cluster; to ⅝″ long; spread at all angles.

Blooms: May.

Elevation: 1,000 to 4,000′.

Habitat: Sandy washes, plains, and canyons.

Comments: Has forked branches resembling deer antlers. New fruits often develop on last year's fruits, occasionally a chain of 2 or 3 is seen. Seventeen species of *Cylindropuntia* in Arizona. Photograph taken in Tucson area, May 12.

WHIPPLE CHOLLA

Cylindropuntia whipplei (Opuntia whipplei)
Cactus Family (Cactaceae)

Height: To 2½′ when shrubby.

Flowers: Pale yellow to lemon yellow; to 1¼″ wide; followed by a yellow, spineless, nearly round to egg-shaped fruit; with shallow cavity at top; to 1¼″ long, ¾″ wide; remaining through winter.

Joints: Green, cylindrical, tubercles to ⅜″ long; joint to 6″ long, ¾″ wide.

Spines: White to pinkish tan, needlelike; 7 to 14 per areole; 1 long spine in cluster to 2″ long.

Blooms: June–July.

Elevation: 4,500 to 7,000′.

Habitat: Plains and grasslands.

Comments: Plants often form mats or grow as low bushes. Seventeen species of *Cylindropuntia* in Arizona. Photograph taken at Wupatki National Monument, June 5.

TURK'S HEAD

Blue Barrel
Echinocactus horizonthalonius
Cactus Family (Cactaceae)

Height: To 12″; usually closer to 8″.

Flowers: Pink, funnel-shaped, opening fully to 2¾″ wide; followed by dry fruit, to 1″ long, ½″ in diameter, covered with woolly hairs.

Stems: Bluish green, with flattened globe to short columnar shape; to 8″ in diameter.

Spines: Gray; central spine curving downward, to 1¼″ long; radial spines, to 1″ long, spreading in all directions.

Blooms: May–June.

Elevation: 3,000 to 3,500′.

Habitat: Desert.

Comments: 8-ribbed. Two species of *Echinocactus* in Arizona. Photograph taken at Tohono Chul Park, Tucson, April 15. Variety *nicholii,* long assigned to the Arizona populations, is in question as being a valid taxa. Found in southwestern Pinal and north-central Pima Counties.

MANY-HEADED BARREL

Echinocactus polycephalus var. *polycephalus*
Cactus Family (Cactaceae)

Height: To 2'.

Flowers: Yellow tinged with pink; to 2" wide; followed by a fruit densely covered with white, woolly hairs to ¾" long; fruit to 1" long, ½" in diameter.

Stems: Roundish to cylindrical, hidden by spines; to 2' long, 10" wide, with 13 to 21 ribs; numerous in dense clumps of up to 30 branches, to 4' in diameter.

Spines: Central spines: reddish to pink with gray, felty, deciduous hair covering; rather flat; 4 per areole (3 near-straight, the fourth, lower spine curving downward); to 3" long. Radial spines: 6 to 8 per areole; to 1¾" long.

Blooms: February–March.

Elevation: 100 to 2,500'.

Habitat: Rocky or gravelly slopes.

Comments: A barrel cactus. Two species of *Echinocactus* in Arizona. Photograph taken in Kofa Mountains, February 21.

CLARET CUP CACTUS

Echinocereus coccineus (Echinocereus triglochidiatus var. *melanacanthus)*
Cactus Family (Cactaceae)

Height: To 6".

Flowers: Scarlet, to scarlet-orange, with bright green stigma in center; to 1½" wide, 2¼" long; occurring at top of stem, followed by a 1"-long fruit with deciduous spines.

Stems: Green to bluish green, to 2½" in diameter; with 9 to 10 ribs forming a crowded clump.

Spines: Grayish, tan, or white; central spine to 2½" long.

Blooms: May–July.

Elevation: 4,000 to 9,000'.

Habitat: Open, rocky hillsides and ledges.

Comments: Two varieties of this cactus. Flowers remain open for several days. Fifteen species of *Echinocereus* in Arizona. Photograph taken near Greer, June 15.

STRAWBERRY HEDGEHOG

Hedgehog Cactus
Echinocereus engelmannii
Cactus Family (Cactaceae)

Height: To 20".

Flowers: Varying shades of magenta, with green stigma in center, cup-shaped; to 3" wide; bloom for several days, followed by green, spiny fruit (red when ripe), to 1¼" long, 1" in diameter.

Stems: Green and cylindrical, to 3" in diameter, with 10 to 13 ribs; forming loose or dense cluster to 3' wide.

Spines: Color varies from whitish to golden yellow to pinkish to black; straight, central spines to 2" long; radial spines to 1" long.

Blooms: March–April.

Elevation: To 5,000'.

Habitat: Sandy and rocky flats and hillsides.

Comments: Fruits are edible; they produce a sugary juice and may be eaten like strawberries. Fifteen species of *Echinocereus* in Arizona. Photograph taken south of Phoenix, March 31.

ARIZONA RAINBOW CACTUS

Echinocereus rigidissimus var. *rigidissimus*
Cactus Family (Cactaceae)

Height: To 1'.

Flowers: Magenta to lavender, funnel-shaped, to 5" wide; 1 to 4 flowers, followed by green to greenish purple, oval fruit to 2½" long.

Stems: Usually single, erect, and columnar, with some red to white spines forming alternating horizontal bands on stem; 15 to 22 ribs; to 4" in diameter.

Spines: Pink to gray to light brown; very dense, flat, comb-shaped, covering the entire stem.

Blooms: June–August.

Elevation: 4,000 to 6,000'.

Habitat: Grasslands, mountains, and limestone hills of southern Arizona.

Comments: Fifteen species of *Echinocereus* in Arizona. Photograph taken at Madera Canyon, June 10.

FISHHOOK BARREL CACTUS

Compass Barrel
Ferocactus wislizeni
Cactus Family (Cactaceae)

Height: To 11′ (usually much less), to 2′ in diameter.

Flowers: Shades of orange to yellow to reddish; cup-shaped; day-blooming; to 2½″ wide; in crown at top of stem; followed by yellow, barrel-shaped, scaly fruit to 1¾″ long, 1⅜″ in diameter.

Stems: Single, massive and cylindrical, with 20 to 28 ribs.

Spines: Grayish to reddish, in dense clusters along ribs; large, sharply hooked, flattened, central spine to 2″ long, surrounded by slender, hairlike spines.

Blooms: July–September.

Elevation: 1,000 to 4,500′.

Habitat: Sandy desert and gravelly slopes in desert or grasslands.

Comments: Drawn toward direct sunlight; faster growth on the cactus's shady side causes barrels to lean in a southerly direction, hence the name "compass" cactus. Instead of water, cactus is filled with a slimy alkaline juice. Yellow fruits persist all year. Three species of *Ferocactus* in Arizona. Photograph taken in vicinity of Mesa, July 13. The central spine on the similar **Coville's Barrel** (*Ferocactus emoryi*) is normally not hooked, and is surrounded by stout, stiff spines.

DEVIL CHOLLA

Club Cholla
Grusonia emoryi (Opuntia stanlyi)
Cactus Family (Cactaceae)

Height: To 6″, rarely to 1′.

Flowers: Yellowish to greenish yellow; to 2″ wide; followed by yellow, spiny fruit to 3″ long, ¾″ in diameter.

Joints: Cylindrical; to 6″ long, 1½″ in diameter.

Spines: Straw-colored to brownish, straight, 18 to 21 per areole; to 2″ long.

Blooms: May–June.

Elevation: 2,500 to 4,000′.

Habitat: Plains and mesas in sandy soil.

Comments: Belongs to the club/mat-forming cholla group. Plants form mats several yards in diameter. There are several varieties of this species. Three species of *Grusonia* in Arizona. Photograph taken at Roper Lake, May 2.

PINCUSHION CACTUS

Fishhook Cactus
Mammillaria grahamii var. *grahamii (Mammillaria microcarpa)*
Cactus Family (Cactaceae)

Height: To 6".

Flowers: Pink to lavender; to 1" wide; forming crown at top of stem; lasting several days; followed by red, smooth, club-shaped fruit to 1" long.

Stems: Cylindrical, with close-set "nipples" obscured by spines; solitary at first, then branching, to 2" in diameter.

Spines: Grayish, dense, in clusters; central spine is dark reddish brown and hooked like an unbarbed fishhook.

Blooms: April–August.

Elevation: To 4,500'.

Habitat: Dry, gravelly areas in deserts; usually under bushes.

Comments: "Mammillaria" refers to nipplelike projections on stems. Pollinated by bees. Nine species of *Mammillaria* in Arizona. Photograph taken in Mesa area, July 21.

BEAVERTAIL CACTUS

Beavertail Prickly Pear
Opuntia basilaris
Cactus Family (Cactaceae)

Height: Grows in clumps 2' high and up to 6' in diameter.

Flowers: Pink to magenta; to 3" wide; at upper end of pad; followed by grayish, oval fruit to 1" long, sparsely covered with glochids.

Stems: Grayish green, with flat joint-pads; oval to spoon-shaped; spineless but with glochids; to 1' long, 6" wide, ½" thick.

Spines: None; has clusters of very fine, brown to reddish brown glochids, to ⅛" long.

Blooms: March–May.

Elevation: 200 to 4,000'.

Habitat: Sandy or gravelly soils of canyons; washes or flats in desert.

Comments: Glochids detach easily upon contact and are difficult to remove. Joint or pad resembles a beaver's tail. Pads root very easily. Native Americans used both fruits and pads for food. Pack rats feed on seeds. Twenty-four species of *Opuntia* in Arizona. Photograph taken at Cattail Cove State Park, March 18.

ENGELMANN'S PRICKLY PEAR

Nopal
Opuntia engelmannii (Opuntia phaeacantha var. *discata)*
Cactus Family (Cactaceae)

Height: To 5'.

Flowers: Yellow, orange, or reddish; to 3¼" wide; on edge of flat pad; followed by smooth, red to purplish, narrow-based, cylindrical fruit, to 3" long, 1½" in diameter.

Stems: Green to bluish green, with circular or oblong pad; to 16" long, to 9" wide; in upright or sprawling chains.

Spines: Ash gray to white, to 3" long, either flattened, curved, or straight. Areoles have brown or yellowish glochids.

Blooms: April–June.

Elevation: 1,500 to 7,500'.

Habitat: Sandy soils of flats, hills, and valleys in desert and grasslands.

Comments: Most common prickly pear in Arizona. A wide, spreading cactus to 15' in diameter. Pollinated by bees. Fruits, called *tunas*, eaten by birds and rodents; also used for jelly and for making red dye. Javelinas eat pads. Stem pulp used to make face cream and water purifier. Glochids difficult to remove from skin. "Itching powder" was made from glochids. Twenty-four species of *Opuntia* in Arizona. Photograph taken at Usery Mountain Recreation Area, April 22.

MOHAVE PRICKLY PEAR

Grizzly Bear Prickly Pear
Opuntia polyacantha var. *erinacea (Opuntia erinacea* var. *erinacea* and var. *ursina)*
Cactus Family (Cactaceae)

Height: To 1'.

Flowers: Reddish to pink to yellow; to 3½" wide, followed by brownish, elliptical fruit, to 1¼" long, ½" wide.

Stems: Bluish green, purplish on tips, elliptical to oblong; to 5" long, 2" wide.

Spines: White to pale gray, straight, long; longest at tip, reduced in size down joint; to 4" long.

Blooms: May–June.

Elevation: 3,000 to 7,000'.

Habitat: Sandy or gravelly soils or rocky hillsides in desert and woodlands.

Comments: In sprawling clumps. Twenty-four species of *Opuntia* in Arizona. Photograph taken near Sunset Crater National Monument, May 31.

PURPLE PRICKLY PEAR

Santa Rita Prickly Pear
Opuntia santa-rita (Opuntia violacea var. *santa-rita)*
Cactus Family (Cactaceae)

Height: To 5′.

Flowers: Pale yellow; to 3½″ wide; followed by red to purplish, smooth, slender fruit, to 1½″ long, ¾″ in diameter.

Stems: Greenish blue to pink to pale violet-purple; thin, flat, almost round; to 8″ wide.

Spines: Few, if any. Reddish brown, to 3″ long when present. Areoles, about 1″ apart, have reddish brown glochids.

Blooms: Spring.

Elevation: Below 4,000′.

Habitat: Sandy or gravelly soils.

Comments: Color of pads varies with drought conditions or lower temperatures. Pads are eaten by rodents and cattle. This cactus is especially vulnerable to attack by tiny, cochineal scale insects, who reproduce and live under patches of sticky, white cottony fuzz, where they suck the juices from cactus pads. If fuzz is crushed a bright carmine-red dye is produced. Twenty-four species of *Opuntia* in Arizona. Photograph taken at Patagonia Lake State Park, May 10.

SENITA CACTUS

Old Man
Pachycereus schottii (Lophocereus schottii)
Cactus Family (Cactaceae)

Height: To 21′.

Flowers: Pale pink, nocturnal; to 1½″ wide; often 2 or more at an areole, within hairlike area; followed by red, egg-shaped fruit to 1¼″ long.

Stems: Green to gray-green, to 5″ in diameter; with clumps to 15′ in diameter; 5 to 9 ribs.

Spines: Upper branches: gray, bristlelike, to 50 per areole, to 3″ long. Lower branches: gray, 8 to 10 per areole, to ⅜″ long.

Blooms: April–August.

Elevation: 1,000 to 2,000′.

Habitat: Sandy soils of desert.

Comments: Concentration of these at Senita Basin at Organ Pipe Cactus National Monument in southwestern Arizona. *Senita* means "old one" in Spanish. When stems age, upper spines become gray and hairlike. One species of *Pachycereus* in Arizona. Photograph taken at Organ Pipe Cactus national Monument, October 23.

DESERT NIGHT-BLOOMING CEREUS

Arizona Queen-Of-The-Night
Peniocereus greggii
Cactus Family (Cactaceae)

Height: Erect or sprawling to 8', but usually less.

Flowers: Waxy white, pointed, perianth segments; numerous white to yellow-tipped stamens; nocturnal, lasting only one night; very fragrant; to 4½" wide, 8½" long; followed by an orangish red, elliptical fruit with short spines, dulling with age, to 3" long, 1½" in diameter.

Stems: Lead-colored and slender, usually 4- to 5-ribbed; to ½" in diameter; unbranched or with up to 12 branches.

Spines: Dark-colored, about 11 to 13 per areole; upper spines to 1⁄32" long, lower ones to ⅛" long; with some whitish color.

Blooms: June–July.

Elevation: 1,000 to 3,500'.

Habitat: Desert flats and washes under trees or shrubs.

Comments: The inconspicuous, apparently dead stems are usually supported by branches of desert shrubs or trees. After dusk the flowers open in spasms, their aroma carrying as far as 100' and attracting moths and other night-feeding insects. The blossoms wilt shortly after sunrise the following morning. The tuberous root is turniplike; 27 roots weighed at the University or Arizona ranged from 1½ to 43 pounds, but roots usually weigh 5 to 15 pounds. Two species of *Peniocereus* in Arizona. Photograph taken at Desert Botanical Garden, Phoenix, June 25.

ORGAN PIPE CACTUS

Pitahaya Dulce
Stenocereus thurberi (Lemaireocereus thurberi)
Cactus Family (Cactaceae)

Height: To 20'.

Flowers: Pale lavender, nocturnal, funnel-shaped; to 3" wide; on sides or tips of stems; followed (in July) by a red, edible, nearly round, 3"-diameter spiny fruit.

Stems: Columnar cactus, free-branching, with stems arising from ground level; 12 to 20 ribs on stem; stems to 8" in diameter.

Spines: Brown to black, 11 to 19 per brown-felted areole; straight, to ½" long; spreading in all directions.

Blooms: May–June.

Elevation: 1000 to 3,500'.

Habitat: Stony desert and rocky hillsides of Organ Pipe Cactus National Monument.

Comments: Resembles pipes of an organ. Organ Pipe Cactus National Monument in southwestern Arizona was established to preserve this species. Tohono O'Odham Indians harvest fruits for syrup, using pulp and seeds for winter food. Sanborn's long-nosed bats are the plant's most important pollinators. One species of *Stenocereus* in Arizona. Photograph taken at Organ Pipe Cactus National Monument, March 30.

GYMNOSPERMS

© RICHARD SPELLENBERG

ARIZONA CYPRESS

Cupressus arizonica
Cypress Family (Cupressaceae)

Height: To 40′ (90′ maximum).

Trunk: To 2′ in diameter (5½′ maximum).

Bark: Variable. Northern Arizona populations tend to have smooth, dark reddish bark with age, while southern and southeastern Arizona plants have rough bark.

Cones: Reddish brown, short-stalked, hard, woody; wedge-shaped scales with a point in center of each; to 1″ in diameter.

Leaves: Pale, bluish green, scalelike; thick, pointed, resinous; to 1⁄16″ long.

Elevation: 3,500 to 5,500′.

Habitat: Canyons and slopes.

Comments: Evergreen. Cones open when mature and remain attached to tree for a number of years. Crown either conical or rounded. One species of *Cupressus* in Arizona.

COMMON JUNIPER

Ground-Cedar
Juniperus communis var. *depressa*
Cypress Family (Cupressaceae)

Height: Shrub to 3′.

Bark: Reddish brown, scaly.

Cones: Bluish, berrylike; to 5⁄16″ in diameter, grow at junction of leaves and branchlets.

Needles: Broad, white band above, shiny, dark green beneath; needle-shaped and concave, sharp-pointed and stiff; to ½″ long; in whorls of 3, spreading at right angles to branchlets.

Elevation: 7,500 to 11,500′.

Habitat: Rocky soils from spruce-fir forests to timberline.

Comments: Prostrate evergreen shrub growing to 10′ in diameter. A valuable erosion fighter. Cones take three seasons to mature. "Berries" of all junipers consist of cone scales that have thickened and grown together—actually cones that never open. "Berries" are eaten by birds and other wildlife, and are used commercially to add flavor to gin. Seven species of *Juniperus* in Arizona. Photograph taken in mountains above Greer, August 14. This juniper is recognizable by its dwarf size and shrubbiness.

ALLIGATOR JUNIPER

Cedro Chino
Juniperus deppeana var. *deppeana*
Cypress Family (Cupressaceae)

Height: To 50′.

Trunk: To 4′ in diameter.

Bark: On older trees the bark's deep fissures are divided into 1 to 2″ squares that resemble an alligator's hide.

Cones: Reddish brown beneath a grayish, waxy coating; hard, 4-seeded; to ½″ in diameter.

Leaves: Bluish green scalelike, pointed; to 1⁄16″ long; dense on branches.

Elevation: 4,500 to 8,000′.

Habitat: In oak and pinyon-juniper woodlands, and lower elevation ponderosa pine.

Comments: Evergreen; largest juniper species in Arizona. Rounded or pyramidal crown. Cones do not mature until second year. "Berries" eaten by wildlife. Cut stumps send forth new shoots. Slow grower; lives 500 to 800 years, with records of 1,100 and 1,400 years. Seven species of *Juniperus* in Arizona. Photograph taken on Mount Graham, April 21. Mature trees recognized by alligator-like bark.

ONE-SEED JUNIPER

Sabina
Juniperus monosperma
Cypress Family (Cupressaceae)

Height: To 25′.

Trunk: To 1½′ in diameter. Has several limbs arising from ground.

Bark: Gray, shreddy, fibrous.

Cones: Coppery green, covered with a bluish, waxy substance; to ¼″ in diameter; usually only 1-seeded.

Leaves: Yellowish green, scalelike; about 1⁄16″ long.

Elevation: 3,000 to 7,000′.

Habitat: Plateaus, plain, foothills, and pinyon-juniper woodlands in the northern reaches of Arizona.

Comments: Evergreen shrub or small tree with very aromatic wood. Usually shrubby, and limbs normally well-hidden by lower branches. Cones mature in one year. Seeds smaller than Utah juniper. Male and female flowers borne on different trees; pollen on male trees and "berries" on female trees. Wood used for fuel and fence posts. Cones eaten by wildlife. Native Americans had many uses for bark and seeds. In the southern two-thirds of Arizona grows a close relative of one-seed juniper; **Roseberry** or **Redberry Juniper** (*Juniperus coahuilensis*). Form is similar in appearance; however, cones are reddish. Seven species of *Juniperus* in Arizona. Photograph taken in Ashurst Lake area, September 6. This juniper is recognizable by its many limbs arising from ground level; limbs are usually hidden by lower branches.

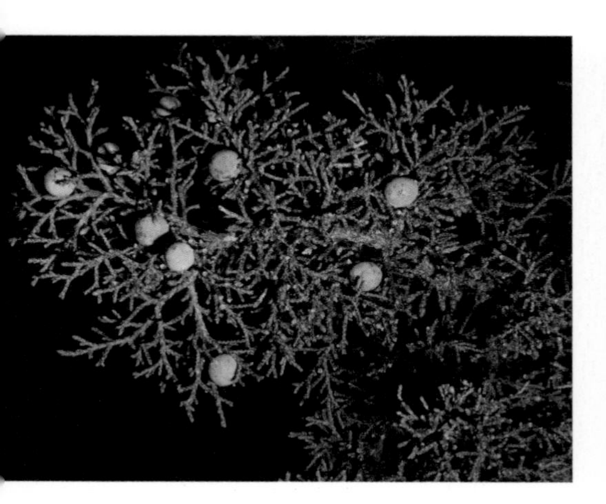

UTAH JUNIPER
Western Juniper
Juniperus osteosperma
Cypress Family (Cupressaceae)

Height: To 20′.

Trunk: To 1½′ in diameter. Has clearly defined trunk.

Bark: Gray, shreddy in long strips, fibrous.

Cones: Reddish brown covered with grayish, waxy coating; 1- to 2-seeded; to ¾″ in diameter.

Leaves: Yellowish green, scalelike; about ¹⁄₁₆″ long.

Elevation: 3,000 to 7,500′.

Habitat: Dry hills, plains, plateaus, and mountains in pinyon-juniper woodlands.

Comments: Most common juniper in Arizona. Has broad, rounded crown. Seeds larger than **One-seed Juniper** (page 419). Male and female flowers on same tree. Cones mature in two seasons. Native Americans use berries for beads and medicine, and wood for firewood, posts, and hogans. Cones eaten by wildlife. Mistletoe often grown in these trees. Seven species of *Juniperus* in Arizona. Photograph taken at Ashurst Lake area, September 6. This juniper is recognizable by its definite trunk.

ROCKY MOUNTAIN JUNIPER
Western Red Cedar
Juniperus scopulorum
Cypress Family (Cupressaceae)

Height: To 40′, but usually about 20′.

Trunk: To 1½″ in diameter.

Bark: Reddish brown or gray, shreddy, fibrous.

Cones: Blue, covered with a grayish, waxy coating; ¼″ in diameter; usually 2-seeded.

Leaves: Grayish green, scalelike, ⅛″ long; young twigs on immature trees have needlelike leaves.

Elevation: 5,000 to 9,000′.

Habitat: Mesas and rocky mountain slopes.

Comments: Wood and leaves have pencil smell when crushed. Branches often droop at ends. Cones mature in two seasons; provide food for wildlife. Wood used for cedar chests, fuel, fence posts, and lumber. Seven species of *Juniperus* in Arizona. Photograph taken at Oak Creek Canyon, June 8. This juniper recognizable by its upright growth and drooping branch tips.

NEVADA JOINTFIR

Mormon Tea
Ephedra nevadensis var. *aspera*
Joint-Fir Family (Ephedraceae)

Height: Usually 3′ to 4′.

Flowers: Tiny, pale yellow, in dense conelike clusters. Male and female flowers on separate plants.

Leaves: Reduced to scales when present; usually 2 per node (joint).

Blooms: February–March.

Elevation: Below 4,500′.

Habitat: Desert and grassland.

Comments: Yellow-green stemmed perennial shrub. Valuable soil binder whose scalelike leaves help conserve moisture. Mormon settlers made tea from dried stems. Native Americans used plant medicinally for treatment of certain diseases. Ten species of *Ephedra* in Arizona. Photograph taken in Superstition Mountains, February 6.

SUBALPINE FIR

Rocky Mountain Fir
Abies bifolia (Abies lasiocarpa)
Pine Family (Pinaceae)

Height: to 90′.

Trunk: To 2′ in diameter.

Bark: Grayish brown, smooth, becoming fissured and scaly with age.

Cones: Dark purplish to almost black, covered with beads of pitch; cylindrical, to 4″ long; stand upright in top part of tree, disintegrate at maturity, leaving behind a vertical core in tree.

Needles: Dark bluish green with 2 silvery lines on surfaces; spreading or in 2 rows, curve upward; flexible, soft, flat, with rounded or notched tip; to 1″ long.

Elevation: 8,000 to 12,000′.

Habitat: Cool, moist spruce-fir forests.

Comments: Evergreen; smallest of true firs. Crown is long, narrow, and comes to sharp point. Branches extend almost to base of tree. Browsed by deer and sheep. Seeds eaten by birds and small mammals. Two species of *Abies* in Arizona. Photograph taken at Hannagan Meadow, August 17. (Mature cone in photograph dropped on us by a careless red squirrel.)

WHITE FIR

Balsam Fir
Abies concolor
Pine Family (Pinaceae)

Height: To 150'.

Trunk: To 3½' in diameter.

Bark: Dark gray, thick, deeply furrowed.

Cones: Grayish green, to 5" long, upright in top part of tree; do not fall after shedding seeds.

Needles: Pale blue-green or silvery; spreading and curved upward; flat, to 3" long.

Elevation: 5,500 to 9,000'.

Habitat: Ponderosa and spruce-fir forests.

Comments: Evergreen. When young, forms a perfect pyramid if growing in open. Pollen-producing cones appear on lower branches in spring and early summer. These soon die after wind currents send the pollen upward to female cones forming near top of tree. Birds and mammals eat seeds; porcupines chew on bark. Two species of *Abies* in Arizona. Photograph taken at Black Canyon Lake area, June 5.

ENGELMANN SPRUCE

White Spruce
Picea engelmannii
Pine Family (Pinaceae)

Height: To 100' (rarely this height in Arizona).

Trunk: To 3' in diameter (rarely in Arizona).

Bark: Purplish to reddish brown; thin and scaly.

Cones: Chestnut brown with papery scales; stiff, rounded and thinner at tip; pendent, to 2½" long.

Needles: Dark green or pale blue-green; slightly curved, flexible, 4-sided in cross section; to 1¼" long.

Elevation: 8,000 to 12,000'.

Habitat: Moist spruce-fir forests.

Comments: Twigs are minutely hairy. Narrow, pointed, conical crown; horizontal to drooping branches nearly to ground. Shallow root system easily uprooted in winds. Wood weak and knotty, used by Native Americans for bows and hoops. Two species of *Picea* in Arizona. Photograph taken near Mexican Hay Lake, July 2. Mainly distinguished from **Blue Spruce** (*Picea pungens*) (page 423) by its shorter cones with scales thinner at tips; brownish, scaly bark; and flexible needles. If you squeeze a spruce branch without hurting your hand, it's this species.

BLUE SPRUCE

Colorado Blue Spruce
Picea pungens
Pine Family (Pinaceae)

Height: To 80′.

Trunk: To 2′ in diameter.

Bark: Dark gray or brown; thick, rough, furrowed into ridges.

Cones: Chestnut brown; scales more or less straight across, and are not thinner at tip; to 4″ long.

Needles: Silvery blue and stiff; protrude in all directions from branch; diamond-shaped in cross section; to 1½″ long.

Elevation: 7,000 to 11,000′.

Habitat: Mixed conifer forests.

Comments: Twigs not hairy. Conical crown of bluish foliage on young trees. State tree of Colorado and Utah. Smaller than Engelmann spruce, and less widely distributed than Engelmann spruce in Arizona. Two species of *Picea* in Arizona. Photograph taken in mountains above Greer, July 2. Mainly distinguished from **Engelmann Spruce** (*Picea engelmannii*) (page 422) by its longer cones with scales of even thickness, dark grayish, furrowed bark, and stiff needles. If you squeeze a spruce branch and say "ouch," it's more than likely this species.

BRISTLECONE PINE

Foxtail Pine
Pinus aristata
Pine Family (Pinaceae)

Height: To 40′.

Trunk: To 2½′ in diameter.

Bark: Whitish and smooth on young trees; reddish brown, scaly, fissured on mature trees.

Cones: Dark purplish brown, hanging; each scale tipped with a stiff, ¼″-long, incurved prickle; to 4″ long.

Needles: Dark green, curved, to 1½″ long; 5 in bundle, crowded, forming brushlike groupings along branch ends. Remain on tree up to 30 years before being shed.

Elevation: 9,500 to 12,000′. (Found only on San Francisco Peaks in Arizona.)

Habitat: Spruce-fir forests up to timberline.

Comments: Resembles a bushy, twisted shrub at timberline. Some bristlecone pines in California are over 4,000 years old. Ten species of *Pinus* in Arizona. Photograph taken at Great Basin National Park, Nevada, July 21.

MEXICAN PINYON PINE

Pinus cembroides (Pinus discolor)
Pine Family (Pinaceae)

Height: To 20′.

Trunk: To 1′ in diameter.

Bark: Light gray, smooth when young; reddish brown and scaly on older trees.

Cones: Dull reddish brown, round or egg-shaped with thick scales; to 2″ wide, 2″ long; open to reveal hard, dark brown seeds to ¾″ long (known as "pinyon nuts").

Needles: Blue-green with silvery lines; 3 in bundle, fine, flexible; to 2½″ long.

Elevation: 5,000 to 7,500′.

Habitat: Dry, rocky slopes with juniper and oaks.

Comments: Slow grower. Can reach age of 350 years. Cones fall in the winter after seeds disperse. Seeds or nuts under cone scales are hard-shelled and flavorful. Pinyon nuts food for Native Americans; also consumed by birds and rodents. Pinyon pitch used as jewelry cement and for waterproofing baskets. Ten species of *Pinus* in Arizona. Photograph taken at Chiricahua National Monument, April 25. This pinyon pine recognizable by bundle of 3 needles.

PINYON PINE

Pinus edulis
Pine Family (Pinaceae)

Height: to 35′.

Trunk: To 30″ in diameter.

Bark: Gray to reddish brown, furrowed into scaly ridges.

Cones: Light brown to yellowish brown; egg-shaped; thick, blunt scales; to 2″ long.

Needles: Dark green, slightly curved; usually 2 per bundle; to 2″ long.

Elevation: 4,000 to 7,000′.

Habitat: Mesas, plateaus, and lower mountain slopes.

Comments: Compact, rounded crown; often with short, crooked trunk. Slow grower. Large, ½″-long brown seeds from cones are oily and edible; known variously as piñones, pinyon nuts, pine nuts, Christmas nuts, and Indian nuts. They are eaten by wild turkeys, pinyon jays, and mammals, and are used commercially (raw and roasted) and in candies. Most drought-resistant of all pines in Arizona. Wood used for fence posts and fuel. Ten species of *Pinus* in Arizona. Photograph taken near Sunset Crater National Monument, September 8. This pinyon pine recognizable by bundle of 2 needles. **Singleleaf Pinyon** (*Pinus monophylla*) is very similar, but its needles occur singly.

APACHE PINE

Pinus engelmannii
Pine Family (Pinaceae)

Height: To 75'.

Trunk: To 30' in diameter.

Bark: Dark brown and lighter brown; deeply furrowed with age.

Cones: Light brown; asymmetrical at base; scales tipped with prickles; conical or egg-shaped; to 5½" long.

Needles: Dark green; usually 3 in bundle; spreading or drooping; to 15" long.

Elevation: 5,000 to 8,200'.

Habitat: Dry, sandy soil in southeastern Arizona.

Comments: Has mostly taproots. Seeds eaten by wildlife. Not a common pine in Arizona. Can live up to 500 years. Ten species of *Pinus* in Arizona. Photograph taken near Cave Creek, Portal, April 23. Recognized by very long, widely spreading or drooping needles.

LIMBER PINE

Pinus flexilis
Pine Family (Pinaceae)

Height: To 50'.

Trunk: To 3' in diameter.

Bark: Young trees: smooth, whitish gray. Mature trees: dark brown to black, split by deep furrows.

Cones: Yellowish brown; columnar, without prickles; thickened, rounded scales, blunt-pointed tip; to 6" long.

Needles: Dark green, long-pointed, with silvery white lines on all surfaces; not toothed; 5 in a bundle; to 3½" long.

Elevation: 7,500 to 10,000'.

Habitat: Spruce-fir forests and ponderosa forests (to a lesser degree).

Comments: Short trunk, widely branched crown with drooping, plumelike, flexible branches. Squirrels feed on seeds. Foliage browsed by elk and deer. Ten species of *Pinus* in Arizona. Photograph taken near Willow Springs Lake, September 14. Distinguished from **Southwestern White Pine** (*Pinus strobiformis*) (page 427) by its wider cones ending in blunt tips, needles with silvery white lines on all surfaces, and drooping ends of branches.

CHIHUAHUA PINE
Pinus leiophylla var. *chihuahuana (Pinus chihuahuana)*
Pine Family (Pinaceae)

Height: To 60′.

Trunk: To 2′ in diameter.

Bark: Dark reddish brown to black; very thick, broad ridges, deep furrows.

Cones: Light brown and shiny; scales tipped with shedding prickles; to 3″ long; three years to mature.

Needles: Pale green to dull gray with white lines; to 4½″ long; 3 in bundle, sheaths around bundles soon shed. Needles often appear randomly in clusters on trunk.

Elevation: 5,000 to 7,800′.

Habitat: Dry, rocky slopes. Scattered in pine forests in mountains of southeastern Arizona.

Comments: Cones stay on tree a long time. Ten species of *Pinus* in Arizona. Photograph taken at Cave Creek, Portal, April 22. Similar to **Ponderosa Pine** (*Pinus ponderosa*) (at right) but has shorter needles, smaller cones, and much darker bark.

PONDEROSA PINE
Western Yellow Pine
Pinus ponderosa
Pine Family (Pinaceae)

Height: To 125′.

Trunk: To 4′ in diameter.

Bark: Young trees: dark brown to almost black; older trees: cinnamon brown to orangish yellow with irregular fissures.

Cones: Light reddish brown, tipped with prickly scales; conical or egg-shaped; to 5″ long.

Needles: Dark green, to 7″ long; usually 3 in a bundle, at times 2 (5 in **Arizona Pine** *Pinus arizonica*, a closely related species found in southeastern Arizona).

Elevation: 5,000 to 9,500′ (6,000 to 9,000′ for Arizona pine).

Habitat: Mountains and higher plateaus.

Comments: The most abundant pine in Arizona. With mainly surface roots, tree is easily blown down in open areas. The wood beneath the bark is twisted, which protects many of these trees from wind damage. Some large trees are 400 to 500 years old. Ten species of *Pinus* in Arizona. Photograph taken at Sunset Crater National Monument, May 31. Mature trees recognizable by large, straight, orangish brown, fissured trunk free of lower branches; existing branches turn upward.

SOUTHWESTERN WHITE PINE

Pinus strobiformis
Pine Family (Pinaceae)

Height: To 80′.

Trunk: To 3′ in diameter.

Bark: Dark gray or dull reddish brown, becoming deeply furrowed and narrowly ridged.

Cones: Yellowish brown, cylindrical; long and slightly thickened cone scales; tip narrow, spreading, and bent back; to 9″ long.

Needles: Bluish green, silvery white lines on inner surface only, to 3½″ long; 5 in bundle, slender, finely toothed near tip.

Elevation: 6,500 to 10,000′.

Habitat: Dry, rocky slopes and canyons.

Comments: Closely related to limber pine. Seeds eaten by wildlife. Ten species of *Pinus* in Arizona. Photograph taken at Greer, August 10. Unlike similar **Limber Pine** (*Pinus flexilis*) (page 425), cones are slimmer and have narrow, bent-back tips, and needles have white lines only on inner surfaces.

DOUGLAS FIR

Pseudotsuga menziesii
Pine Family (Pinaceae)

Height: To 200′ in Pacific Northwest; 100 to 130′ in Arizona.

Trunk: To 6′ in diameter in Pacific Northwest, but less in Arizona.

Bark: Dark reddish brown, very thick, deeply furrowed.

Cones: Reddish brown, thin with rounded scales; long, distinctive 3-pointed papery bracts extending from between scales; to 3″ long, hanging from branches.

Needles: Dark bluish green; protrude in all directions from branch; narrow, flat, and soft, rounded at apex; to 1⅛″ long.

Elevation: 6,500 to 10,000′, down to 5,000′ in canyons.

Habitat: Mixed with ponderosa pines or with spruce-firs.

Comments: Not a true fir. The largest tree in Arizona. Valuable for its timber. Compact, conical crown with drooping side branches. Seeds and foliage eaten by wildlife. One species of *Pseudotsuga* in Arizona. Photograph taken in Willow Springs Lake area, September 15. The cones' 3-pointed, paper bracts are unique to this species.

FERNS

FORKED SPLEENWORT

Asplenium septentrionale
Spleenwort Family (Aspleniaceae)

Description: Dark green, shiny, grasslike leaves, grows to 6" long, ⅛" wide.

Elevation: 7,700 to 9,000'.

Habitat: Crevices in rocks.

Comments: Evergreen. Nine species of *Asplenium* in Arizona. Photograph taken near Willow Springs Lake, September 9.

MAIDENHAIR SPLEENWORT

Asplenium trichomanes
Spleenwort Family (Aspleniaceae)

Description: Dark green, evergreen, dainty, rounded. Pinnae: ¼" long, toothed at tips. Stipe and rachis: dark purplish brown and brittle. Grows to 7" tall, ½" wide. Sori: few, elongated, often overlapping each other.

Elevation: 6,000 to 9,000'.

Habitat: Moist cracks under overhanging rock ledges.

Comments: A tiny fern. Nine species of *Asplenium* in Arizona. Photograph taken at Woods Canyon Lake, July 7.

BRACKEN

Brake

Pteridium aquilinum var. *pubescens*

Bracken Fern Family (Dennstaedtiaceae)

Description: Dark green, very coarse texture, thick; bipinnate to tripinnate; broadly triangular, edges of segments turned under. Stipe: smooth, stiff, about same length as leafy part; green at first, turning dark brown with age. Frond to 3' long, 3' wide. Grows to 4' high.

Elevation: 5,000 to 8,500'.

Habitat: Meadows, open woodlands, pine forests, and burned-over areas.

Comments: Most common fern; weedy. Often found in large colonies. Killed by first frost. One species of *Pteridium* in Arizona. Photograph taken at Black Canyon Lake, June 4.

LADY FERN

Athyrium felix-femina

Shield Fern Family (Drypoteridaceae)

Description: Green, not evergreen; delicate, pinnately cleft. Pinnae with pointed tips, cleft, and toothed. Lower pinnae project forward. Rachis: smooth and slightly grooved. Grows to 3' tall, 8" wide at widest section. Sori: dark brown, curved, in 2 rows on underside of each pinnule.

Elevation: 7,000 to 9,500'.

Habitat: Shaded areas along streams and springs in rich soil.

Comments: Fronds form a vase-shaped, circular cluster. One species of *Athyrium* in Arizona. Photographs taken at Lee Valley Reservoir in mountains above Greer, July 2.

FRAGILE BLADDER FERN

Cystopteris fragilis
Shield Fern Family (Dryopteridaceae)

Description: Bright green to dark green; stalk very brittle, pinnules fan-shaped and very variable in toothing. Stipe: black to dark brown. Rachis: smooth, green or straw-colored. Grows to 10" tall, 3" wide. Sori: brown.

Elevation: 5,000 to 12,000'.

Habitat: Rich, moist soil among rock ledges and springs, in shade.

Comments: Four species of *Cystopteris* in Arizona. Photograph taken at Woods Canyon Lake, August 9.

MALE FERN

Shield Fern
Dryopteris filix-mas
Shield Fern Family (Dryopteridaceae)

Description: Dark green above, lighter green beneath, semi-evergreen, leathery; pinnules parallel-sided and blunt-tipped. Grows to 18" tall, 8" wide; widest at center of frond. Sori: large, whitish, located toward midvein.

Elevation: 6,500 to 10,000'.

Habitat: Rock crevices in rich soil, cool forests and along streams.

Comments: The drug aspidium is derived from this fern. Four species of *Dryopteris* in Arizona. Photograph taken on San Francisco Peaks, June 4.

FLOWER CUP FERN

Woodsia plummerae
Shield Fern Family (Dryopteridaceae)

Description: Light green, thin; pinnules fringed and wavy-edged. Stipe: dark reddish brown; undersides of fertile fronds have large dark brown patches covering the sori. Pinnae to 1" long. Grows to 7" long, 2" wide.

Elevation: 2,000 to 9,000'.

Habitat: In the shade of cliffs and rock ledges.

Comments: Mostly rock-inhabiting ferns. Six species of *Woodsia* in Arizona. Photograph taken at Woods Canyon Lake, August 3.

WESTERN POLYPODY

Polypodium hesperium (Polypodium vulgare)
Polypody Fern Family (Polypodiaceae)

Description: Dark green above, lighter green beneath, evergreen; pinnately cleft, 4 to 14 pairs of pinnae with rounded tips. Grows from 4 to 15" tall, 1¾" wide. Sori: brown, round, in 2 rows on underside of each pinnule.

Elevation: 7,000 to 9,000'.

Habitat: Moist slopes in canyon and conifer forests.

Comments: Creeping, scaly rhizomes. One species of *Polypodium* in Arizona. Photograph taken at Woods Canyon Lake, July 7.

COCHISE CLOAK FERN

Narrow Cloak Fern
Astrolepis cochisensis (Notholaena cochisensis)
Cloak Fern Family (Pteridaceae)

Description: Olive green above, brownish scales beneath; tall narrow fronds. Stipe: round, brownish. Rachis: reddish brown, very hairy, to 8" long. Pinnae: 1 or 2 pairs of lobes, roundish to oval, hairy, to ¼" long, 3⁄16" wide. Frond and stem to ½" wide, 9" long.

Elevation: 1,000 to 7,000'.

Habitat: Dry, rocky slopes and canyons.

Comments: Poisonous to livestock. Six species of *Astrolepis* in Arizona. Photograph taken at Saguaro National Park West, April 17.

WAVY CLOAK FERN

Astrolepis sinuata (Notholaena sinuata)
Cloak Fern Family (Pteridaceae)

Description: Olive green above, brownish scales beneath; tall narrow fronds. Stipe: round, brown with whitish scales, to 4" long. Rachis: very scaly, to 16" long. Pinnae: 3 to 6 pairs of lobes, wavy-edged, to ¾" long. Front and stem to 1¼" wide, 2' long.

Elevation: 1,000 to 7,000'.

Habitat: Dry, rocky slopes.

Comments: Often found in limestone areas. Six species of *Astrolepis* in Arizona. Photograph taken at Tortilla Flat, December 10. A similar looking species, **Golden Lipfern** *(Cheilanthes bonariensis)*, is discussed on page 435.

GOLDEN LIPFERN

Cheilanthes bonariensis
Cloak Fern Family (Pteridaceae)

Description: Similar in appearance to Wavy Cloak Fern, the leaflets of this species are densely woolly beneath, not scaly.

Elevation: 4,000 to 7,000'.

Habitat: Golden lipfern is found in rock crevices in the oak woodlands of mountains in much of southern Arizona.

Comments: The range of this species is huge, extending south to Argentina, and includes the islands of Jamaica and Hispaniola. Photograph taken in the Santa Rita Mountains, May 11.

BEADED LIP FERN

Fairy Sword
Cheilanthes wootonii
Cloak Fern Family (Pteridaceae)

Description: Golden green above, thick cinnamon-brown scales beneath, tripinnate, beadlike segments. Stipe: dark brown, scaly, woolly haired. Grows to 6" at lower elevations, 11" in mountains.

Elevation: 2,000 to 8,000'.

Habitat: Dry slopes among rocks.

Comments: Grows in long rows. Seventeen species of *Cheilanthes* in Arizona. Photograph taken on Mount Graham, April 21.

CALIFORNIA CLOAK FERN

Notholaena californica (Cheilanthes deserti)
Cloak Fern Family (Pteridaceae)

Description: Dull green and very glandular above, yellowish beneath; rough, star-shaped fronds. Frond to 1½" wide, 1½" long. Stipe: round, chestnut brown. Grows to 5" tall.

Elevation: 1,000 to 3,000'.

Habitat: Crevices on dry, rocky slopes; in canyons.

Comments: Six species of *Notholaena* in Arizona. Photograph taken in Superstition Mountains, January 28.

STAR CLOAK FERN

Standley's Cloak Fern
Notholaena standleyi
Cloak Fern Family (Pteridaceae)

Description: Dark green and shiny above, covered with golden wax beneath; symmetrical, star-shaped fronds. Stipe: round, reddish brown. Front to 4" wide. Grows to 8" high.

Elevation: 1,000 to 6,500'.

Habitat: Dry banks and rock ledges.

Comments: Loses more than 50 percent of its water content during drought, forming a dusty-brown curl. Six species of *Notholaena* in Arizona. Photograph taken at Tortilla Flat, December 10.

WRIGHT'S CLIFF BRAKE
Pellaea ternifolia var. *wrightiana*
Cloak Fern Family (Pteridaceae)

Description: Bluish green. Stipe: round, grooved, shiny, very dark chestnut brown to almost black. Blade: narrowly triangular, bipinnate. Pinnae: slightly wavy-edged with margins rolled under. Grows to 15" tall.

Elevation: 4,000 to 8,000'.

Habitat: Rocky hillsides and crevices.

Comments: Evergreen rock fern. Eight species of *Pellaea* in Arizona. Photograph taken on Mount Graham, April 21.

SPINY CLIFF BRAKE
Pellaea truncata (Pellaea longimucronata)
Cloak Fern Family (Pteridaceae)

Description: Bluish green, triangular-shaped frond; bipinnate, up to 10 pairs of oval leaflets. Stipe: shiny, chestnut brown, hairless, grooved, stiff. Grows to 15" high.

Elevation: 2,000 to 6,000'.

Habitat: Rocky crevices and cliffs.

Comments: Eight species of *Pellaea* in Arizona. Photograph taken in Superstition Mountains, February 4.

GLOSSARY

Acaulis, -e Stemless.

Acerosus, -a, -um Stiff-needled; sharp.

Achene (ah-KEEN) A small, dry fruit that does not split open; one-seeded.

Acris Biting or acrid.

Acuminate Tapering to a point.

Acute Sharp-pointed.

Aduncus, -a, -um Hooked or bent backward.

Alatus, -a, -um Winged.

Albicans Whitish.

Alternate Not opposite each other (as leaves on a stem).

Ambiguus, -a, -um Uncertain (as in a plant's identity).

Angled Sided (as in the shape of fruits).

Angularis, -e Angled.

Angustifolius, -a, -um Narrow-leaved.

Annual A plant that completes its life cycle in one season.

Annuus, -a, -um Annual.

Anther The part of a stamen bearing pollen.

Arenarius, -a, -um Growing in sand.

Areole A raised area in a cactus from which spines develop.

Argenteus, -a, -um Silvery.

Arvensis, -e Of the fields.

Asper, -a, -um Rough.

Asymmetrical Irregular; not divided into like and equal parts.

Aureus, -a, -um Golden.

Axil The angle where the leafstalk or flower stalk joins the stem.

Banner The upper petal in a pea flower, or the three erect petals of an iris flower.

Barbatus, -a, -um Bearded or barbed.

Barbed With a backward-facing tip; resembling a fishhook.

Basal At or near the base (as leaves).

Berry A fleshy or pulpy fruit that does not usually split open and that has one or more seeds.

Biennial A plant that completes its life cycle in two years, growing vegetation in the first year and flowers and seeds in the second, then dying.

Biflorus, -a, -um Double-flowered.

Bifolius, -a, -um Double-leaved.

Bipinnate Describes compound leaves having secondary leaflets that are also pinnate.
Blade The portion of a leaf not including the stalk.
Bloom White, powderlike coating.
Bract A modified leaf at base of flower head or fruit.
***Bracteatus**, -a, -um* Bearing bracts.
***Brevicaulis**, -e* Short-stemmed.
Brevipes Short-stalked.
Bristle A stiff, hairlike structure.
Bud A developing stem, leaf, or flower.
Bur A spiny or prickly fruit or seed.
Calyx Collectively, the group of sepals that encircle flower parts.
***Campaniflorus**, -a, -um* Having bell-shaped flowers.
Canescens Having grayish white hairs.
Capsule A dry fruit divided into two or more seed compartments that split longitudinally or, uncommonly, around the circumference.
***Cardinalis**, -e* Scarlet.
Catkin A long cluster of tiny, petalless flowers, often all one sex, also called an "ament."
***Caudatus**, -a, -um* Tailed.
Cell A chamber or compartment (as in a seed).
Chaparral An area dense with leathery-leaved, evergreen shrubs.
Chlorophyll A green, photosynthetic substance found in plants.
Clasping leaf A leaf with leafstalk wrapped around a stem.
Cleft Cut about halfway (as in a leaf).
***Coccineus**, -a, -um* Scarlet.
Column A center structure in a flower formed by united stamens, or stamens united with style or stigma.
Complete Describes flowers with petals, sepals, pistils, and stamens.
Composite A flower that is a member of the Sunflower Family
Compound Describes a leaf divided into two or more smaller leaflets.
Cone A dry fruit with overlapping, woody scales.
Conifer An evergreen, needled, cone-bearing tree or shrub.
***Cordatus**, -a, -um* Heart-shaped.
Corm An enlarged base of a stem, resembling a bulb.
Corolla Collectively, all the petals of a flower.
***Crassifolius**, -a, -um* Thick-leaved.
***Cristatus**, -a, -um* Crested or comblike.

Crown The top branches, twigs, and leaves of a tree.

Cyme A broad, flat inflorescence on which the flowers open first in the center and successively outwards toward the perimeter.

Deciduous Describes a plant that sheds its leaves or other parts at a certain time or season.

Deltoides*, *-deus*, *-a*, *-um Triangular.

Demissus*, *-a*, *-um Weak or low-hanging.

Densiflorus*, *-a*, *-um Densely flowered.

Dentation The toothed edges of leaves or leaflets.

Dentatus*, *-a*, *-um Toothed.

Depressus*, *-a*, *-um Flat.

Dicot or Dicotyledon A plant having two seed leaves (cotyledons); one of the two major divisions of flowering plants.

Digitatus*, *-a*, *-um Fingerlike.

Dioecious (dye-EE-shus) Describes a species with female and male flowers on separate plants.

Disk flowers Tiny, tubular flowers, often forming the center "button" on a composite flower.

Dissected Finely cut or divided into many, narrow segments (as in a leaf).

Diversifolius*, *-a*, *-um With different leaf shapes.

Drupe A fleshy fruit with a hard stone or nut in its center.

Edulis*, *-e Edible.

Elatus*, *-a*, *-um Tall.

Elegans Elegant.

Elliptical Wider in the center and tapering at base and tip (as in a leaf).

Entire Without teeth or notches on margins (as leaves and petals).

Ephemeral Describes a plant or flower that lasts a very short time.

Erectus*, *-a*, *-um Upright.

Escapee A plant that escaped from cultivation and now reproduces on its own.

Evergreen A plant having green leaves all year; describing such a plant.

Fasciculatus*, *-a*, *-um In a tight cluster.

Female flowers Flowers with pistils but no functional stamens.

Filament The thin stalk of a stamen supporting an anther at its tip.

Filifolius*, *-a*, *-um Having fine or threadlike foliage.

Floret A single, small flower.

Flower The reproductive parts of a seed plant.

Follicle A dry, many-seeded fruit opening along one side only.

Frond A fern leaf.

Fruit A mature ovary, containing one or more seeds.

Fruticosus, *-a*, *-um* Shrubby.

Giganteus, *-a*, *-um* Large.

Glaber, *-ra*, *-rum* Hairless, smooth.

Glandular Producing tiny globules of sticky or oily substance (as in a leaf or stem).

Glaucus, *-a*, *-um* Having a white or grayish powder or bloom.

Globosus, *-a*, *-um* Globe-shaped.

Glochids Barbed bristles on cacti.

Gracilis, *-e* Slender and graceful.

Grandiflorus, *-a*, *-um* Large-flowered.

Habitat The environment where a plant lives.

Herbaceous Fleshy-stemmed; not woody.

Herbs Fleshy, nonwoody plants.

Hip A fleshy, berrylike fruit (as in some members of the Rose Family).

Hirsutus, *-a*, *-um* Hairy.

Hispidus, *-a*, *-um* Rough-haired.

Hooked Curved at the tip.

Host A plant furnishing nourishment to a parasite.

Hybrid The offspring of cross-fertilization between two different species.

Imperfect Describes a flower having stamens or pistils, but not both.

Inferior Describes an ovary located low in a flower, with petals and calyx lobes attached above.

Inflorescence Referring to flowers when they do not occur singly, and their arrangement on the flower stem.

Integer, *intergra*, *-um* Whole, with no cut.

Involucre A circle of bracts or leaves supporting a flower head.

Irregular Describes a flower that is not radially symmetrical, with parts of unequal size or shape.

Joint Segment of a stem (as in a cactus), or a plant node (as where leaves join the stem).

Keel The fused lower petals of a member of the pea family, or in some species, the single lower petal; the ridge on some seeds or fruits.

Key A dry, one-seeded fruit with a wing or wings, also called a "samara" (as in maple or ash seeds).

Lanceolate Lance-shaped, or narrow and tapering at the tip (as a leaf shape).

Lanceolatus, *-a*, *-um* Lance-shaped.

Latifolius, *-a*, *-um* Having wide leaves.

Leaflet One segment of a compound leaf.

Leafstalk An appendage attaching a leaf to the plant's main stem.

Legume A seed pod (as in the pea family) that splits along two sides; a member of the pea family.

Linear Long and narrow with parallel margins (as a leaf shape).

Linearifolius, -a, -um Having narrow, parallel-sided leaves.

Lip The upper or lower segment of an irregular (or asymmetrical) flower.

Lobatus, -a, -um Divided into lobes.

Lobed Describes leaves with marginal indentations that dissect less than halfway, and their associated rounded projections.

Longiflorus, -a, -um Long-flowered.

Longifolius, -a, -um Long-leaved.

Luteus, -a, -um Yellow.

Macrocarpus, -a, -um Having large seed pods or fruits.

Macropetalus, -a, -um Large-petaled.

Macrophyllus, -a, -um Large-leaved.

Maculatus, -a, -um Blotched or spotted.

Male flowers Flowers with stamens, but no functioning pistils.

Margin The edge of a leaf or petal.

Marginatus, -a, -um Pertaining to margins or edges.

Micranthus, -a, -um Small-flowered.

Microcarpus, -a, -um Having tiny seed pods or fruits.

Micropetalus, -a, -um Small-petaled.

Microphyllus, -a, -um Small-leaved.

Mollis, -e Soft-haired.

Monocot or Monocotyledon A plant having one seed leaf (cotyledon); one of the two major divisions of flowering plants.

Monoecious Describes a species with separate male and female flowers on same plant.

Naturalized Describes plants from another region that have established themselves with the native flora.

Needle The long, narrow leaf of some conifers.

Nervosus, -a, -um Having prominent nerves (veins).

Node The point on a stem where a leaf or bud sprouts.

Nut A hard-shelled fruit with one seed.

Nutans Nodding.

Nutlet A small nut.

Odorus, -a, -um Fragrant.

Opposite Paired or opposite each other (as leaves on a stem).

Oval Broadly elliptical or egg-shaped.

Ovary The base of a pistil where female germ cells develop into seeds after fertilization.

Ovatus, -a, -um Egg-shaped.

Palmately compound Describes leaves divided or lobed from one point (as fingers growing from palm of hand).

Palustris, -e Growing in wet areas.

Pappus A scale hair, or bristle on the tip of an achene (as on seeds of a dandelion).

Parasite An organism obtaining nourishment from a host organism.

Parviflorus, -a, -um Small-flowered.

Patulus, -a, -um Spreading.

Pedicel A flower stalk.

Perennial A plant living more than two years.

Perfect Describes a flower with both male and female organs.

Perianth Collectively, the calyx and/or corolla.

Petal One of the segments of the corolla, usually colored.

Petiole A leaf stem.

Phyllary One of the bracts below the flower head in members of the sunflower family.

Pictus, -a, -um Painted or variegated.

Pinna A compound leaf division or leaflet (plural: "pinnae").

Pinnately compound Describes leaves divided or lobed along each side of a leafstalk, resembling a feather.

Pinnatus, -a, -um Divided or lobed.

Pinnule A subleaflet of a fern frond or other twice-compound leaf.

Pistil The female organ of a flower, composed of an ovary, a slender style, and a stigma or stigmas at tip.

Pistillate Describes a female flower bearing one or more pistils but no functional stamens.

Pod A dry fruit that splits open when mature.

Pollen Spores, borne by the anthers, that contain the male germ cells.

Pollination The transfer of pollen from anther to stigma.

Pome A fleshy fruit having several seeds (as an apple).

Prickle A weak or rigid outgrowth from bark or epidermis.

Procumbens Trailing or lying down.

Prostrate Lying horizontally.

Pubescens Having hairy-soft, short hairs.

Pumilus, -a, -um Small.

Purpureus, *-a*, *-um* Purple.

Pusillus, *-a*, *-um* Small.

Raceme An unbranched flower stem with stalked flowers, the newest flowers forming at its top.

Racemosus, *-a*, *-um* Having flowers in racemes.

Rachis That part of a fern front stem bearing the leaflets, also called the axis; also, the main stalk of a flower cluster or of a compound leaf.

Radial spines Spines emerging from the edges of an areole.

Radiatus, *-a*, *-um* Raylike.

Ray flowers Flat, straplike flowers on a member of the composite family.

Reflexus, *-a*, *-um* Turned back on itself.

Regular Describes a flower with petals or sepals all of equal size and shape; radially symmetrical.

Repens Creeping.

Reticulatus, *-a*, *-um* Netted.

Ribs The raised rows on stems of cacti; the primary veins of leaves.

Rosette A cluster of leaves in a circular arrangement at the base of a plant.

Rotundifolius, *-a*, *-um* Round-leaved.

Rugosus, *-a*, *-um* Wrinkled.

Samara Dry fruit with wings that do not open when mature (as maple and ash seeds).

Saprophyte A plant living off dead or decaying organic matter and usually lacking chlorophyll.

Scaber, *-ra*, *-rum* Rough to touch.

Scale A greatly reduced leaf or outgrowth on skin.

Seed A mature ovule that has been fertilized.

Sepal A calyx segment.

Sessile Without a stalk or stem.

Shrub A small, woody plant with several stems.

Simple Describes a leaf with one part, not divided into leaflets.

Sori Clusters of spore sacs on a fern frond.

Spike A long flower cluster with each flower attached to the stalk, either directly or nearly so.

Spine A sharp, stiff outgrowth on a plant stem.

Spinosus, *-a*, *-um* Spiny.

Spore The reproductive cell of a non-flowering plant.

Spur The hollow, tubular projection from the base of a petal or sepal, often producing nectar; a short side twig on a tree.

Stamen The male organ of a flower, composed of a slender stalk (filament) tipped with a pollen-producing anther.

Staminate Describes a male flower having one or more stamens, but no functional pistils.

Stellatus*, -*a*, -*um Star-shaped.

Stem The part of the plant above ground where leaves and flowers appear.

Stigma The tip of the pistil that receives pollen at pollination.

Stipe That part of a fern front stem below the rachis (below where leaflets are attached); the "leaf stem" of fern fronds.

Stipules Small, leaflike, paired projections at the base of a leafstalk.

Style The slender stalk of a pistil joining ovary and stigma.

Succulent Fleshy, juicy, and thickened; a plant with those characteristics.

Superior ovary An ovary high in a flower, above the joining point of stamens, sepals, and petals.

Tendril The coil of a modified stem or leaf, often used for support.

Terminal At the end of branch or stem.

Thorn A short, stiff, sharp-pointed branch.

Tomentosus*, -*a*, -*um Covered with soft, matted, flat hairs.

Triflorus*, -*a*, -*um Three-flowered.

Trifoliatus*, -*a*, -*um Three-leaved.

Tripinnate Divided three times.

Tubercle A knoblike projection (as on a cactus joint).

Tubular Describes a flower with united petals forming a tube.

Twining Climbing by coiling around something.

Umbel An umbrella-shaped flower cluster with all flower stalks evolving from same point.

Umbellatus*, -*a*, -*um Having flowers in umbels.

Undulatus*, -*a*, -*um Wavy.

Uniflorus*, -*a*, -*um One-flowered.

United Describes petals fused together.

Variegatus*, -*a*, -*um Having markings of a different color than the basic color.

Versicolor Having variable colors.

Villosus*, -*a*, -*um With long, loose hairs.

Viridiflorus*, -*a*, -*um With green flowers.

Viscosus*, -*a*, -*um Sticky.

Vulgaris*, -*e Common.

Whorl Three or more leaves or flower parts radiating outward from a stem node.

Wing A thin, paperlike flap on a seed capsule, stem, or flower.

Woolly Having soft, wool-like hairs.

PARTS OF A FLOWER

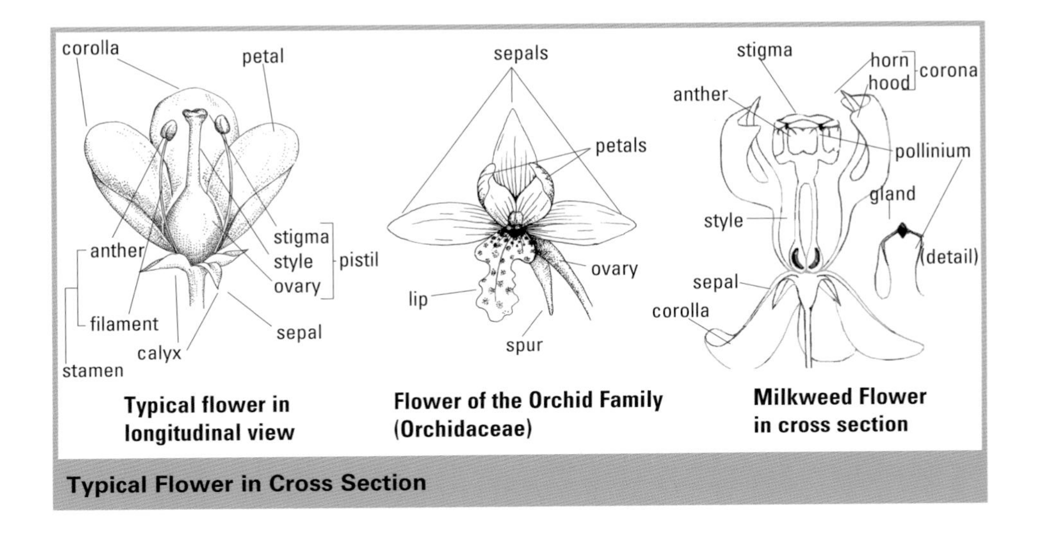

Typical Flower in Cross Section

REFERENCES

Anderson, Edward F. *The Cactus Family*. Portland, OR: Timber Press, 2001.

Anderson, J.L. "Vascular Plants of Arizona: Anacardiaceae," in *CANOTIA: An Arizona journal publishing botanical and mycological papers*, vol. 3, issue 2, pp. 1–10.

________. "Vascular Plants of Arizona: Ericaceae," in *CANOTIA: An Arizona journal publishing botanical and mycological papers*, vol. 4, issue 2, pp. 1–10.

Arizona-Nevada Academy of Science Editorial Committee, eds. A New Flora for Arizona in Preparation. Vol. 26(1), 1992, pp. 22–28, 36–41; vol. 27(2), 1994, pp. 164–168, 190–194, 222–236, 241–245; vol. 29(1), 1995, pp. 6–12, 13–14, 29–38, 39–62; vol. 30(2), 1998, pp. 61–83, 84–95; vol. 32(1), 1999, pp. 1–21, 32–47, 52–54; vol. 33(1), 2001, pp. 50–57, 58–64, 69–72.

Arnberger, Leslie P. *Flowers of the Southwest Mountains*. Globe, AZ: Southwestern Monuments Association, 1962.

Baerg, Harry J. *How to Know the Western Trees*. Dubuque, IA: C. Brown Company Publishers, 1973.

Bair, A., M. Howe, D. Roth, R. Taylor, T. Ayers, and R.W. Kiger. "Vascular Plants of Arizona: Portulacaceae," in *CANOTIA: An Arizona journal publishing botanical and mycological papers*, vol. 2, issue 1, pp. 1–22.

Baker, Marc. "A multivariate study of morphological characters for *Echinocactus horizonthalonius* and *E. texensis* (Cactaceae)" (first draft report, School of Life Sciences, Arizona State University, Tempe, AZ, 2007). www.fws.gov/southwest/federal_assistance/PDFs/Nichol'sTurk'sheadcactusmorphologyAZE-6-9.pdf Feb. 25, 2009.

Bates, S.T., F. Farruggia, E. Gilbert, R. Gutierrez, D. Jenke, E. Makings, E. Manton, Chiang, F., and L. R. Landrum. "Vascular Plants of Arizona: Solanaceae Part Three: *Lycium* L. Wolfberry, Desert Thorn," in *CANOTIA: An Arizona journal publishing botanical and mycological papers*, vol. 5, issue 1, pp. 1–10.

Benson, Lyman. *The Cacti of Arizona*. Tucson, AZ: University of Arizona Press, 1969.

Benson, Lyman, and Robert A. Darrow. *Trees and Shrubs of the Southwest Deserts*. Tucson, AZ: University of Arizona Press, 1981.

Bernard, Nelson T. *Wildflowers Along Forest and Mesa Trails*. Albuquerque, NM: University of New Mexico Press, 1925.

Bowers, Janice Emily. *100 Desert Wildflowers of the Southwest*. Tucson, AZ: Southwest Parks and Monuments Association, 1989.

________. *100 Roadside Wildflowers of Southwest Woodlands*. Tucson, AZ: Southwest Parks and Monuments Association, 1987.

Christie, K., M. Currie, L. S. Davis, M-E. Hill, S. Neal, and T. Ayers, "Vascular Plants of Arizona: Rhamnaceae," in *CANOTIA: An Arizona journal publishing botanical and mycological papers*, vol. 2, issue 1, pp. 1–24.

Coombes, Allen J. *Dictionary of Plant Names*. Portland, OR: Timber Press, 1990.

Craighead, John J., Frank C. Craighead Jr., and Ray J. Davis. *A Field Guide to Rocky Mountain Wildflowers*. Boston, MA: Houghton Mifflin Company, 1963.

Crittenden, Mabel. *Trees of the West*. Millbrae, CA: Celestial Arts, 1977.

Crittenden, Mabel, and Dorothy Telfer. *Wildflowers of the West*. Millbrae, CA: Celestial Arts, 1975.

Desert Botanical Garden Staff. *Arizona Highways Presents Desert Wildflowers*. Phoenix, AZ: Desert Botanical Garden, 1988.

Dodge, Natt N. *100 Desert Wildflowers in Natural Color*. Globe, AZ: Southwestern Monuments Association, 1963.

________. *Flowers of the Southwest Deserts*. Globe, AZ: Southwestern Monuments Association, 1969.

________. *100 Roadside Wildflowers of Southwest Uplands in Natural Color*. Globe, AZ: Southwestern Monuments Association, 1967.

Elias, Thomas S. *The Complete Trees of North America*. New York: Van Nostrand Reinhold, 1980.

Elmore, Francis H. *Shrubs and Trees of the Southwest Uplands*. Tucson, AZ: Southwest Parks and Monuments Association, 1976.

Flora of North America Editorial Committee, eds. *Flora of North America: North of Mexico*. Vols. 1–5, 7, 8, 19–27. New York: Oxford University Press, 1993–2010.

Florists' Publishing Company. *New Pronouncing Dictionary of Plant Names*. Chicago: Florists' Publishing Company, 1990.

Foxx, Teralene S., and Dorothy Hoard. *Flowers of the Southwestern Forests and Woodlands*. Los Alamos, NM: Los Alamos Historical Society, 1984.

Gentry, Howard Scott. *Agaves of Continental North America*. Tucson, AZ: University of Arizona Press, 1982.

Hickman, James C., ed. *The Jepson Manual: Higher Plants of California*. Berkeley: University of California Press, 1993.

Integrated Taxonomic Information System (ITIS). "ITIS Advanced Search and Report." Accessed December 2, 2010. www.itis.gov/advanced_search.html.

Jaeger, Edmund C. *Desert Wild Flowers*. Stanford, CA: Stanford University Press, 1940.

The Jepson Herbarium, University of California, Berkeley. "The Jepson Manual II: Vascular Plants of California." Last modified October 1, 2009. http://ucjeps.berkeley.edu/jepsonmanual/review/.

Kearney, Thomas H., and Robert H. Peebles. *Arizona Flora*. 2nd ed. Berkeley, CA: University of California Press, 1951.

Kodela, Phillip G., and Peter G. Wilson. "New combinations in the genus *Vachellia* (Fabaceae: Mimosoideae) from Australia." *Telopea* 11, no. 2 (2006): 233–44.

Lamb, Edgar, and Brian Lamb. *Colorful Cacti of the American Deserts*. New York: Macmillan Publishing Company, 1974.

Lamb, Samuel H. *Woody Plants of the Southwest*. Santa Fe, NM: Sunstone Press, 1977.

Lehr, J. Harry. *A Catalogue of the Flora of Arizona*. Phoenix, AZ: Desert Botanical Garden, 1978.

Lenz, L. W., and M. A. Hanson. "Typification and change in status of *Yucca schottii* (Agavaceae)." *Aliso* 19, no. 1 (2000): 93–8.

Little, Elbert L. *The Audubon Society Field Guide to North American Trees* (Western Region). New York: Alfred A. Knopf, 1980.

Martin, William C., and Charles R. Hutchins. *Fall Wildflowers of New Mexico*. Albuquerque, NM: University of New Mexico Press, 1988.

________. *Summer Wildflowers of New Mexico*. Albuquerque, NM: University of New Mexico Press, 1986.

McDougall, W.B. *Grand Canyon Wild Flowers*. Flagstaff, AZ: Museum of Northern Arizona, 1964.

Mickel, John T. *How To Know the Ferns and Fern Allies*. Dubuque, IA: William C. Brown Publishers, 1979.

Miller, Howard A., and Samuel H. Lamb. *Oaks of North America*. Happy Camp, CA: Naturegraph Publishers, Inc., 1985.

Mohlenbrock, Robert H. Wildflowers: *A Quick Identification Guide to the Wildflowers of North America*. New York: Macmillan Publishing Company, 1987.

Munz, Philip A. *California Desert Wildflowers*. Berkeley, CA: University of California Press, 1962.

________. *A Flora of Southern California*. Berkeley, CA: University of California Press, 1974.

Newcomb, Lawrence. *Newcomb's Wildflower Guide*. Boston, MA: Little, Brown & Company, 1977.

Newton, D., and L. R. Landrum. "Vascular Plants of Arizona: Solanaceae Part Two: Key to the genera and *Solanum* L.," in *CANOTIA: An Arizona journal publishing botanical and mycological papers*, vol. 5, issue 1, pp. 1–16.

Niehaus, Theodore F., Charles L. Ripper, and Virginia Savage. *A Field Guide to Southwestern and Texas Wildflowers*. Boston, MA: Houghton Mifflin Company, 1984.

Orr, Robert T., and Margaret C. Orr. *Wildflowers of Western America*. New York: Alfred A. Knopf, 1974.

Parker, Kittie F. *An Illustrated Guide to Arizona Weeds*. Tucson, AZ: University of Arizona Press, 1986.

Patraw, Pauline Mead. *Flowers of the Southwest Mesas*. Globe, AZ: Southwestern Monuments Association, 1959.

Pesman, M. Walter. *Meet the Natives*. Denver, CO: Pruett Publishing, 1988.

Phillips, Arthur M. *Grand Canyon Wildflowers*. Flagstaff, AZ: Grand Canyon Natural History Association, 1979.

Rickett, Harold W. *Wild Flowers of the United States*. Vol. 4 (in 3 parts). New York: McGraw-Hill Co., 1973.

Rico Arce, M. L., and S. Bachman. "A taxonomic revision of Acaciella (Legumionosae, Momosoideae)." *Anales del Jardín Botánico de Madrid* 63, no. 2 (July–December 2006): 189–244.

Seigler, David S., John E. Ebinger, and Joseph T. Miller. "The genus *Senegalia* (Fabaceae: Mimosoideae) from the new world." *Phytologia* 88, no. 1 (June 2006): 38–93.

Southwest Environmental Information Network (SEINet). "SEINet Research Checklist: Arizona." Accessed March 4, 2011. http://swbiodiversity.org/seinet/checklists/checklist.php?cl=1&proj=1.

Spellenberg, Richard. *The Audubon Society Field Guide to North American Wildflowers* (Western Region). New York: Alfred A. Knopf, 1979.

United States Department of Agriculture. "Natural Resources Conservation Service (NRCS): Plants Database." Accessed December 4, 2010. http://plants.usda.gov/java/.

Venning, Frank D. *Wildflowers of North America: A Guide to Field Identification*. New York: Golden Press, 1984.

Vines, Robert A. *Trees, Shrubs and Woody Vines of the Southwest*. Austin, TX: Texas Press, 1976.

Ward, Grace B., and Onas M. Ward. *190 Wildflowers of the Southwestern Deserts in Natural Color*. Palm Desert, CA: Best-West Publications, 1978.

Wilken, D.H., and J.M. Porter. "Vascular Plants of Arizona: Polemoniaceae," in *CANOTIA: An Arizona journal publishing botanical and mycological papers*, vol. 1, issue 1, pp. 1–37.

Yatskievych, G., and M.D. Windham. "Vascular Plants of Arizona," in *CANOTIA: An Arizona journal publishing botanical and mycological papers*, vol. 4, issue 2, pp. 1–3.

———. "Vascular Plants of Arizona: Polypodiaceae Polypody Family," in *CANOTIA: An Arizona journal publishing botanical and mycological papers*, vol. 5, issue 1, pp. 1–5.

INDEX

Each genus and scientific (Latin) family name appears in **BOLDFACE** type.

D